T0146333

Applications of 3D Measurement from Images

edited by

Professor John Fryer

Emeritus Professor, Surveying and Photogrammetry, University of Newcastle, NSW, Australia

Dr Harvey Mitchell

Senior Lecturer, School of Engineering, University of Newcastle, NSW, Australia

and

Dr Jim Chandler

Senior Lecturer, Department of Civil and Building Engineering, Loughborough University, Loughborough, UK

Whittles Publishing

CRC Press
Taylor & Francis Group

Published by
Whittles Publishing,
Dunbeath,
Caithness KW6 6EY,
Scotland, UK
www.whittlespublishing.com

Distributed in North America by
CRC Press LLC,
Taylor and Francis Group,
6000 Broken Sound Parkway NW, Suite 300,
Boca Raton, FL 33487, USA

© 2007 J.G. Fryer, H.L. Mitchell & J.H. Chandler

ISBN 10: 1-870325-69-9
ISBN 13: 978-1-870325-69-1
USA ISBN 10: 1-4200-5486-4
USA ISBN 13: 9781-4200-5486-6

Typeset by Compuscript Ltd., Shannon, Ireland

Printed in Poland

Contents

CD contains: supporting imagery, colour versions of the figures and some short animations. These illustrate and expand upon the applications described and provide an opportunity to demonstrate some exciting and innovative final products.

Acknowledgments

Stuart Robson and Mark Shortis wish to acknowledge the following individuals and organisations who have supported and facilitated the case studies cited in Chapter 4: Geraint Jones and Graham Worthington of Airbus UK, Robert Brade of UKEA who supplied the textual content, and EFDA JET who supplied the images. Richard Pappa, Tom Jones, Al Burner, Paul Danehy, Mark Hutchinson and Keith Belvin of NASA LaRC, Vaughn Behun of Swales Aerospace, Joe Blandino of James Madison University, Gary Robertson of ShapeQuest Inc., and former students Louis Giersch and Jessica Quagliaroli who were all involved in the development of photogrammetry methods for the Gossamer spacecraft structures. They also wish to acknowledge the support of George Johnson of Midcoast Metrology, and Bill Goad of NASA Langley Research Centre.

John Fryer wishes to acknowledge several people for their input and guidance in the presentation of the material for Chapter 5. In particular, he is grateful for the assistance of Tim Clarke of the Optical Metrology Centre, UK in preparing the case studies on motor car racing; the assistance of Clive Fraser with some of the images; and the assistance of Joe Wenta in sourcing some interesting forensic examples.

Jim Chandler, the principal author of Chapter 6, has been involved with most of the work described in this chapter. However, he particularly wishes to acknowledge the work by Jan Walstra in quantifying the evolution of the Mam Tor landslide, and that of Stuart Lane in writing the second case study.

Harvey Mitchell wishes to acknowledge that much of the description within Chapter 7 derives from earlier collaborative studies and writing by him and Ian Newton, formerly of the academic staff of the University of Newcastle upon Tyne, UK. He greatly appreciates the contribution made by Ian Newton.

Contributors

Jean-Angelo Beraldin   National Research Council Canada, Institute for Information Technology, 1200 Montreal Road, Building M-50, Room 352, Ottawa, ON K1A 0R6
ph: +1 (613) 990-6871
fax: +1 (613) 952-0215
email: Jean-Angelo.Beraldin@nrc-cnrc.gc.ca

Dr J. H. Chandler   Department of Civil and Building Engineering, Loughborough University, Loughborough, LE11 3TU
email: J.H.Chandler@lboro.ac.uk

Dr Albert K. Chong   School of Surveying, University of Otago, New Zealand
email: chonga@albers.otago.ac.nz

Dr Sabry F. El-Hakim   National Research Council Canada, Institute for Information Technology, 1200 Montreal Road, Building M-50, Room 358, Ottawa, ON K1A 0R6
ph: +1 (613) 991-6381
fax: +1 (613) 952-0215
email: sabry.el-hakim@nrc-cnrc.gc.ca

Professor Clive S. Fraser   Department of Geomatics, University of Melbourne, Victoria 3010 Australia
ph: +61-3-8344-4117
fax: +61-3-9347-2916
email: c.fraser@unimelb.edu.au
web: www.geom.unimelb.edu.au

Professor John Fryer   Emeritus Professor, Surveying and Photogrammetry, University of Newcastle, NSW, Australia
email: John.Fryer@newcastle.edu.au

Professor Stuart N. Lane   Professor in the Department of Geography, University of Durham, Durham, DH1 3LE
ph: +44 (0)191 33 41818
fax: +44 (0)191 33 41801
email: s.n.lane@durham.ac.uk

Dr Harvey Mitchell   Senior Lecturer, School of Engineering, University of Newcastle, NSW, Australia
email: harvey.mitchell@newcastle.edu.au

Cliff Ogleby

The University of Melbourne, Department of Geomatics, Victoria 3010, Australia
ph: +61-3-8344-6806
fax: +61-3-9347-2916
email: clogleby@unimelb.edu.au

Professor Petros Patias

Aristotle University of Thessaloniki, Department of Cadastre, Photogrammetry and Cartography, Univ.Box 473, GR-54124 Thessaloniki, Greece
ph: +30-2-310-99.6116
fax: +30-2-310-99.6128
email: patias@auth.gr

Dr. Stuart Robson

Department of Geomatic Engineering, University College London, Gower Street, London, WC1E 6BT
ph: +44 (0)207 679 2726
fax: +44 (0)207 380 0453
email: S.Robson@ge.ucl.ac.uk
web: www.ge.ucl.ac.uk

Professor Mark Shortis

Professor of Measurement Science, Science, Engineering and Technology Portfolio, RMIT University, GPO Box 2476V, Melbourne 3001, Australia.
ph: +61 3 9925 9628
email: mark.shorts@rmit.edu.au
web: www.geomsoft.com/markss

Dr Jan Walstra

Department of Languages & Cultures of the Near East & North Africa Ghent University, Sint-Pietersplein 6, B-9000 Ghent, Belgium
ph: +32 (0)9 264 3544
email: Jan.Walstra@UGent.be

1 Introduction

John G. Fryer

1.1 Definition

It is widely recognised that the world's maps have been generated primarily by measurement of aerial imagery for about 60 years. What is not so well known is that for approximately 150 years, some surveyors, architects, medical practitioners and others have been determining the size and shape of a large array of objects by analysing images, recorded on film or, more recently, on electronic media. This book describes many diverse examples of application areas for the extraction of 3D information on objects from images. The process is both an art and a science. The rigorous science required for the correct imaging and analysis must be supplemented by some art, some flair and some inventiveness in order to produce successful outcomes. When studying this book, it is easy to underestimate the skills of the professionals whose work is described in each chapter. However, with some training, experience and understanding it is possible for new users to extract spatial data of an acceptable quality for their purposes. This book has been drafted to help develop the required understanding and encourage the use of the techniques described.

A major theme in this book is the large variation in the scale of the projects which are suited to 3D imaging, a feature not associated with other measurement technologies. In most industrial and scientific processes, the scale of a project poses limitations. Scale often controls what can and cannot be done, what is economical and what is not. But, with the extraction of 3D information from imagery, variations in scale have little impact on the mathematics and techniques employed. The imaging methods, the requirements for control points, the calculation processes and the methods of displaying results are independent of the scale of the project. To illustrate and enforce this special feature, the applications in each chapter have been specially chosen to illustrate the diversity of object sizes which can be investigated.

Recognition must be made at the outset of a previous book which explored some application areas of imaging, *Close Range Photogrammetry and Machine Vision* (Atkinson, 1996). In the eleven years since that book explained and illustrated the basic principles of photogrammetry, there have been rapid and continuing developments in imaging, image processing and a massive reduction in the cost of computer hardware. Much of the theory in that book remains valid today; readers desirous of a more detailed theoretical background to photogrammetry and 3D imaging are encouraged to access it.

This book has the aim of concentrating on new and exciting developments in 3D imaging applications, almost all of which have occurred in recent times. 3D imaging techniques have reached a stage of maturity, thanks to digital cameras and computerised solutions, that enables practitioners from many professions to consider their usefulness. There are chapters in this book which should capture the imagination and inspire archaeologists, biologists, engineers, medical professionals, forensic investigators, physical geographers, field- and office-based scientists and sports medicine clinicians.

3D imaging can be seen as another step in mankind's inevitable progression from manual to automated data capture. New applications for 3D imaging are continually being reported at conferences and in the scientific literature. Appreciating its scale independence is a key to understanding its appeal to those of us who have spent a working lifetime fascinated by the amount of data contained in images. Unlocking that data and converting it to useful information is the "holy grail" pursued by researchers and practitioners in this field.

Imaging is a modern term, photogrammetry being the more narrowly focused one it is replacing in the twenty-first century. The principles are the same: there are certain fundamental physical phenomena which must be acknowledged and some geometric constraints which define the boundaries of application areas. That light travels in straight lines is an obvious example of the former while a requirement for multiple images to overlap is a self-evident example of the latter.

Some fundamental considerations remain, with the term photogrammetry in the following quote being used to include other forms of modern imaging. Professor E.H. Thompson, a famous English photogrammetrist, wrote in 1962 "… that photogrammetry is often the method chosen after all else fails". The advantages he listed for photogrammetry included its use as a measurement tool when:

- the object is inaccessible or difficult to access;
- the object is not rigid and its instantaneous dimensions are required;
- it is not certain that measurements will be required at all;
- it is not certain, at the time of measurement, what measurements are required;
- the contours of the surface of the object are required; and,
- the object is very small, especially when it is microscopically small.

Thompson also suggested that "photogrammetry can be useful only if direct measurement is impossible, impractical or uneconomical" (Thompson, 1962).

This statement may have been especially true in the days of strict stereo photogrammetry before computers allowed for the rapid analysis of convergent photographs, but it is worth remembering that there is no universal panacea for all measurement tasks. However, the challenge established by Thompson's statement may have been the catalyst which has generated so many innovations in recent years.

A concern for all imaging scientists, no matter how spatially aware they may be, relates to the extremely low level of awareness of this science by other scientists, engineers, biologists, lawyers, medical practitioners, archaeologists and politicians. Until this lack of recognition in the wider scientific community is addressed, 3D imaging as a real option for the measurement of a vast range of objects will remain the province of a small group of specialists.

1.2 A brief history

It is not the purpose of this book to deal too much with the past. Potted histories of photogrammetric imaging can be found in Atkinson (1980, 1996), Slama (1980) and McGlone (2004). It is obvious, however, that photogrammetry, as a scientific measurement tool, did not commence until imaging techniques were established by the middle of the nineteenth century, although many of the concepts upon which photogrammetry flourished were well known for centuries prior to the first photographic image. Stereoscopic drawings had been made by artists from the 1600s and prototype "pin-hole" cameras were devised in advance of lens developments.

That the initial applications of photogrammetry were of a terrestrial nature is not surprising. Reliable airborne platforms, apart from kites and balloons which were often at the mercy of the elements, did not appear until the twentieth century, and so mapping of city-scapes from roof-tops and architectural studies of buildings were early areas for fruit-ful study. In 1858 Albrecht Meydenbauer, a German architect, commenced surveys of historical monuments, churches and buildings for recording their condition, before and after their reconstruction after damage. In 1885 he established a state institute in Berlin, and so laid the foundations for the recording of cultural heritage.

Other application areas for 3D imaging also started in the second half of the nineteenth century. After the American Civil War in 1863, the American physician Oliver Holmes used photogrammetry as a technique for gait analysis, so that prosthetic devices could be better fitted to limbless soldiers. Thus medical applications were born, but he could have scarcely dreamt that the same geometrical principles, using modern cameras, video technology and computers, would be applied to all parts of the human body, both external and internal, within the next 140 years.

Scientific imaging received formal recognition when the International Society for Photogrammetry was formed in 1910 and technical commissions to concentrate study into specific areas commenced in 1926. Terrestrial, architectural and engineering photogrammetry were represented as study groups. Later, the name was changed to the International Society for Photogrammetry and Remote Sensing to reflect advances in satellite and airborne sen-sors. Since that time, large congresses have attracted thousands of attendees every four years. Authors of chapters in this book have been regular attendees at many of these congresses and smaller specialist group meetings, held every two years, which cover the application areas cited in this book. The case studies reported in their chapters reflect this international collaboration and involvement.

1.3 A growing maturity

The analysis of 3D images was an extremely rigorous and geometrically constrained process until the 1970s. The analogue or optical-mechanical stereoplotting equipment used to take mea-surements, follow contour lines and draw plans, was large, expensive, heavy, yet fragile, and needed a specially trained operator. Easy access to computers heralded a new era. Analytical photogrammetry became possible, although the mathematics involved with the implementation of analytical photogrammetry was complex, involving the solution to sets of non-linear equa-tions.

The collinearity equations which had been derived in the 1930s, but rarely applied as they required a very time consuming iterative solution, suddenly became the focus of attention for photogrammetrists. These equations, which could take days to solve by hand, formed the basis of most new applications but it took almost 30 years before the authors of scientific papers in imaging were mature enough just to reference them rather than fill up the introductory section of their paper with an exposition of their formulation! In an analytical stereoplotter, this calcu-lation of the collinearity equations (or their complementary form, the coplanarity equations) has to be performed as a background task approximately a hundred of times per second to allow three-dimensional viewing of the 3D model that is jitter-free.

Just as even the most traditional and conservative mapping agencies had concluded that the computer age had really arrived and were replacing pen and ink draftspersons with large and costly flatbed plotters, the 1990s saw the dawn of the era we know as digital photogrammetry. The revolution in techniques continued. The large, cumbersome, stereoplotting machines, some of which had been retro-fitted with electric motors, encoders and analytical photogrammetric software were finally sentenced to death. For some, this was commuted to a life sentence as a curiosity in a technology museum, but for many others it was a trip to the scrap metal dealer for recycling.

The early digital era relied on the technology of electro-optical or video cameras, frame-grabbers and computers. True digital technology, where there were no analogue-to-digital conversions, only arrived this century when breakthroughs came in the production of image sensing arrays. Often referred to as charge coupled devices (CCDs), these sensors allowed digital cameras to become so cheap that film-based cameras lost their pre-eminent place in imaging and fashion conscious teenagers refused to use mobile phones that didn't have integrated megapixel cameras!

As in each era of developments in imaging, the burst of new digital inventions has not directly translated into ready acceptance of photogrammetric principles and mass acceptance of imaging as a measurement tool. The real-time extraction of planimetric and altimetric features is moving from science fiction to reality (Toutin and Beaudoin, 1995) but some experts caution that it still has a long way to go. In 2005, Professor W. Förstner, the President of the International Society's Technical Commission dealing with Photogrammetric Computer Vision and Image Analysis stated "the potential of photogrammetric applications could be increased by at least two orders of magnitude if the basic photogrammetric problems, namely calibration and orientation were solved for cameras in the hand of the normal consumer". He went on to explain the need for software to assist and guide users and awaits the development of the "micro compass in the camera or a GPS receiver in the watch of the user, the camera being linked to a wearable computer" (Förstner, 2005).

Cameras have not been the only item of recent digital developments. Other measurement technologies to capture the size and shape of objects include automated total stations (electronic theodolites), airborne and terrestrial laser scanners and user-friendly image processing software packages. These devices are making inroads into areas traditionally the preserve of photogrammetrists. The ability of new equipment to capture the coordinates for thousands of data points in a short time, known as *point clouds*, has opened new avenues for research and applications. Laser scanners can capture tens of thousands of points per second, although it is now often forgotten that photogrammetric image correlation software can extract similar quantities of data at rates that are only a fraction slower.

The analysis of the data in point clouds and converting them into meaningful information about the size and shape of objects is a major challenge for the first part of the twenty-first century. Other challenges include: how to segment the point clouds into geometric surfaces; how to automatically classify what particular shapes represent; how to combine point clouds taken from different viewpoints; how to detect, eliminate and intelligently fill occluded sections on the object; how to use the intensity of a reflected laser signal; and how to develop strategies for general problems related to the type of surface such as semi-transparency (glass), texture-free sections and specular reflections. In short, the new technologies offer much, but bring with them their own peculiar constraints.

Notwithstanding new difficulties faced by users of the modern technologies mentioned above, we are entering, just as the pioneers of imaging did 150 years ago, a new and exciting era. Cartographers have known for centuries that image presentation is all important and "a picture is worth a thousand words". Software manufacturers are producing cheap, user-friendly packages which allow image processing to be done on personal computers. The production of an animated image sequence representing the spatial world is now within the grasp of anyone prepared to read a computer manual.

1.4 To the future

Most chapters in this book detail the special mathematical considerations needed for specific applications. Engineering is a diverse field of endeavour which relies heavily on measurement techniques for the efficient construction and subsequent monitoring of structures. The concept of 3D imaging not being constrained by scale is presented in Chapters 3 and 4 which describe applications involving bridges, radio telescopes and major infrastructure facilities. The 3D measurements are not an end in themselves: they must be combined with CAD or other models to demonstrate compliance to manufacturing tolerances, amounts of wear, reliability under test conditions and so on. The 3D imaging professional is an essential part of a team: he or she is a provider of a service and their involvement in a project must always be recognised.

The use of images for the gathering of evidence for criminal or other legal investigations has been termed forensic photogrammetry (see Chapter 5). It may be considered an even more diverse field of application than those mentioned above. Not only do trained photogrammetrists take sets of images of crime or accident scenes, but very often key items of evidence emerge from casual snapshots taken by persons passing by. These fortuitous images usually lack the control points and the rigour associated with the work of professionals, but they are often the only images which display objects vital to the solution of the investigation.

The evolution of the earth's landscape has been a subject of philosophical discussion, and later study, for a long time. Remote sensing from high and low airborne platforms has given us the ability to further investigate the mechanisms that are slowly, but sometimes spectacularly, at work in our environment. Terrestrial and close range imaging and other surveying measurements can complement aerial sensing and Chapter 6 explains the methodologies presently being explored and developed to assist understanding of the basic processes shaping the earth's evolution.

When the American physician Oliver Holmes used photogrammetry to design prosthetic devices after the American Civil War, little did he realise that his scientific approach would set the scene for continuing research into medical applications of imaging. Chapter 7 on medical photogrammetry describes the special features associated with this form of imaging, a primary one being concern for the patient: how to most expeditiously use non-invasive imaging techniques in ways which do not alarm, restrain, or infringe the rights of the individual, yet provide cost-effective solutions. The population of the earth is still growing, and individuals are growing older than ever before in history. New problems associated with ageing are outstripping the resources of the developed world to cope. Automated techniques using 3D imaging will not be the panacea to this dilemma, but might just help analysis and diagnosis in certain areas. Chapter 7 also describes how sport-related imaging has grown from a low base in the last two decades to the stage where specialised equipment is available for the analysis of both individual and team performance in most high profile sports. Has this application area reached a saturation point and maturity? Given the developments of the last 140 years in biostereometrics, one can only guess what inroads 3D imaging will yet make to sporting achievements.

Photogrammetric applications involving biology and zoology are considered in Chapter 8. Just as preserving ancient monuments is important to our cultural heritage, the recording and analysis of our plants and land and sea animals is equally important to the understanding of our biological make-up. The chapter has an emphasis on underwater creatures, not only because of the extra difficulties working in such an environment, but because more and more public emphasis is being placed on marine resources and their fragility when faced with fishing and other maritime operations. The special challenges posed by working through air–water interfaces or underwater are explored and explained.

Architectural photogrammetry commenced in the mid-nineteenth century and as Chapter 9 explains, its uses are expanding and evolving. Modern techniques have been embraced in the study of our oldest artifacts and monuments. Perhaps this is the field of application which has used fly-throughs and animations the most to illustrate the earth's culture and heritage. As population pressures on the world's treasures grow, the very items people wish to experience are under increasing threat of destruction. Many major tourist attractions such as the pyramids of Egypt, Stonehenge in England and Machu Picchu in Peru are already deteriorating due to their popularity. Ready access to international travel allows millions of people to inspect our antiquities. The need to record and then display them in a "virtual museum" has never been greater and may be either the key to their sustainability or, paradoxically, seal their ultimate fate.

This book concludes with a chapter which looks at the new and emerging developments which may or will influence the future applications of 3D imaging. Laser scanners and their integrated use with digital images are reviewed. The development of computer models is explained along with the use of software to exploit these models in short animation video clips. Some applications illustrate this approach to the presentation of 3D information.

It is a difficult task to look into the future. Without being too prescriptive, it is obvious that the ubiquitous computer will be central to future developments. We can surely look forward to greater integration of imaging tools with position measuring devices. For example, a digital camera which not only records the spectral intensity of each pixel but also measures the exact distance from the camera to each point imaged is under development. If the camera has a GPS receiver, levelling sensor and micro-compass fitted, then those points will be automatically coordinated on the earth's reference frame as well. Automatic feature extraction would follow and wireless transmission back to a base computer would be a matter of course. This guesswork will stop here, as futurists usually have bad track records.

A DVD containing supporting imagery and colour versions of most of the figures in the book forms part of this publication. These include some short animations to illustrate and expand upon the applications described in some of the chapters. This DVD has been designed so that it should operate in most reasonably modern PCs. The concept of including a DVD with this publication seemed appropriate, given the level of computerisation of 3D imaging and the sophistication of its output products. No longer will a simple plan suffice many clients, a fly-through, animated sequence or movie is needed to convince everyone from a local authority planner to a judge and jury. This DVD provides the ideal opportunity to demonstrate some of these exciting and innovative final products.

1.5 A caveat

Despite all the exciting possibilities for 3D imaging techniques, users of this technology must keep in mind some fundamentals which will form the bases for future applications, just as they do for present day imaging. Light will continue to travel in straight lines and an understanding of geometry will be just as valid tomorrow as it is today. The need for an appreciation of spatial analysis, call it geomatics or surveying principles, is still a paramount criterion for the extraction of geometrically strong 3D objects from images.

1.6 References

Atkinson, K.B. (Ed.), 1980. *Developments in Close Range Photogrammetry-1*, Applied Science Publishers, London, UK.

Atkinson, K.B. (Ed.), 1996. *Close Range Photogrammetry and Machine Vision*, Whittles Publishing, Scotland.

Förstner, W., 2005. ISPRS Comm. III – Contribution to ISPRS Highlights. Highlights magazine of *International Society for Photogrammetry and Remote Sensing*, 10(3), 15–17.

McGlone, J.C., 2004. *Manual of Photogrammetry*, American Society for Photogrammetry and Remote Sensing, Bethesda, Maryland, USA.

Slama, C.C. (Ed.), 1980. *Manual of Photogrammetry*, 4th edn., American Society of Photogrammetry, Falls Church, Virginia, USA.

Thompson, E.H., 1962. Photogrammetry, *The Royal Engineers Journal* 76(4), 432–444 and reprinted in *Photogrammetry and Surveying, A Selection of Papers by E.H. Thompson*, 1910–1976, Photogrammetric Society, London, 1977, 242–255.

Toutin, T. and Beaudoin, M., 1995. Real-time extraction of planimetric and altimetric features from digital stereo SPOT data using a digital video plotter. *Photogrammetric Engineering and Remote Sensing*, 61(1), 63–68.

1.7 Further reading

Al-Garni, A.M., 1995. The fourth dimension in digital photogrammetry. *Photogrammetric Engineering and Remote Sensing*, 61(1), 57–62.

Carbonnell, M., 1984. Comité International de Photogrammétrie Architecturale. *International Archives of Photogrammetry and Remote Sensing*, 25(A5), 151–155.

Deville, E., 1895. *Photographic Surveying*, Government Printing Bureau, Ottawa, Canada.

El-Hakim, S.F. and Barakat, M.A., 1989. A vision-based coordinate measuring machine (vcmm). In Grün, A. and Kahmen, H. (Eds.), *Optical 3D Measurement Techniques*, Wichmann Verlag, Karlsruhe, Germany, 216–218.

Faig, W., 1975. Calibration of close-range photogrammetric systems-mathematical formulation. *Photogrammetric Engineering and Remote Sensing,* 41(12), 1479–1486.

Fourcade, H.G., (1903). On a stereoscopic method of photographic surveying. *Transactions of the South African Philosophical Society*, 14(1), 28–35 (also published in 1902. *Nature*, 66(1701), 139–141).

Gates, H.W.C., Oldfield, S., Forno, C., Scott, P.J. and Kyle, S.A., 1982. Factors defining precision in close-range photogrammetry. *International Archives of Photogrammetry*, 24(5/1), 185–195.

Grün, A.W. 1993. A decade of digital close range photogrammetry – achievements, problems and prospect. *Photogrammetric Journal of Finland*, 13(2), 16–36.

Haggren, H., 1993. 3D video digitising. *Photogrammetric Journal of Finland*, 13(2), 37–45.

Harley, I.A., 1967. The non-topographical uses of photogrammetry. *The Australian Surveyor*, 21(7), 237–262.

Herron, R.E. (Ed.), 1983. *Biostereometrics '82, SPIE Proceedings* 361, Society of Photo-optical Instrumentation Engineers, Bellingham, Washington, USA.

Hottier, P., 1976. Accuracy of close-range analytical restitution: practical experimentation and prediction. *Photogrammetric Engineering and Remote Sensing*, 42 (3), 345–375.

Karara, H.M. (Ed.), 1979. *Handbook of Non-topographic Photogrammetry*. American Society of Photogrammetry, Falls Church, Virginia, USA.

Konecny, C., 1981. *Development of Photogrammetric Instrumentation and its Future*. Finnish Society of Photogrammetry, Helsinki, Finland, 50th Anniversary Publication, 21–48.

Laussedat, A., 1898, 1901, 1903. *Recherches sur les Instruments, les Methodes et le Dessin Topographiques*, Gauthier-Villars, Paris (Vol. 1, 1898; Vol. 2(1), 1901; Vol. 2(2), 1903).

Lenz, R., 1989. Image data acquisition with CCD cameras. In Grün, A. and Kahmen, H. (Eds.), *Optical 3D Measurement Techniques*, Wichmann Verlag, Karlsruhe, Germany, 22–34.

Meydenbauer, A., 1894. Ein deutsches Dankmälerarchiv (Monumenta Germaniae). *Deutsche Bauzeitung*: 629–631.

Novak, K., 1990. Integration of a GPS-receiver and a stereovision system in a vehicle. *International Archives of Photogrammetry and Remote Sensing*, 28(5/1), 16–23.

Welch, R., 1985. Cartographic potential of SPOT image data. *Photogrammetric Engineering and Remote Sensing*, 51(8), 1085–1091.

Wheatstone, C., 1838. Contribution to the physiology of vision. – Part the first. On some remarkable, and hitherto unobserved, phenomena of binocular vision. *Philosophical Transactions of the Royal Society of London for the year MDCCCXXXVIII*, Part II, 371–394.

Wong, K.W. and Ho, W.H., 1986. Close-range mapping with a solid state camera. *Photogrammetric Engineering and Remote Sensing*, 52(1), 67–74.

2 Fundamentals of photogrammetry

Harvey Mitchell

2.1 Basic concepts: measurement with a camera

To most people, a camera is a device for creating images, for personal, technical, artistic or any other reasons. But to photogrammetrists and other specialists involved in extracting three-dimensional information from images, a camera is a measuring instrument, able to be used to determine the shapes of objects. The coordinates obtained by photogrammetry can be remarkably accurate, especially if the cameras and other equipment are of appropriate quality, if good software is utilised, and if the survey is well designed.

Photogrammetry is the use of the measurements taken from two-dimensional images to obtain three-dimensional coordinates of points on those objects. The theory and practical procedures for achieving these outcomes by photogrammetry are outlined in this chapter. This introduction provides only the key concepts of photogrammetry. It has been written to be general and simple, with the aim of covering only what is necessary for non-photogrammetrists to comprehend the subsequent series of chapters, which might enable them to initiate photogrammetric measurement using available software. Readers will find further detail which can provide a full understanding of the theory and practice of photogrammetry, enabling them to make the most of photogrammetric measurement, by referring to a photogrammetry text or reference book, such as Wolf and Dewitt (2000), Mikhail *et al.* (2001) or McGlone (2004); for a detailed theoretical coverage, see also Cooper and Robson (1996).

The basic principles of measurement with a camera can be understood by visualising the camera simply as a positive lens with a film or other suitable sensor placed in the focal plane of that lens. When an image is obtained with such a camera, many objects might be seen on the image, but this explanation of photogrammetric concepts starts with a consideration of just one point which is imaged by the camera. From that point, there is one light ray which travels in a straight line, passing through the centre of the camera lens to finally reach the imaging plane of the camera. Although the image of any point will be formed by many light rays, which pass through the various regions of the lens to be concentrated onto a single point on the imaging plane, the ray which is used in photogrammetric deductions is the one particular ray (sometimes called the chief ray) which travels in a straight line from the object, through the centre of the lens to the image plane.

It is helpful to imagine that the chief ray actually travels in the reverse direction, starting at the imaging plane, travelling outwards through the lens into space, towards the object which we want to measure. This ray can then be envisaged as a line radiating for an unknown distance from the front of the camera, in a direction which is fixed relative to that camera, as in Figure 2.1.

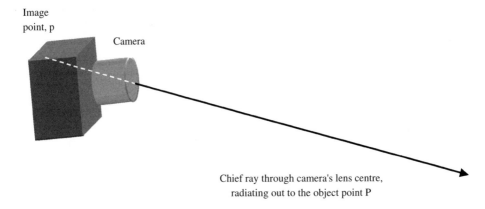

Image
point, p

Camera

Chief ray through camera's lens centre,
radiating out to the object point P

Figure 2.1 *Using the camera to define a direction in space.*

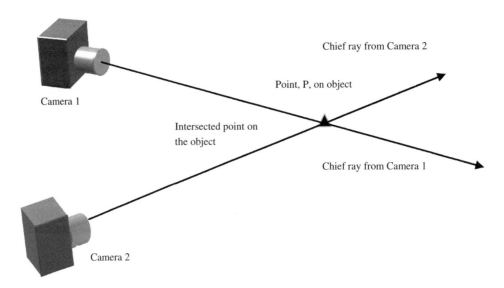

Chief ray from Camera 2

Camera 1

Point, P, on object

Intersected point on
the object

Chief ray from Camera 1

Camera 2

Figure 2.2 *Defining the position of an object point using intersecting chief rays.*

If the position and orientation of the camera are known, the spatial location of the chief ray can be established. Then, if two images of that same object have been obtained from two different positions, two chief rays to the one object point can be reconstructed. The point from which the chief rays originated can be located at the intersection of those rays, as shown in Figure 2.2.

This procedure can be replicated for other object points, as in Figure 2.3. In this way, it should be possible to deduce the spatial position of enough points to depict the complete shape of the object. Of course, the number of images need not be restricted to two, but it is easier to begin by thinking of only two images.

In the next section of this chapter, the procedure for mathematically reconstructing the ray paths to find their intersection point is explained; in later sections, practical aspects in the implementation of the procedure are introduced.

2.2 Mathematical concepts

The task of deducing the positions of the intersection of chief rays using the points found on the imagery is, of course, achieved mathematically. When using contemporary software, it is not crucial to understand the mathematics beyond the concepts explained in Section 2.1. Some readers may not want to read this section in detail. However, an understanding of the mathematical solution is advantageous in order to make the most of the options and to overcome problems.

2.2.1 Mathematical model

Various quantities are involved in a mathematical reconstruction of any photogrammetric situation. Some of the quantities will be regarded as *unknown* and obtaining their value is the goal of the mathematical solution. Those unknown quantities which are of greatest interest are generally the coordinates of the points on the object. Other quantities will be regarded as *known*, in the sense that they have known numerical values. Of these, it is necessary to distinguish between those that are rigidly fixed in the solution (the focal length of the camera being a typical case) and those that are regarded as *measurements*. The crucial distinguishing characteristic of measurements is that they are recognised as including a small amount of unavoidable error, and so may not be perfectly correct.

All quantities, known and unknown, are related in a set of equations which model the ray intersection problem which was described above, and which is depicted in Figure 2.3. These equations are introduced here using the terminology and notation which has been adopted by the international photogrammetric community.

A fundamental aspect of the mathematical modelling is the adoption of a number of different sets of coordinates: first, each image has its own image coordinate system. It is a three-dimensional Cartesian coordinate system, denoted (x,y,z), (and usually with subscripts which identify the image). The origin of this coordinate system is at the centre of the lens, the perspective centre of the image. The z axis is along the lens axis, the positive direction being from the lens towards the imaging plane. The point of intersection of the lens axis and the imaging plane is called the principal point of the image. (This point is normally very close to the centre of the film or digital sensor array, and many users simply assume that this is so). Although the x and y axes are in the plane of the lens, the x and y coordinates

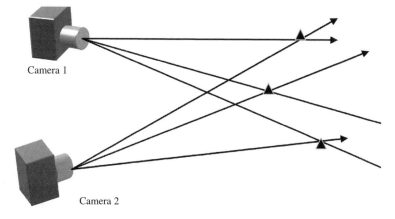

Camera 1

Camera 2

Figure 2.3   *Defining the position of a number of object points by using multiple chief rays.*

can in practice be quite simply measured in the plane of the image, from the principal point. Furthermore, the x and y coordinates should be measured on the "negative" image which is "upside-down" in the camera, but they can in practice be measured on a positive image, as in Figure 2.4. The figure shows that the x axis "across" the positive image, to the right, and the y axis "up" this image, assuming the camera and its image can be regarded as having an apparent upward direction. Because all image points lie on the image plane, they have the same z coordinate. Using this conventional coordinate system is convenient, but it is not crucial to the theory.

Secondly, an object coordinate system (sometimes alternatively called the ground coordinate system), denoted (X, Y, Z), is adopted for the object points. Often, this three-dimensional Cartesian system has its X and Y axes horizontal and the Z axis vertical. For fixed objects, it is usually based on an existing permanent state, national or global coordinate system. (Some ground coordinate systems refer to east, north and height coordinates, but these can easily be thought of as X, Y and Z). If the object is movable, the coordinate system may be arbitrarily positioned, and the origin and orientation may be selected for convenience; in that case, the (X, Y, Z) coordinate system need not even be oriented horizontally and vertically.

To enable the image and object coordinate systems to be related, the position of the lens (being the origin of the image coordinate system) of camera i is given coordinates (X_i, Y_i, Z_i) within the (X, Y, Z) system.

In addition, the image coordinate system will generally not be parallel to the object coordinate system, so allowance must be made for the orientation of the image coordinate system relative to

Figure 2.4 *Conventional image coordinate system in relation to an image.*

the object coordinate system. To do this, the angles of rotation of a camera's image coordinate (x_i, y_i, z_i) system relative to the (X, Y, Z) object coordinate system are denoted by three angles, ω_i, φ_i and κ_i. Specifically, the camera axes are normally assumed by the photogrammetric community to have rotated from a position in which the x, y and z axes are parallel to the X, Y and Z axes: firstly by an angle ω about the X axis, then by an angle φ about the Y axis, and finally by an angle κ about the Z axis. The signs of the three angles are given by conventional rules relating to right-handed Cartesian coordinate systems. Fortunately, this level of detail is irrelevant to most software users, being more relevant to software programmers, but Figure 2.5 shows the general concept of the angles of rotation.

It can now be seen that the path of a chief ray can be summarised as follows: the ray emanates from the image point p with measured image coordinates (x_p, y_p) on image i, travelling undeviated through the centre of the lens which has object coordinates (X_i, Y_i, Z_i), to meet the object at a point P which has coordinates (X_P, Y_P, Z_P). The relationship between these various elements—the coordinates of points on the images, the coordinates of the camera lenses, and the measured coordinates on the images—can be mathematically modelled using the equations of a straight line in space. As photogrammetry textbooks and reference books show (e.g. Wolf and Dewitt, 2000; Mikhail *et al.*, 2001; McGlone, 2004; Cooper and Robson, 1996), two equations can be written to represent the chief ray in terms of the various elements defined above:

$$x_p = c_i \frac{r_{11i}(X_P - X_i) + r_{12i}(Y_P - Y_i) + r_{13i}(Z_P - Z_i)}{r_{31i}(X_P - X_i) + r_{32i}(Y_P - Y_i) + r_{33i}(Z_P - Z_i)} \qquad (2.1a)$$

$$y_p = c_i \frac{r_{21i}(X_P - X_i) + r_{22i}(Y_P - Y_i) + r_{23i}(Z_P - Z_i)}{r_{31i}(X_P - X_i) + r_{32i}(Y_P - Y_i) + r_{33i}(Z_P - Z_i)} \qquad (2.1b)$$

where the quantities r are elements of a rotation matrix, and involve ω_i, φ_i and κ_i.

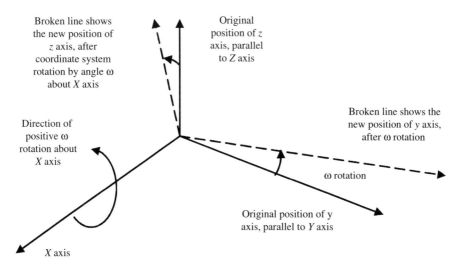

Figure 2.5 *Defining the angles between the coordinate systems. This figure illustrates that a rotation ω about the X axis rotates the y and z axes.*

Because all chief rays pass through the centre of the lens, the distance from the perspective centre to the image plane, along the z axis, is an important quantity. This distance is called the principal distance, denoted c, (or more specifically c_i for camera i), and it is involved in Equations (2.1). The principal distance will be similar to that quoted as the focal length of the lens, but focusing can cause the principal distance to differ by small amounts from the lens's focal length.

Equations (2.1) are called the collinearity equations, being derivable by assuming that the image point, lens centre and object point are co-linear, along the path of the chief ray. For each point on each image on which a point appears, there will be two collinearity equations of this form: one equation for x_p and one equation for y_p. Thus, if 15 object points appear on three images, there will be $15 \times 3 \times 2$ (or a total of 90) equations to be solved simultaneously to find the unknown quantities.

So far, it has been assumed in this explanation that $X_i, Y_i, Z_i, c_i, x_i, y_i, \omega_i, \varphi_i$ and κ_i, are known, but in practice the camera position X_i, Y_i, Z_i, and orientations $\omega_i, \varphi_i, \kappa_i$, are usually not known. To overcome this problem, "control points", whose (X, Y, Z) coordinates are known and whose image coordinates are also measured, are included in the solution (see Section 2.3.5).

2.2.2 Mathematical solutions

A determination of the unknowns can be obtained provided that enough quantities are known and/or measured to permit a valid mathematical representation of the intersecting rays. The geometrical meaning can be envisaged: a valid solution for a point of interest will be achieved if that point is intersected by at least two rays, which come from different cameras whose location and orientation can be determined owing to the control points.

The solution of Equations (2.1) to find the object coordinates and other unknowns is invariably calculated by software, whose processes are apparent to users. Solving Equations (2.1) generally requires that the equations be first transformed in some way. Most commonly, they are linearised to make them suitable for least squares estimation of the solution. This is usually achieved by Taylor's expansion, which requires that estimates of the final values of the unknown quantities be provided. Readers should be made aware of the creation of linearised equations by an alternative approach, the "direct linear transformation" (DLT) approach which was developed by Abdel-Aziz and Karara (1971). This solution is widely used and frequently referred to in publications because it enables a solution to be found without needing initial estimates of the unknowns. Some traits of the DLT solution have been remarked upon by Cooper and Robson (1996, p. 33). Understanding these transformations is not crucial to executing photogrammetric solutions by software users, who are generally oblivious to them; interested readers will find the processes described in photogrammetry textbooks such as those referred to above.

The most common goal of a photogrammetric survey is to find the coordinates of the object points in the (X, Y, Z) coordinate system. Therefore the most typical solution seeks values of X_P, Y_P, Z_P, for each point (P), using given values of $x_{Pi}, y_{Pi}, X_i, Y_i, Z_i, c_i, \omega_i, \varphi_i$, and κ_i. When there are more than two images, not all points will appear in all images, but provided that a point of interest is imaged from at least two positions, a solution can be obtained for its coordinates. When all the chief rays are adjusted in one big solution for a number of cameras or images, it is described as a "bundle adjustment", suggestive of the bundles of light rays from each camera in the solution.

Once the equations have been formed, the solution may proceed in two stages: first, there may be a "resection" stage, in which the positions and orientations of the cameras are determined

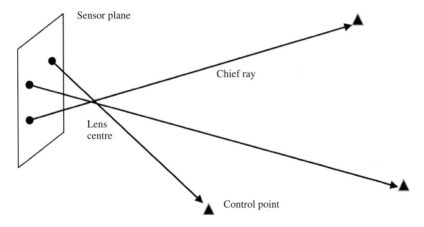

Figure 2.6 *The concept of resection. If the positions of the control points (shown as triangles) are known, the location and orientation of the camera sensor plane can be calculated.*

using some of the equations, especially those involving the control points; and, secondly, there may be an "intersection" stage in which the coordinates of points on the object are determined using the rest of the equations. The intersection process has already been explained: it involves finding the point in space where the rays from the image points pass through the camera lenses to meet on the object. The resection stage can be perceived as the reverse: the camera locations and orientations are determined using the rays from given points on the object (Figure 2.6). There must be three given, or control, points in order to perform a resection. In practice, the resection and intersection may be carried out simultaneously by software, but appreciating that there are two distinct mechanisms in the solution can be useful.

It should be noted that the value of the principal distance, c, may or may not be the same for all images used in the survey. The principal distance varies with focusing, so the principal distances may vary from one image to another, especially if a zoom lens is used. Moreover, if scenes are being measured from multiple simultaneous images, the use of different cameras may mean the introduction of different principal distances.

In most cases the number of rays involved in the solution is more than necessary, i.e. redundant observations are utilised in typical photogrammetric solutions. The existence of superfluous observations is advantageous, as the surplus data can contribute to an improved precision and accuracy for the calculated parameters. Redundancies can also enable certain parameters to be sought in the solution instead of being measured. Solving for an uncertain principal distance is a good example of utilising redundant information to advantage. However, solving for greater numbers of parameters reduces the number of redundancies which in turn weakens the precision, so unnecessarily solving for parameters in this way is not necessarily wise.

2.3 Practical implementation

2.3.1 Stages of a photogrammetric solution

Photogrammetry as a measuring technique offers a myriad of options to the user, and many different approaches may achieve the same desired result. The photogrammetric user can choose the options, after carrying out a cost/benefit analysis. Even so, the operation for acquiring results by photogrammetric measurement involves a number of generally standard stages:

(1) Photography: There are photographic considerations in all photogrammetry, as quite obviously the imagery must be of suitable clarity and must provide appropriate coverage of the object. It is generally true that the better the image quality (and the larger the object appears within that image), the easier it is to define the image points and hence the more precise the final results will be.

The issue of camera choice is discussed below in Section 2.5. At this stage, it is probably useful to be clear that special cameras designed for photogrammetric use are not necessarily required, and, in particular, they are not essential for many of the applications discussed in this book.

Selecting the camera locations can also be crucial to obtaining not only coverage of the object but also the required accuracy (see Section 2.4).

(2) Image formatting: It may be necessary to convert images into a format suited to the subsequent processing, whether from film to digital format or from one digital format to another to suit the chosen photogrammetric software.

(3) Image coordinate measurement: For all those images in which points of interest appear, the x and y image coordinates must be measured. There are so many ways of achieving this, some manual, some automatic, that the matter is discussed separately in Section 2.3.3. For many of the applications discussed in this book, the required (x,y) image coordinates would be determined using digital images in conjunction with software on a personal computer.

(4) Processing: These computations typically have two distinct stages even though they may not be apparent to the users of the software. The first stage, interior orientation, recognises that the image coordinates as measured in practice are not necessarily related to the true principal point. The measurements may be based on an arbitrary and temporary coordinate system, and must be converted to a system based on the principal point so that they comply with assumptions made in the derivation of the collinearity equations. Moreover, it has to be recognised that the origin of coordinate system as defined by the apparent centre of the film or the electronic sensor, does not necessarily correspond to the principal point defined by the lens axis. The definition of the principal point position relative to the centre of the sensor can be determined by calibration processes. Locating the position of the true principal point is sometimes assisted by *fiducial marks* whose positions relative to the principal point are determined in the calibration, and which are superimposed onto the exposed image (see Figure 2.7).

Other corrections may also be made to the raw coordinates at this stage: to account for lens distortion, non-orthogonality of the sensor axes, differential scales on the x and y axes, and so on, if the appropriate corrections have been determined, by camera calibration.

The second stage, the object point coordinate calculation, involves the intersection and resection processes referred to in Section 2.2.2, and culminates in the provision of the required coordinates of the object points. In photogrammetric textbooks, this solution stage will often be identified as "exterior orientation". The process is typically apparent to the users of most software, but, like interior orientation, it is crucial to the underlying theory.

An analysis of the results, including such tasks as checking for satisfactory point distribution, gross faults, and assessment of precisions, should be seen as an essential component of this stage. Proficient software will automatically estimate the precisions of the determined parameters at this time (see Section 2.4).

Figure 2.7   *Corner portion of an aerial photograph showing in the top left corner a fiducial mark, whose position relative to the true principal point is known.*

(5) Output: Finally, the results of the photogrammetric computation need to be transformed into a form of output suited to the user. The final photogrammetric product may be supplied to the client in many formats, ranging from a simple text-file listing of the coordinates of identified points, to dynamic three-dimensional computer visualisation fly-throughs, created by draping the imagery across movable perspective or orthographic views of surfaces. Paper or digital maps, in which perspective and other errors have been eliminated, and in which the third dimension may be shown via contours, can still be surprisingly useful means of representing topographic surfaces. A combination of display options can be informative (see Figure 2.8).

It is worth defining the meanings of some common terms related to the representation of a surface in terms of Z coordinates for a series of (X,Y) coordinates. A *digital terrain model* (DTM) is characterised by the fact that points depicting the terrain are on an evenly spaced grid. Such a model can be depicted using perspective views, as in Figure 2.9. The expression *digital elevation model* (DEM) is also used. The term *digital surface model* (DSM) is sometimes used to define those DTMs which show only the natural surface of the earth, excluding buildings and other human-made features, and also excluding vegetation. These definitions are not universal, and not all groups of users may mean the same thing as given here by DEM, DTM and DSM. In a *triangulated irregular network* (TIN) the points are not evenly spaced. Such a network cannot be viewed using a perspective view of the sort shown in Figure 2.9, but of course it can be shown with the assistance of shading and other graphical visual assistance techniques.

Rectification is a procedure which was carried out with aerial photographs to remove the effects of the aeroplane's tilts at the time of photography. It had the effect of creating a photograph that had various errors removed. Some positional errors arising from the perspective effect of the topography remained. Today, the same sort of procedure is carried out digitally on aerial and space imagery, and, moreover, other positional errors (including the relief effect mentioned above) can be removed digitally so that the final image is correct like a map product would be, but it looks like an image. The product of this procedure, sometimes known as *digital rectification*, is an *orthoimage*.

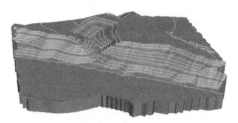

Figure 2.8 *Expressive three-dimensional surface representation by a combination of contours, perspective and shading (most noticeable on cliff face). Reproduced with permission of Dr S. Buckley and Dr J. Mills, University of Newcastle upon Tyne, UK.*

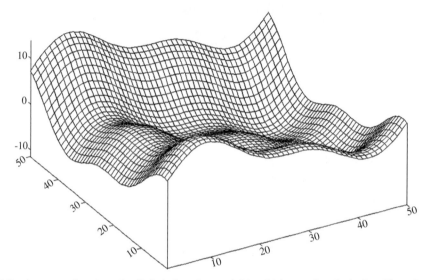

Figure 2.9 *A perspective view of a digital elevation model in which a surface is depicted by points on an evenly spaced grid. Figure created using Surfer software version 7.0 (Golden Software, Golden, Colorado, USA).*

2.3.2 Software

Photogrammetric software which operates on an ordinary desktop computer is now widely available to help someone carry out all the photogrammetric stages described in Section 2.3.1, i.e. to:

- select suitable image points or assist in their selection;
- measure the image coordinates;
- carry out the interior orientation;
- carry out the exterior orientation;
- calculate the object point coordinates; and,
- create the desired output for the client.

Prices vary, but it is certainly fair to say that much of it is affordable. Some of it is certainly cheap. It is probably true that the availability of inexpensive but competent photogrammetric software on the market has made photogrammetry accessible to non-photogrammetrists over the last decade or two. Examples, but not an exhaustive list, include the Australis (Photometrix,

2005), Leica Photogrammetry Suite (Leica Geosystems, 2005), iWitness (Photometrix, 2005), PhotoModeler (PhotoModeler, 2005), Rolleimetric (Photarc, 2005), ShapeCapture (ShapeCapture, 2005), 3DM Analyst (ADAM Technology, 2005) and VirtuoZo (Supresoft, 2005) software packages. These packages differ in their capabilities and prices. Some require manual operations to detect corresponding points, but others are capable of image matching by software and are thus able to carry out automated contouring across surfaces. The most complex and highly capable software (not listed here) may be operated on a dedicated computer, in which case the system may be regarded as a "digital photogrammetric workstation".

Some of the functions enumerated in Section 2.3.1 are considered in greater detail below, and it is intended to highlight the role of the different software packages in these stages.

2.3.3 Point selection

It will be recognised that a crucial element of the photogrammetric procedure is the selection of appropriate object points on the image, whether digital or film. The tasks of point selection and point measurement can be strongly interrelated, but it is worth recognising during planning that there is normally a choice in the type of point that is measured. The points to be coordinated by this photogrammetric procedure may be chosen in various ways:

- The points may be selected (especially by the software, but perhaps by an operator) at an approximately regular spacing.
- Points may be chosen on the basis of the likelihood of recognising the same points on different images, and/or the likelihood of being able to measure their coordinates accurately; this can depend on the location of good visual detail or rich image texture.
- Distinct points of interest may be chosen manually on the object (such as the corners of a house window).
- Sometimes, the points may be marked with an identifiable target prior to imaging (see Figure 2.10).

Software generally divides itself into two clearly distinct categories, depending whether the software is intended for either use with targets, or the measurement of unmarked points. These two different cases tend to be applied in contrasting circumstances, as suggested by Table 2.1.

What is important is that the chosen points provide adequate coverage across the surface to depict its shape to the required quality.

2.3.4 Image point measurement

The measurement of the required (x,y) image coordinates of chosen points of interest, whether control points, targeted points or nondescript surface points, may be accomplished manually (perhaps aided by human stereoscopic vision) or automatically; it may be done utilising one, two or more images at the same time. Indeed, a crucial attribute of any software is the level of automation offered to users to assist in the measurement of the image coordinates. Whichever case applies, it is likely that, for many of the applications discussed in this book, the measurement would be at least assisted by software, and the coordinates will be provided in pixels, rather than distances as they are with film images. Various types of assistance can be identified.

The measurement of manually selected points may be aided by digital image enhancement. The software may or may not utilise a stereoscopic procedure, if optical instrumentation permits, to assist the manual location of corresponding points. Some software can automatically locate the same point on multiple images, provided that the image texture in the vicinity of that

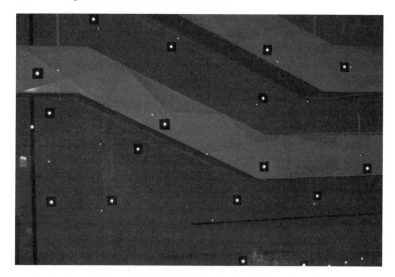

Figure 2.10 *Retro-reflective targeted points used in photogrammetric measurement.*

Table 2.1 Differences between measuring surfaces and points in photogrammetry

Aspect	Surface shapes	Points (usually marked targets)
Application examples in close range measurement	Manufactured shapes: glass, metal sheets Medical (e.g. backs, faces) Topography	Industrial quality control in large built items, e.g. deformation studies Architecture
Output type	Graphical views, orthoimages, rendered views, perhaps contours	Coordinate lists
Measurement in digital photogrammetry	Surface matching of points chosen by the software, often on a grid covering the area	Targets can often be detected by software, and even uniquely identified by software if suitable targets are used
Number of points	Many (any part of the surface with suitable texture can be selected by the software as a suitable point)	Limited number (targets placed manually)
Precision	Lower (1:1,000–1:20,000)	Higher (1:10,000–1:200,000)
Note:	Not all surfaces are suited to photography, or may not be suited to point identification; texture may be added by projected light or temporary painting or marking, especially on small objects	Targets may have to be placed, which can be laborious, especially if access is difficult. Strictly, the technique is then no longer non-contact

point is sufficient, as suggested in Section 2.3.3. The outcome is that, by doing this repeatedly for many points, the software can automatically measure almost continuously across a surface. Software limitations are caused by two camera axes which are highly convergent, as this may cause the various images of the same region not to appear similar, which can disrupt fully automated surface mapping. The techniques used for automated matching are not relevant, but are named here because they can appear in case studies and software descriptions: area-based matching involves a simple correlation of image texture, while feature-based matching involves some analysis of the imagery to find patterns.

Target measurement may occur at a number of levels:

- Targets may be identified and measured manually, using a cursor on the computer screen.
- Highly retro-reflective targets, such as those seen in Figure 2.10, are often recognised automatically by software. An interesting specialised case is the use of highly reflective targets, imaged by the use of a flash and a small camera aperture, regardless of the light conditions. This enables the required targets to be visible on the image, to aid automatic detection and measurement (see Figure 2.11).
- The capacity for software to recognise distinctive targets having a defined and suitable pattern (e.g. by correlating an area of image with a "template" image of the target) is available; (e.g. the Australis software (Photometrix, 2005)).
- Some software will recognise a number of different targets, making it possible to introduce automated identification (see Figure 2.11). (A few specialists have even built the target recognition into the camera, to enable very fast processing (Fraser, 1997; Ganci and Handley, 1998).

Automated processing of digital imagery can be very fast (often over ten thousand points can be identified and measured per minute, compared to less than ten points per minute when selected

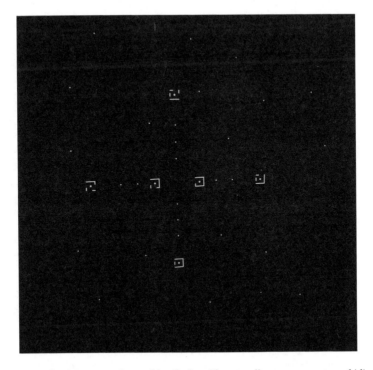

Figure 2.11 *Highly reflective targets imaged by flash, with a small camera aperture, hiding background objects having normal reflection properties. Six of the targets are coded to enable them to be automatically identified by suitable software. Reproduced with permission of Dr S. Robson, University College London, UK.*

manually). It can also be accurate, as point measurement on digital images is normally to sub-pixel accuracy. Note that all automated results warrant checking for gross errors, by human examination of the imagery or at least by human perusal of the final results.

2.3.5 Control

Control points are simply points which are on or in the vicinity of the object, whose coordinates (some or all three of them) in the object coordinate system have been determined by a surveying method separate from the photogrammetric survey. The image coordinates of the control points are included in the solution along with any other image coordinates, but in the solution their (XYZ) coordinates will be fixed rather than treated as unknown. Control points are essential if the results are to be tied to a specified ground coordinate system. This is a typical requirement in topographic mapping, when landscape is to be mapped in a national coordinate system for engineering and scientific purposes.

Control points may be specially marked with targets so that they are easily identified or they may be points selected after the imagery has been examined.

Alternative forms of control, notably fixed distances between two identifiable points provided by a bar of known length for example, as in Figure 2.12, may also be used as control, but users should be aware that not all software will handle this relatively uncommon form of control.

Choosing control can create a difficulty for the inexperienced user of photogrammetry. Specifically, selecting the appropriate number, distribution and accuracy of control points which will provide the required accuracy of the object points (see Section 2.4) depends on theory which is well beyond the scope of this introduction. Moreover, such an accuracy prediction

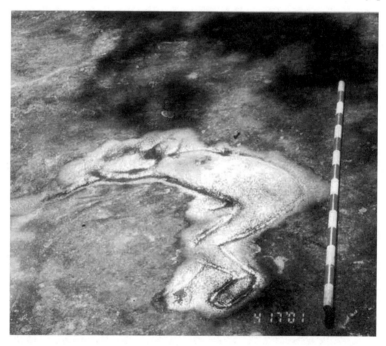

Figure 2.12 *Photogrammetric control in the form of a rod with known length intervals. The location and orientation of the rod is not known, so it does not enable the object to be tied to a ground coordinate system, but it can scale the object. Photograph of rock engraving, reproduced with permission of Professor J. Fryer, University of Newcastle, New South Wales, Australia.*

capacity is beyond the capability of many programs; software capable of predicting network accuracy can be regarded as superior. However, a good general rule is that control points should surround the object of interest, so avoiding extrapolation beyond the control field.

Control points are a crucial element of photogrammetric recording, but establishing their ground coordinates can create novel problems for some users who are not familiar with ground surveying techniques.

Sometimes, superfluous control may be added to deliberately strengthen the mathematics of a solution, especially one which is made weak by a large number of unknowns in the solution: the greater the number of redundancies, the higher the precision of the final results.

2.3.6 Stereoscopic vision

Much historical photogrammetry is built around the measurement of image points by a human operator. The process was aided in many cases by optical instruments which made use of human stereoscopic vision acuity to identify these points. The role of stereoscopic vision deserves some comment, because it is often seen as a central element of photogrammetry. While historically stereoscopic vision was essential for photogrammetric processing, its role now is often not crucial. It lies only in providing assistance, such as during the identification of common points on photographs, provided that they come in pairs and provided that the configuration of the two photographs is such that it tends to mimic normal human vision. This situation is commonly found with aerial photography, where the photographs are parallel and have suitable overlap, and the object is relatively plane and distant. Indeed, the process of creating maps and other data extracted from aerial photography is still labour-intensive because of the need to ensure that results are of high accuracy in an era when fully digital processing is not yet flawless. In other cases, however, stereoscopic vision often need not be utilised because software can carry out the required recognition of conjugal points on images. The chosen image geometry may make stereoscopic vision impossible anyway.

Stereoscopic viewing using simple viewers (Figure 2.13) remains a useful aid to studying photographs which have been taken with almost parallel camera axes and a significant fraction of overlap to provide coverage of common objects.

Figure 2.13 *Table-top stereoscopic viewer.*

2.3.7 Single image solutions

Measurement from a single image is also possible, although, as may be obvious, three-dimensional solutions by ray intersection are not possible. However, if assumptions can be made about objects in the imagery, such as background terrain being level, or the walls and floors of the building being mutually perpendicular, three-dimensional results can nevertheless be produced (see Figure 2.14). While these computations are not necessarily feasible with all photogrammetric software, some software has been specifically written for situations involving single images where additional intelligence provided by perpendicular walls and floors, can be exploited. For example, the ShapeCapture software (ShapeCapture, 2005) utilises this sort of information so that by using a high level of operator input three-dimensional results can be obtained from a single two-dimensional image.

2.4 Data quality considerations

2.4.1 Error sources

In reality, the photogrammetric process is subject to many sources of error, and it is crucial to recognise that computed results cannot be perfect. The errors can be considered in three groups:

(1) Measurement errors, which are randomly distributed, thus having equal likelihood of a positive or negative sign; they typically affect the image coordinates, and they may affect the coordinates of control points.
(2) Systematically occurring errors which are caused by imperfections in the adopted value of a quantity which appears fixed in the calculations (such as the principal distance)

Figure 2.14 *An example of single photograph geometry from a security camera during a robbery, showing a case with numerous vertical and horizontal planes and parallel lines, and many points between which the distances can be measured easily. Faces have been deliberately blurred in the interest of privacy. Reproduced with permission of Professor J. Fryer, University of Newcastle, New South Wales, Australia.*

or errors, omissions or inadequacies in the assumed geometry which is the basis of the collinearity equations, such as:

- errors in the pixel geometry of the digital sensor;
- a lack of flatness in the plane of the digital sensors or film;
- film distortion during wind-on or processing;
- errors in the assumed position of the principal point;
- lens distortion;
- non-perpendicularity between the sensor plane and the lens axis; and
- refraction (especially over longer distances).

(3) Genuine mistakes, "blunders" or "gross errors" in the measurement and processing, such as:

- wrongly identified and measured control points;
- falsely matched points during automated point measurement; and
- using inappropriate units, for example: entering the focal length in millimetres when metres are required.

Systematic errors in the geometry are ignored in the conventional and simplest forms of the collinearity equations, as given by Equation 2.1, but they can be incorporated into these equations if desired. Systematic errors can also be limited by calibrating equipment (determining the principal distance, and determining the differential scale and the angle between the rows and columns in the sensor array), and inserting the appropriate corrected values into the equations. Sometimes, in the mathematical bundle solution, parameters describing the errors may be sought as unknown quantities, termed *additional parameters*.

2.4.2 Measures of quality

Because results achieved in any photogrammetric process cannot be perfect, assessing the quality of results is an important undertaking following any photogrammetric solution. Quality in relation to errors may be specified by precision and/or accuracy, and, less easily, by reliability. These terms may have different definitions in different scientific communities; here *precision* describes the variability of a quantity, without regard to how close it is to the correct value, and is best indicated by a standard deviation for normally distributed errors. *Accuracy* can be defined as the difference between the determined quantity and the true value, and is generally the desired statistical indicator of the quality of results. However, accuracy can only be assessed by comparing computed values with "known" ones. If such test or "check" points with known coordinates are available, then the achieved level of accuracy can be assessed. It is usually indicated by a root mean square (rms) error, which is derived from the square root of the mean of the squared differences between the determined and true coordinates at the check points.

Precision is derived most commonly in the solution during the least squares estimation of parameters in the collinearity equations. The precision of the required quantities depends on such factors as:

- the precision of the measurements of the (x,y) image coordinates on the image;
- the geometry of the intersecting rays, and specifically the angles between them; and
- the precision of control point coordinates.

Estimates of precision are important as a substitute for accuracy, because accuracy is usually not available unless there are test points, whereas precision has the advantage that it can be calculated without the availability of test points. However, it is an indicator of accuracy only in the absence of systematic errors. Because some systematic errors must be expected, a precision value may be an optimistic indicator of accuracy.

The capacity to critically assess the overall quality of the results is an important feature of photogrammetry software. Proficient software packages provide the ideal statistical output: a set of precision estimates for each of the object space coordinates (i.e. *X, Y* and *Z* separately) for each of the points sought in the photogrammetric survey, and even their covariances. Some packages provide just a single standard deviation value as a precision for the entire network. Other software packages provide only rudimentary indicators of precision, but the various types of indicators which they use are beyond the scope of this introductory discussion.

Reliability relates to the ease with which gross errors, blunders or mistakes can be detected. Means of estimating and improving reliability are relatively complex and are not treated further here.

The topic of error statistics in photogrammetry is almost limitless. The interested reader will find that the subject has been comprehensively reviewed by Cooper and Cross (1988).

2.4.3 System design and testing

It is obvious by now that a consequence of the existence of error sources is that each photogrammetric project has a design challenge: to select the equipment, imaging configuration and processing which will achieve the required accuracy, presumably at minimum cost.

Some basic planning is needed to ensure that all object points are imaged from the desired number of cameras, and from at least two. However, it is also necessary to ensure that the chosen configuration leads to the desired accuracies. Planning the optimum camera configuration prior to a survey is a most difficult aspect; a few software packages offer mathematical design assistance. Sometimes, experience can allow photogrammetric users to have confidence that accuracies will be achieved or exceeded. Otherwise, design is aided by adopting standard or recommended procedures, and users should realise that many photogrammetric surveys are successfully executed, and are not prohibited by a need for design. In the simplest case, users can resort to a rule-of-thumb: because precision reflects the geometry of the chief ray intersections and the measurement precision of the image coordinates, the object coordinate precision is indicated by the image precision multiplied by the scale number of the imagery, so that a 0.1 mm image measurement precision on a photograph at 1:100 scale could translate into about 10 mm precision on the object. And, in general, the more images which are obtained, and hence the more measurements which are made, the more precise and reliable will be the result. The point is simply that errors cannot be ignored, and in the following chapters, sources of error and design considerations are highlighted when they are relevant to the given applications. A detailed study of network design has been provided by Fraser (1996).

It is necessary to recognise the benefits of a testing stage with new photogrammetric arrangements, primarily to test whether the accuracies of the results meet the requirements. Tests also serve to confirm other aspects of the solution. Accuracy confirmation can be done by using objects of known dimensions.

2.5 Sensors

2.5.1 "Non-metric" cameras

For most cases discussed within the realm of this book, "non-metric" cameras, i.e. cameras not specifically constructed by the maker for measurement purposes will be used, and it can be expected that the cameras will be digital. Certainly, computer-based processing of digitally collected images could be anticipated. Digital cameras provide faster turnaround, and the costs of digital cameras and small format scanners are not high. Digital cameras can permit the examination of images before leaving the site, unlike film (in some applications cameras are linked directly to computers to provide "real-time" results). The disadvantages of lower accuracy due to the low spatial resolution of earlier digital cameras are being overcome. While the accuracies achievable with digital cameras using simple techniques may be as low as 1:10,000, this may be adequate in many cases. Moreover, some measurement with digital cameras can achieve accuracies of around 1:1,000,000, especially if targets are used.

To quantify those items that can be expected to remain unchanged, such as focal length, lens distortion characteristics, principal point location, and film or sensor distortion (Section 2.5.3), camera calibration can be used. It can even compensate for lower optical and sensor qualities. Calibration is normally carried out by photographing objects with known characteristics, perhaps sets of parallel lines or more complex pre-surveyed targets. As explained in Section 2.2, it is possible to evaluate some fixed camera parameters during the photogrammetric solution, if there are redundant observations. Methods for achieving this include the use of test-fields of control points, or self-calibration during bundle adjustments within any photogrammetric project. Full accounts of calibration techniques have been given by Fryer (1996) and Fryer and Brown (1986).

Although the use of a cheaper camera may be compensated for by a careful calibration, the requirement for calibration may be seen as unnecessarily onerous and avoidable, so the approach of using a more expensive and higher quality camera may be preferable.

Scanning of film is still used as a means of creating digital imagery. Users should be aware that the scanning can, of course, introduce some additional geometric errors into the imagery.

2.5.2 Photogrammetric cameras

There is on the market a small number of cameras made specifically for photogrammetric measurement purposes. These "metric" cameras usually have a fixed focus (at or near infinity), so that the principal distance will not change, and can be determined exactly by calibration. Of course, metric cameras boast high quality components, notably low distortion lenses and, in the past, the use of glass plates to carry the photographically sensitive emulsion. Demand for them has diminished and is restricted to certain specialised high accuracy applications (e.g. Geodetic Systems Incorporated, 2005). For lower accuracy applications, the historical advantage of high quality lenses and other characteristics can now often be overcome by the careful calibration of the characteristics of an inferior lens.

2.5.3 Characteristics of digital imagery

A detailed discussion of digital imagery has been provided by Dowman (1996) and Shortis and Beyer (1996), but a few fundamental concepts are reproduced here, if only to help users understand the jargon.

The most important aspects of the sensor are the geometric errors in a digital sensor which can affect photogrammetric accuracy directly:

- the sensor plane may not be perfectly flat;
- the rows and columns of pixels may not be perpendicular to each other; or
- the sensor plane may not be perpendicular to the lens axis.

Users should also note that it cannot be assumed that the individual pixels have the same height as width. This is not a fault, but it can contribute to error in the results if it is not recognised in calculations via the collinearity equations.

Other characteristics of digital imagery affect only the recorded intensity. However, the intensity values can affect the interpretation of the image. Digital sensors are subject to random *noise*, which can be attributed to electronic limitations. The level of the noise in a sensor is often compared with the range of intensities that the sensor can detect, in a *signal-to-noise ratio* (SNR). The SNR indicates the maximum contrast that can be achieved. The SNR should be distinguished from the *dynamic range* of a digital sensor, which is effectively the range of discernable intensity values that the sensor can record, as a multiple of the level of background noise which occurs even when the sensor is exposed to no light.

2.6 Aerial photography

Mapping from aerial photography deserves individual attention because it constitutes photogrammetry's most widespread and most commercially active application. It is a highly specialised field; aerial survey cameras (see Figure 2.15) are of extremely high quality, and are correspondingly expensive, with values of the order of a million US dollars (2005).

Aerial film has a resolution which can only be described as remarkable (see Figures 2.16 and 2.17).

Digital aerial cameras are only now beginning to match film resolution but can provide a superior dynamic range.

Figure 2.15 *Highly precise Leica RC30 aerial camera. Reproduced with permission of Leica Geosystems.*

Figure 2.16   *Sample of aerial photography, taken from an altitude of 1200 m with a camera having a focal length of 300 mm. Photography reproduced with permission of AAMHatch, Melbourne, Australia.*

Figure 2.17 *Extracted portion of the aerial photograph shown in Figure 2.16, when scanned so that each pixel represents 60 mm on the ground (or 0.015 mm on the original photograph). The high resolution is readily apparent. (The building in this figure appears just to the right of centre of Figure 2.16). Photography reproduced with permission of AAMHatch, Melbourne, Australia.*

Near-vertical aerial photographs for mapping are normally taken in a series of consecutive overlapping exposures (the overlap suiting human stereoscopic vision), providing a geometry which simplifies processing and aids consistent mapping accuracy. Historically, maps were made using large and highly specialised optical-mechanical instruments designed for stereoscopic viewing by human operators, which effectively converted image coordinates to ground coordinates in real time but using an analogue procedure. These instruments were later computer controlled. Now computer-based software is capable of being used for all operations, including the selection of image points, substituting for the human operator's stereoscopic vision. However, this latter aspect is not yet reliable in all circumstances, so that mapping equipment which employs human stereoscopic observation for various levels of assistance in the plotting operations remains in widespread use in map production (see Figure 2.18). Human checking of computed results remains a crucial operation in aerial mapping. In terms of its wide commercial usage, the expense of the equipment used, and its continued use of human vision, aerial mapping can be contrasted with many other applications which are encountered in this book.

Satellite imagery is occasionally used for photogrammetric mapping, especially as higher resolution imagery has become available in recent years, and because spacecraft provide recurring coverage. However, the resolution of the best satellite imagery, at about 1 m × 1 m per pixel, is not yet comparable with aerial photography which can be obtained at low altitudes (see

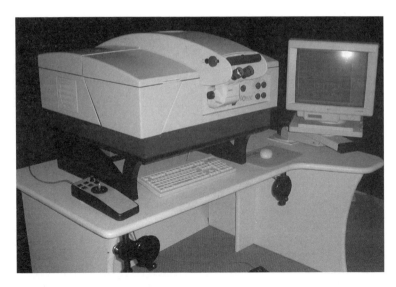

Figure 2.18 *ADAM Technology PROMAP System, a precision analytical stereoplotter, for use with an operator able to observing pairs of large format (240 mm × 240 mm) aerial photographs in stereoscopic mode. Reproduced with permission of ADAM Technology.*

Figure 2.17), and is not necessarily cheaper to purchase. Users of satellite imagery must also recognise that satellite scanners do not provide the perspective geometry of optical cameras, and consequently the software needs to be suitable for processing this type of imagery.

2.7 Alternatives to photogrammetry for three-dimensional measurement

There are some non-photogrammetric options which are available for three-dimensional surface shape measurement.

Laser scanners using distance measurement

There are two types of laser scanner. The first measures a distance and direction to unmarked and unidentified points on the object. The technology is such that point density is usually high, and measurement is very fast, 5000–100,000 points being measured per second, as the laser spot ranges across the surface to gather what is often termed a "point cloud" of data. Range capabilities vary with differing instrumentation, from a metre to hundreds of metres, with accuracies being commensurate: from about 0.1 mm to perhaps 25 mm (see Figure 2.19 and Chapter 10 for more details).

Aerial laser scanners also use this principle. The laser beam which emanates from a scanner in an aircraft is reflected from the ground, and the time difference between the emission and reflection indicates the distance. The direction of the laser is also recorded. This information is combined with accurate information about the position and orientation of the aircraft during flight, and the three-dimensional coordinates of points across the area of ground being scanned may be calculated. GPS and inertial navigation systems are normally used to position the aircraft. Typical characteristics of the scanning are up to 20 scan lines every second, and up to 250 measurements on each scan line. For a flying height of 1 km, the scan width may be about 500 m, with a vertical accuracy of each individual laser measurement less than 150 mm.

Figure 2.19 *I-Site brand laser scanner based on distance measurement. Reproduced with permission of Professor J. Fryer, University of Newcastle, New South Wales, Australia.*

Laser scanners using triangulation

The second type of laser scanner does not measure laser distances directly but uses a camera and projected light, separated at a fixed base distance, in what is known as a triangulation technique (see Figure 2.20). The image from the camera is used to determine the position of a projected spot, line or perhaps multiple lines. Because of the need to detect the projected light, this scanner is usually restricted to measurement indoors and/or at ranges of only a few metres. Moreover, as the range increases, a fixed short base distance introduces greater uncertainty in point precision. However, at close range its accuracy matches that of the distance scanner. Small triangulating scanner instruments in which the laser source and sensor are on a movable wand are also available, e.g. fastSCAN by Polhemus (Polhemus, 2005).

It will be realised that the concepts of ray intersection used in laser triangulation are really identical to those used in photogrammetry, as described in Section 2.1, and the distinction in theory between laser triangulation and photogrammetry is slight. Moreover, if two cameras are used to observe the projected line or spot, then the technique is even more similar to photogrammetry. However, the distinction is probably unimportant, and is related more to the theory than the execution. What is important here is that the photogrammetry techniques of this book are directed at users of cameras who collect images of the objects they wish to measure, and who are likely to analyse the images themselves, with the assistance of computer software. Triangulating laser scanners could thus be regarded as a narrow and specific category of photogrammetric measurement.

Mechanical probes

Various types of mechanical devices, based on the determination of the position of a probe, are in existence on the commercial measuring market and in research laboratories for measuring small objects. These vary from coordinate measuring machines which move automatically across a fixed object placed under the probe to others which can be guided manually across the surface of an object, being usually suitable for objects which are more complicated (see Figure 2.21).

Figure 2.20 *Vivid 910 triangulating laser scanner. Reproduced with permission of Konica Minolta.*

Figure 2.21 *Coordinate measuring machine: Roland Picza PIX-30 motorised profiler measuring a steel plate.*

Figure 2.22 *Use of surveying equipment for close range measurement of rock engravings which could plausibly be executed by photogrammetry. Reproduced with permission of D. Lambert, New South Wales National Parks and Wildlife Service, Australia.*

Surveying instruments

Fairly obviously, conventional surveying equipment, which is relatively low cost and commonly available, can be used to comprehensively measure objects which are also suited to measurement by photogrammetry. Deserving particular mention is the "total station" instrument, an automatic distance measuring electronic theodolite, some versions of which are capable of programmed self-driving to selected locations. Programmable models which measure distances from a few metres to hundreds of metres without a special reflector and to millimetre accuracy are capable of performing automated measurement across the sorts of objects which suit photogrammetric measurement, such as the rock engravings shown in Figure 2.22, and the choice between measurement by surveying equipment and cameras may need to be based on a careful comparison. The world of surveying is clearly too large to be given further attention here.

2.8 Gaining experience

This chapter is not intended to promote photogrammetry so much as to introduce it. Photogrammetry will be seen to be simple in geometrical concept, even if the mathematics programmed into software is quite intricate. However, many users of modern photogrammetric software will find that the mathematical detail is irrelevant. Numerous hardware and software options are commercially available to enable a user to reach measurement goals. An important lesson to be learnt from this chapter is the need to plan photogrammetric surveys and to optimise resource usage while obtaining required surface coverage and measurement accuracies. An understanding of the geometric concepts can assist in this crucial stage.

So how does an interested user get started, if, as the following chapters suggest, measurement from imagery is the tool which can produce the three-dimensional information that is required? The purchase of the camera is probably not a significant challenge. The main consideration is probably to purchase the most appropriate software required for the style of task, and to gain experience with it. Once the user has the images, it should be possible to simply try it.

2.9 References

Abdel-Aziz, Y.I. and Karara, H.M., 1971. Direct linear transformation from computer coordinates into object space coordinates. In *ASP Symposium on Close Range Photogrammetry*, American Society of Photogrammetry, Bethesda, Maryland, USA, 1–18.

ADAM Technology, 2005. See http://www.adamtech.com.au/ (accessed 23 August 2005).

Cooper, M.A.R. and Cross, P.A., 1988. Statistical concepts and their application in photogrammetry and surveying. *Photogrammetric Record*, 12(71), 637–663.

Cooper, M.A.R. and Robson, S., 1996. Theory of close range photogrammetry. In Atkinson, K.B. (Ed.), *Close Range Photogrammetry and Machine Vision*, Whittles Publishing, Scotland, 9–51.

Dowman, I.J., 1996. Fundamentals of digital photogrammetry. In Atkinson, K.B. (Ed.), *Close Range Photogrammetry and Machine Vision*, Whittles Publishing, Scotland, 52–77.

Fraser, C.S., 1996. Network design. In Atkinson, K.B. (Ed.), *Close Range Photogrammetry and Machine Vision*, Whittles Publishing, Scotland, 256–281.

Fraser, C.S., 1997. Innovations in automation for vision metrology systems. *Photogrammetric Record*, 15(90), 901–911.

Fryer, J.G.,1996. Camera calibration. In Atkinson, K.B. (Ed.), *Close Range Photogrammetry and Machine Vision*, Whittles Publishing, Scotland, 156–180.

Fryer, J.G. and Brown, D.C., 1986. Lens distortion in close range photogrammetry. *Photogrammetric Engineering and Remote Sensing*, 52(2), 51–58.

Ganci, G. and Handley, H.B., 1998. Automation in videogrammetry. *International Archives of Photogrammetry and Remote Sensing*, 32(5), 53–58.

Geodetic Systems Incorporated, 2005. See http://www.geodetic.com/ (accessed 23 August 2005).

Konica Minolta, 2005. See http://kmpi.konicaminolta.us/eprise/main/kmpi/content/ISD/ISD_Category_Pages/3dscanners (accessed 2 November 2005).

Leica Geosystems, 2005. See http://www.gis.leica-geosystems.com/products/ (accessed 2 November 2005).

McGlone, J.C., 2004. *Manual of Photogrammetry*, 5th edn., American Society of Photogrammetry, Falls Church, Virginia, USA.

Mikhail, E.M., Bethel, J.S. and McGlone, J.C., 2001. *Introduction to Modern Photogrammetry,* Wiley, New York, USA.

Photarc, 2005. See http://www.photarc.co.uk/rollei.htm (accessed 2 November 2005).

Photometrix, 2005. See http://www.photometrix.com.au (accessed 23 August 2005).

PhotoModeler, 2005. See http://www.photomodeler.com (accessed 2 November 2005).

Polhemus, 2005. See http://www.polhemus.com/fastscan.htm (accessed 1 April 2005).

ShapeCapture, 2005. See http://www.shapecapture.com/shapecape_2002.htm (accessed 23 August 2005).

Shortis, M.R. and Beyer, H.A., 1996. Sensor technology for digital photogrammetry and machine vision. In Atkinson, K.B. (Ed.), *Close Range Photogrammetry and Machine Vision*, Whittles Publishing, Scotland, 106–155.

Supresoft, 2005. See http://www.supresoft.com (accessed 1 November 2005).

Wolf, P.R. and Dewitt, B.A., 2000. *Elements of Photogrammetry: with Applications in GIS*, 3rd edn., McGraw-Hill, Boston, MA, USA.

3 Structural monitoring

Clive Fraser

3.1 Introduction

The application of photogrammetric measurement techniques to structural monitoring has increased markedly since the development of automated digital close range photogrammetry systems in the mid-1990s. As a deformation measurement tool, photogrammetry displays a number of beneficial attributes, including: the non-contact nature of the technique; the speed of the image recording phase; the nowadays easy access to suitable digital cameras; the availability of highly automated easy-to-use data processing systems; and, the fact that the technology is applicable to objects of virtually any size, and the flexibility of the technique in the important area of measurement accuracy.

The aim of this chapter is to highlight the accuracy, reliability, flexibility and high productivity of digital close range photogrammetry as a tool for monitoring structural deformation. The term vision metrology is also often used to describe the technology when it is applied to higher accuracy 3D measurement tasks. Selected case studies will be examined, with each demonstrating one or more of the significant practical merits of the photogrammetric approach. There are a number of operational modes for digital photogrammetric systems, ranging from on-line, real-time target tracking systems, such as those used in human motion capture applications, to off-line networks involving tens of images and a thousand or more object points of interest. The case studies to be considered, which were all structural monitoring projects carried out in recent years, cover a wide range of measurement designs and configurations and it is therefore important to first present a brief review of photogrammetric project design principles as they relate to measuring deformation.

3.2 Network design considerations

Network design is an important aspect of structural monitoring to ensure that the required high accuracies will be achieved. A full outline has been given by Fraser (1996).

Figure 3.1 shows a hypothetical case of the monitoring of structural deformation of an object of interest via photogrammetric means. In this case the object was first photographed in its "unloaded" state using a network of six overlapping images. This measurement produced (X,Y,Z) coordinates for the 14 surface monitoring points indicated by solid circular dots on the object, as well as the four points not subject to displacement, which are shown mounted on two tripods. A load was applied to the object causing it to deform to a new steady state indicated by the dashed outline in the figure. The (X,Y,Z) coordinates of the displaced surface points, shown as open dots were again measured using another network of six images. In the first instance, the deformation

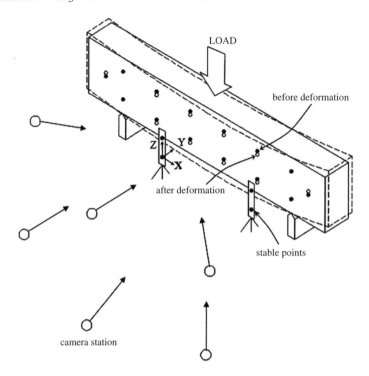

Figure 3.1 *Network geometry for hypothetical deformation survey.*

of the object being monitored is quantified by the geometric relationship between the Cartesian coordinates of monitoring points determined in the two photogrammetric measurements.

Once the two measurement networks are brought into a common reference coordinate system, utilising the shape-invariant transformation being afforded by the four stable points, both absolute and relative displacements can be determined. The resulting analysis of structural deformation generally utilises this point displacement data as the fundamental measurement information from which the parameters of the displacement function, initially, and the deformation response model, ultimately, are quantified. Put simply, in order to accept the hypothesis that the measured point displacements are a statistically significant consequence of deformation and not the result of random or systematic photogrammetric measurement errors, the (X, Y, Z) monitoring point coordinates for each epoch must be determined to a given level of accuracy and reliability. This means that the user of the vision metrology approach must be able to ascertain the likely precision and accuracy of measurement at the project planning stage.

The precision of any optical triangulation system is essentially a function of the angular measurement resolution, the geometry of the intersecting rays at each target point and the mean sensor-to-object distance, or image scale. Angular measurement resolution in photogrammetry is quantified as the angle subtended by the standard error of image coordinate measurement over the principal distance/focal length of the camera. Measurement resolution is directly dependent upon the digital image processing algorithm used to determine image coordinates, and in absolute terms is also a function of the pixel size of the sensor.

The range of angular resolution values for modern digital cameras employed in close range photogrammetry is considerable, since pixel sizes nowadays vary from about 2–10 μm, with the trend in consumer-grade cameras being to higher resolution charge coupled device (CCD) arrays of smaller pixels. Consider two modern digital cameras, for example the 6 megapixel Nikon D100 SLR-type camera with an 18 mm lens (field of view of $67° \times 48°$) and the 5 megapixel consumer-grade Canon IXUS50 with an integrated 5.8–17.4 mm zoom lens. Under the assumption that the deformation survey being conducted has high contrast, artificial targets for monitoring points, and the image measurement accuracy for these is 0.03 pixel, the angular measurement resolution of the D100 camera, which has 8 μm pixels, will be 2.8 seconds of arc. For the IXUS50 camera, which has a so-called 1/2.5 inch CCD array with 2.8 μm pixels, the corresponding angular measurement resolution will vary from 3 seconds of arc for a focal length of 5.8 mm to 1 second for a 17.4 mm focal length.

What must always be kept in mind, however, is that at the highest angular resolution, when the lens is fully zoomed in, the IXUS50 has a field of view of only about $23° \times 19°$, whereas when the lens is fully zoomed out, the field of view is $64° \times 53°$. A field of view for the camera of about 40–80° is generally required for close range photogrammetric measurement in industrial and engineering applications. Although from a network design standpoint the longer the focal length the better, so as to optimise angular resolution, the choice of focal length is typically constrained by the requirement to have a working field of view. Thus, there is invariably a practical trade-off, with one compromising a little on angular resolution or imaging scale in order to achieve a desirable field of view.

In designing a photogrammetric network it should be recalled that object point triangulation precision varies linearly in accordance with variations in angular measurement resolution, as is indicated by the following expression:

$$\bar{\sigma}_c = \frac{q}{k^{1/2}} S \sigma = \frac{q}{k^{1/2}} \, d \, \sigma_a \tag{3.1}$$

where $\bar{\sigma}_c$ is the rms value of (X,Y,Z) object point coordinate standard errors; S the scale number, given as the mean object distance d divided by the camera principal distance c; σ the image coordinate standard error; σ_a the corresponding angular measurement standard error; q a design factor expressing the strength of the basic camera station configuration; and k the average number of exposures at or near each station. Another way of expressing k is as the number of camera stations divided by the number required to form a generic network. A "generic" network, as described below, is a camera station configuration displaying optimal ray intersection geometry. For the case of the network shown in Figure 3.1, the generic network could be considered to comprise the four corner camera stations and thus the value of k could be taken for practical purposes as either 1 or 1.5. The design factor q is determined by the geometry of the ray intersections in the photogrammetric network. For a given camera station geometry, therefore, only one value of q will apply. For a sensor configuration comprising a single image at each station, the mean standard error value $\bar{\sigma}_c$ is simply the dimensional equivalent of the angular standard error σ_a over the distance d in instances where the q value is close to unity. As the geometry of the ray intersection improves, so the variance of triangulation decreases to a finite minimum. As it happens, a q value of 1 might be expected for a weakly convergent network of three to four camera stations, but q will rarely assume a value below 0.5, even in networks with optimal ray intersection geometry.

It is helpful when conducting a photogrammetric network design to adopt the notion of a generic sensor configuration that displays optimal geometric strength. The addition to this network of further stations which do not result in a broader diversity of convergence angles will not lead to a "stronger" intersection geometry *per se*, even though the added stations will enhance precision and reliability since they represent further sets of redundant observations. Once the basic configuration or generic network of camera stations is in place, the standard error of triangulation will improve in proportion to the square root of the number of exposure stations. This fact accounts for the presence of k in Equation 3.1, where it can be seen that to double measurement accuracy, i.e. to halve the value of $\bar{\sigma}_c$, four exposures need to be taken for each of the stations of the generic network. The additional images do not have to be at or near those of the basic geometry, for as long as the strength expressed by convergence angle diversity is not added to, the net improvement in precision can be expected to be the same.

Strictly, the empirical network design equation (Equation 3.1) provides a measure of precision, which is only an initial indication of the anticipated accuracy of triangulated object point coordinates within a convergent, multi-station photogrammetric network. It must be remembered (Section 2.4.2) that precision and accuracy are only equivalent if the system is free from all systematic errors, which for close range photogrammetry necessitates a fully calibrated metric camera with stable interior orientation. The design formula will be referred to for each of the case studies of structural monitoring by vision metrology, which are now considered. All these applications employed automated image measurement of artificial, high contrast targets. Further details and references to other relevant publications can be obtained from the single reference for each project provided at the end of the chapter.

3.3 The Puffing Billy Railroad Bridge

As a first example of the application of digital close range photogrammetry to structural monitoring, the case of a survey to determine the deformation of a 100-year old rail bridge under the loading of a locomotive will be considered (Johnson, 2004). In many respects this application highlights the potential of modern photogrammetric techniques, designed for use by non-specialists, to derive accurate, automated 3D measurements using off-the-shelf technology and inexpensive software systems. Previously this would have been considered a specialist and complex measurement task.

The Puffing Billy Tourist Railway is a major tourist attraction in the Australian state of Victoria. Trains first operated on this narrow gauge railway line in 1900 and for almost the last 50 years the railway has been under the control of a preservation society which runs tourist trains through the attractive, heavily forested Dandenong Ranges. One wooden trestle rail bridge, built between 1899 and 1900 remains essentially in its original form, though it underwent restoration in 1975. This bridge is shown in Figure 3.2, where the major structural components can also be seen.

The piles are some 450 mm in diameter, and spanning the 6.1 m gap between the piles are four longitudinal beams measuring 200 mm in thickness and 450 mm in height. Forming the deck of the bridge are planks with a 100 mm × 100 mm cross-section. The rails rest on sleepers on the decking. After some hundred years of operation it was deemed appropriate to investigate the current structural characteristics of this bridge. Specifically, a deformation survey of the spans between support piles was required, in which the displacement of the decking and support beams under the load of one of the historic locomotives was to be quantified. This required the 3D measurement of the section of the bridge indicated in Figure 3.2, first in the absence of the train, and secondly with the engine parked on the bridge, as shown in the figure. The accuracy required was 2 mm in the vertical direction.

As is often the case with largely volunteer organisations such as the Puffing Billy Preservation Society, there was a compelling engineering need for the deformation survey, but little budget to allocate to it. Thus, the structural monitoring of the Monbulk Creek Bridge became a student project within the Department of Geomatics, University of Melbourne, Victoria, Australia.

Figure 3.2   *Survey of railway bridge deformation.*

From a reasonably cursory analysis, making use of the empirical accuracy formula of Equation 3.1, it was apparent that by employing an SLR-type digital camera, high contrast targeting and the close range photogrammetric software system iWitness (Photometrix, 2005), it would be possible to determine bridge deflection data to the required accuracy in a two-epoch deformation survey. The camera available was the previously mentioned Nikon D100 with an 18 mm lens, which could cover the full deck area of interest, namely two spans, from a set-back distance of 15 m, thus generating an image scale of close to 1:800. An image measurement accuracy of 0.1 pixel or better was anticipated. With the adoption of a convergent imaging geometry and camera self-calibration, there was every confidence that a value of the empirical design factor of $q = 0.7$ could be achieved, and substituting these values into Equation 3.1 yielded a likely object point accuracy of 0.5 mm from a camera station configuration comprising 6–8 stations, even though the steep site was not conducive to an optimal network design.

Following the installation of targets, which required abseiling down each of the piles from the bridge deck, photography was recorded. The camera station configuration for the "loaded" survey is shown in Figure 3.2. The images were then measured semi-automatically and (X,Y,Z) object point coordinates were computed for each of the "loaded" and "unloaded" cases. The resulting mean coordinate standard error in the direction of most interest, namely the vertical, was 0.4 mm in each of the two photogrammetric networks. Less than 30 minutes per network was required for the image measurement and data processing and this could be carried out in the field, thus ensuring an additional level of quality control. An initial assumption was then made that the lower three targets on each of the three piles were not subject to movement and would therefore

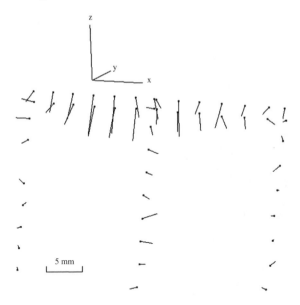

Figure 3.3 *Point displacements due to locomotive loading.*

provide the required stable, common reference point system for the two measurement epochs. The coordinates from the "loaded" measurement were transformed via a rigid-body transformation into the reference system established by these nine points and the point displacements were quantified, as indicated by the plot of the displacement vectors in Figure 3.3. These displacements were obtained to an accuracy averaging 0.7 mm. The maximum vertical deflection on the deck of the wooden bridge was 5.6 mm, which was within the allowable engineering tolerance.

Overall, the deformation survey of the Monbulk Creek rail bridge was a relatively straightforward process, and it was carried out using inexpensive off-the-shelf digital camera technology and a data processing system designed for non-specialist users. Yet, reliable non-contact 3D coordinate measurements were made to an accuracy averaging 1:35,000 of the size of the target array, all within a matter of an hour or two. It is not so many years since the time when such a photogrammetric measurement would have required specialist cameras and operator expertise. Nowadays this practical and powerful 3D measurement tool is available to anyone with a good quality consumer-grade camera and an appropriate, easy-to-use data processing system capable of automated image measurement and triangulation via a self-calibrating bundle adjustment. Moreover, the Puffing Billy bridge survey illustrates how the accuracy of the final (X,Y,Z) coordinate data can be planned for and predicted prior to the imagery being recorded, such that the actual photogrammetric survey becomes a verification of the network design and planning stage, in which Equation 3.1 and the generic network concept proved very useful.

3.4 The Hobart Radio Telescope

3.4.1 Project requirements

Unlike 100-year old wooden trestle bridges, parabolic antennas constitute structures which are relatively frequently measured by close range photogrammetry. The Hobart Radio Telescope, shown in Figure 3.4, is a parabolic antenna of 26 m diameter, the main reflector of which

Figure 3.4 *The 26 m diameter Hobart Radio Telescope.*

comprises 252 aluminium panels. The antenna's gain performance is a function of the integrity of the shape of the parabolic reflector. As part of an overall survey of the reflector surface, to accurately determine both its shape and shape changes associated with different elevation settings, it was required that the 3D positions of panel centres be measured at three elevation settings, namely horizontal, 45° and vertical. The chosen 3D measurement technology was vision metrology (see Fraser *et al.*, 2005).

When performing antenna measurements via image based measurement, to the generally required accuracy of 1:200,000 or better, an effective angular measurement resolution for the imaging rays of a second of arc or so is required. For this to be achieved for a single image from a digital camera, exceptionally accurate image measurement would be required. For example, the camera selected for use in the Hobart Radio Telescope deformation monitoring project was an INCA 4.2 from Geodetic Systems (Geodetic Systems Incorporated, 2005). The 4k × 4k CCD array of this camera, which has an 18 mm focal length lens, has a pixel size of 9 μm. Thus, an angular measurement resolution of 1 second implies an image measurement accuracy of 0.09 μm or 1/100th of a pixel, which is 2.5 times the optimal centroiding accuracy of 1/40th of a pixel generally associated with single image measurement of a retro-reflective target.

The surface measurement accuracy required for the Hobart antenna was 0.065 mm (rms, one standard deviation precision) or 1:400,000 of the diameter of the reflector, which would allow surface point displacements to be determined to 0.1 mm accuracy. This was a very tight tolerance indeed, and to effectively enhance the angular measurement resolution to the required level, the concept of hyper-redundancy was employed. The reader will recall that a generic network design is one where optimal multi-ray intersection geometry is obtained with as few camera stations as practicable. Hyper-redundancy is a concept whereby once the generic network is in place, many additional images are recorded, with the beneficial impact upon object point precision being equivalent to the presence of multiple exposures at each camera position within the generic network. The effective number of images per station within a hyper-redundant network might well be in the range 10–20 or more. As is apparent when it is considered that a hyper-redundant network may comprise hundreds of images, the concept is only applicable in practice to fully automatic vision metrology systems. Here it proves to be a very effective means of enhancing measurement accuracy at the cost of minimal additional work in the image recording phase. In reference to Equation 3.1, in a network displaying hyper redundancy, k will assume a large value of, say, 5–10 or more.

3.4.2 Network configurations for the radio telescope

Photogrammetric measurement of the radio telescope had to take into account a number of network design constraints. These related to practical limitations in positioning the camera, and included an image scale constraint, a field of view constraint and constraints related to incidence angle and visibility. For the case of the telescope oriented vertically upwards, the image scale constraint arose because the crane obtained for the photography had a finite operating height limit of 40 m, thus allowing a maximum camera station height above the centre of the dish of only about 14 m. Allowing for convergent imaging with an average angle of incidence to the horizontal of 50°, this led to a mean object distance of 18 m. The field of view and incidence angle constraints followed directly from this, since it was now apparent that only a portion of the target array, averaging 100 panel centre points, would be imaged from each camera station. The visibility constraint arose from the visual obstruction caused by the quadripod structure (see Figure 3.4).

Through a rearrangement of Equation 3.1, and under the assumption of homogenous object point accuracy, the number of images needed for each target point to yield the required accuracy could be calculated, approximately, as follows:

$$k = \left(qS\sigma/\bar{\sigma}_c \right)^2 \tag{3.2}$$

For values of $q = 0.6$, $\sigma = 0.2$ μm (1/40th of a pixel), $S = 18$ m/18 mm $= 1000$ and $\bar{\sigma}_c = 0.065$ mm, k assumes a value of 4 when rounded up to the nearest integer. In practice, constraints in the positioning of the camera platform, coupled with the need to ensure that each target is covered by a sufficient number of rays displaying the necessary geometry, led naturally to the tendency to capture as many images as practicable. As mentioned, this hyper-redundant coverage has no impact upon data processing time and relatively little extra time is needed for the photography. In the case of the Hobart Radio Telescope measurement, the recording of all required images from a single crane set-up took 20 minutes, which was less time than that needed to establish the crane in a given position. The positioning limitations imposed on the camera platform from a single crane set-up position also applied when the radio telescope was pointing both in the horizontal direction and at a 45° elevation angle. Thus, it was decided to also maintain an average object distance of 17 m for these two networks, as this would enhance the prospects of achieving the same level of surface measurement accuracy.

Figure 3.5 shows the final camera station geometry for the hyper-redundant network for the vertical orientation, which comprised 148 images. The actual locations were dictated by manoeuvrability limitations of the man-box of the crane, rather than by concerns for optimal geometry. Figure 3.6 illustrates the hyper-redundant intersection geometry for one target location at the 45° orientation. In this particular network, each target appeared in an average of 45 images, the lowest number being 15 and the highest 75. The average number of points per image was 80, with the lowest and highest being 27 and 135, respectively.

The desired surface measurement accuracy of 0.065 mm was achieved in all three photogrammetric surveys, which were processed using the *Australis* vision metrology software

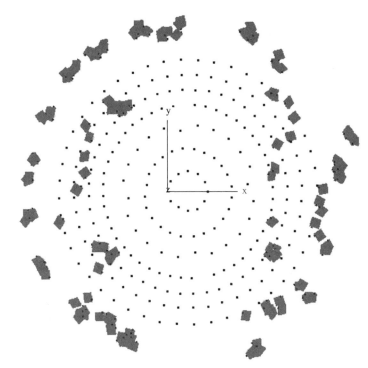

Figure 3.5   *Hyper-redundant camera station network for the vertical case.*

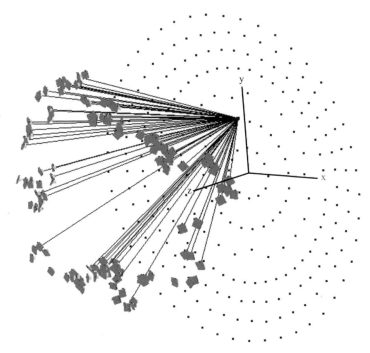

Figure 3.6   *Sample multi-ray intersection geometry within the 45° network.*

(Photometrix, 2005). The results validated the assumptions adopted in the network design process. As well as enhancing accuracy, the hyper-redundancy concept greatly improves internal reliability in the final bundle adjustment, leading to better gross error detection. The number of rejected observations in each network was close to 100, out of some 19,000 or more, with there being a maximum of four for any one image. The high level of internal consistency in each network can be appreciated by the fact that image coordinate measurements were rejected if their associated residuals were greater than 0.15 pixels.

3.4.3 Deformation analysis

The analysis of reflector surface deformation for the Hobart Radio Telescope had two components: an evaluation of the conformance of the reflector surface to a true parabola of revolution, and a determination of the change in shape of the reflector as it was re-oriented from the vertical through 45° to the horizontal. A best-fitting paraboloid surface was determined for each telescope orientation, with the resulting rms discrepancy between the measured and design surfaces being close to 1.5 mm in each case. Significant surface departures from design were relatively local in nature, being confined to areas of several panels, generally at the rim of the dish. The maximum measured surface departure of any panel centre point was 8.8 mm.

An illustration of the extent of surface deformation associated with antenna re-orientation is provided in Figure 3.7, which shows the individual surface point movements that occurred

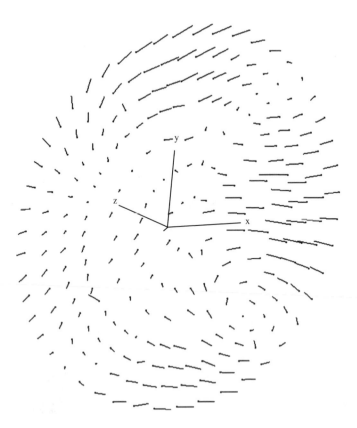

Figure 3.7   *Point displacements between the vertical and horizontal positions.*

between the vertical and horizontal settings. Here, the overall rms discrepancy vector length in the telescope axis direction (Z coordinate) was 0.6 mm. Most of the deformation occurred in the outer panels where the maximum gravitationally induced point displacement was 2.6 mm. For the vertical to 45° case, where deformation was less, the corresponding value was 0.3 mm.

The Hobart Radio Telescope deformation monitoring project highlighted both the very high accuracy attainable with vision metrology, and the flexibility afforded by the hyper-redundancy concept for enhancing measurement accuracy. With modern fully automatic vision metrology systems, the only cost involved in adopting the hyper-redundancy approach is the generally small additional time required to record the images; the cost in additional data processing time is a matter of seconds or a few minutes at worst.

3.5 The North Atrium of Federation Square

3.5.1 Atrium construction and monitoring requirements

Federation Square was conceived in 1996 as a celebration of Australia's Centenary of Federation. It is as big as a city block and houses about a dozen buildings, different yet linked and coherent. One of the principal architectural structures in Federation Square is the Atrium, which is an open, galleria-like public space comprising two distinct elements. One of these, the North Atrium, forms a covered forecourt and is made up of galvanised structural frames in a triangular geometry, developed into a folded 3D structure glazed on the outside. The framework structure, which evolved from the triangular geometry of the façades, was designed with a cantilever system that allows the final 14 m section to be suspended in free space, as indicated in Figure 3.8.

Figure 3.8 *North Atrium showing 14 m cantilevered section.*

Due to the geometric complexity of the North Atrium structure, its lack of redundancy and the importance of preventing any possible interaction between the superstructure and the intricate façade skin of glass, an accurate structural analysis was essential prior to and during construction. Particular emphasis was placed on an evaluation of the deformation characteristics in the cantilever section of the atrium, where significant overall deflections were expected both in the de-propping process during construction, and in the glazing. An independent deformation monitoring system was required and automated digital close range photogrammetry was selected to be the measurement tool to quantify structural deformation, which could then be compared with the point deflections anticipated from theoretical predictions.

The superstructure frame for the North Atrium is fabricated from 200 mm square hollow sections and comprises two skins of in-plane frames separated by a gap of approximately 1.5 m, as can be seen in Figure 3.9. The frames consist of four- to five-sided irregular polygons connected by in-plane diagonals based on a "pinwheel" configuration allowing a sense of randomness whilst preserving geometrical definition which was essential for the attachment of the glazed façade system. The roof structure consists of a series of 2.0 m deep trusses which

Figure 3.9 *Two skins of in-plane frames.*

have the dual function of supporting the roof and facilitating the walkways and catwalks required for maintenance work, while the roof structure over the cantilever section is far more complex with a structural system similar to that adopted for the walls.

Construction of the North Atrium was a complex operation, which called for a comprehensive understanding of how the structure would deform over time. In addition to determining the structural performance at the final permanent state, continuous evaluation of deflections, both absolute and differential, during the construction phase was required. This allowed the designers to develop acceptable façade connection details to facilitate the predicted movements. Staged second-order elastic analysis to predict such movements was carried out for this assessment, concurrently with the construction program. The critical construction activities included the de-propping of the cantilever section and the installation of the glazing which dominates the contribution towards the overall dead load. Figure 3.10 illustrates the structural deflection predicted from finite element analysis, with the maximum deflections at any node point in the cantilevered section being of the order of 50 mm.

In view of the complexity of the Atrium structure and the specific requirement for accuracy in determining deflections of the structure, an independent monitoring program was sought to validate the theoretical predictions and to provide some "peace of mind" feedback for the owners. A five-epoch monitoring program was established for the North Atrium construction, with the 3D coordinates of 71 structural node points being determined at each epoch to 0.3 mm accuracy using an automated vision metrology system, V-STARS (Geodetic Systems Incorporated, 2005) in this case (see Fraser *et al.* (2003)). The measured coordinates then allowed both absolute and differential node point deflections to be computed to the required accuracy of 0.5 mm. The five monitoring epochs were: an initial condition before de-propping, three during de-propping and one during the glazing process.

The deformation monitoring took place during what was a very busy and involved construction process and minimal disruption to construction activity was an absolute imperative for the monitoring surveys. The fact that vision metrology provides a very fast, automated, economical, reliable and highly accurate (proportional accuracy of close to 1:100,000) technique, which can operate efficiently in spite of the cluttered construction environment, made it the obvious and possibly only suitable 3D measurement choice.

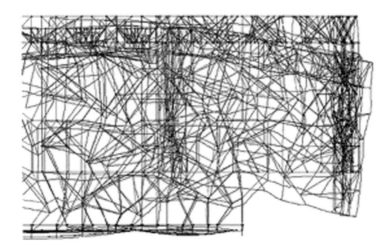

Figure 3.10 *Deflected shape due to dead load in cantilever section (not to scale).*

3.5.2 Five-epoch survey program

The V-STARS system, with an INCA 4.2 digital camera, was employed to measure the 71 node points signalised with 2 cm retro-reflective targets. Some 100 coded targets were employed to effect automatic exterior orientation, point labelling and to provide network tie points. Although the photogrammetric approach could accommodate most of the constraints imposed by the construction site, one difficulty was visibility. Node points on both the inner and outer frames were to be monitored, which gave rise to some targets being visually obstructed from the desired camera station positions. The approach adopted to overcome this problem was simply to add more camera stations in accordance with the hyper-redundancy concept. Thus, the nominal 32-station network shown in Figure 3.11 usually produced between 100 and 140 images. Photography was captured from a telescopic boom (Figure 3.12), with the operator occupying usually three to five different viewing positions at a nominal location in order to minimise visual obstructions. As with the Hobart antenna project, the added effort of recording a network with 100, 120 or even 150 images was measured in minutes, both at the image acquisition stage and within the final photogrammetric orientation, which was fully automatic. The recording of the images at each epoch took about 25 minutes, whereas the bundle adjustment and subsequent computation of deflection vectors required only 5 minutes.

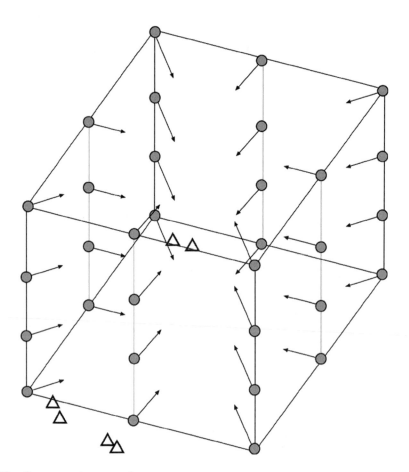

Figure 3.11   *Camera station network.*

Figure 3.12 *Recording of images.*

In order to provide a time-invariant reference system, six targets were established as stable "datum" points. These were positioned on lower sections of the Atrium framework, though not upon an absolutely stable structure since this was not feasible. Some small movement of the datum points, which are indicated by triangles in Figure 3.11, was thus anticipated, though the magnitude of their deflections was predicted to be of limited consequence for the determination of accurate overall structural deformation.

Consistent with expectations from the empirical accuracy formula of Equation 3.1, an accuracy of 0.15 mm (rms, one standard deviation) or better was achieved in node point positioning in all five photogrammetric surveys. The self-calibrating bundle adjustments involved about 170 object points (nodes and coded targets) and 90–140 images, with image scales generally being in the range 1:500–1:1000. The combination of retro-reflective targeting and precise image centroiding yielded an image coordinate measurement accuracy of 1/40th of a pixel (approx. 0.2 μm). Accurate absolute scale was not a requirement for the deformation monitoring, but redundant scale information was nevertheless used as it provided a further external accuracy check. At the first measurement epoch, two 2.5 m long precise invar scale bars were included in the network, with the discrepancy in the measured lengths of these known distances being only 0.05 mm.

3.5.3 Measured versus predicted deformation

The primary aim of measuring node point deflections arising from the de-propping and installation of the glazing was to compare the actual versus predicted deformation of the North Atrium structure and thus both validate the results of a second-order finite element frame analysis and allow a refinement, if necessary, of the structural model. As a first step in this process, all five photogrammetric networks required shape-invariant transformation into the chosen reference system, which was provided by the six datum points at epoch 1. Node point deflections where then computed from coordinate differences, with the results being shown

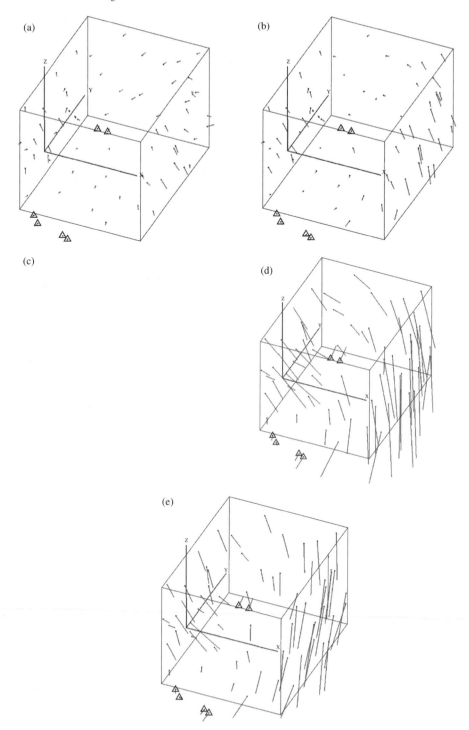

Figure 3.13   *Computed node point displacements between measuring epochs; (a)–(c) effect of de-propping; (d) de-propping and (e) glazing (total dead load).*

in the plots in Figure 3.13. The maximum node deflection between two epochs was 22 mm, which was less than predicted.

Figure 3.13 shows that the six datum points did not remain stable throughout the monitoring period and it is apparent that as the structure was de-propped the increased loading upon the lower frames induced small lateral movements of 5 mm or less at the maximum loading (epoch 5). It was fortunate that these movements were horizontal as they then had little effect on the computed values of the settling of the frames of the cantilevered section.

Significant deformation accompanied the de-propping of the cantilevered section of the North Atrium structure, as shown in Figure 3.13. The maximum node point deflection at the first de-propping (Figure 3.13a) was 2.5 mm, with the corresponding deflections for the second (Figure 3.13b) and third (Figure 3.13c) de-propping being 5.3 mm and 8.5 mm, respectively. With the significant load accompanying the installation of the exterior glazing, the deflections sharply increased, especially vertical displacements on the glazed north and west walls where a maximum deflection of 22 mm from initial condition occurred (Figure 3.13d). Of this overall 20 mm vertical settling, approximately 14 mm could be attributed to the weight of the glazing (Figure 3.13e). In general, the behaviour of the structure was in good agreement with theoretical predictions based on the Atrium's design, even though a number of the nodes had deflections that were less than predicted.

The successful conduct of the multi-epoch deformation monitoring of the complex North Atrium structure highlighted the applicability of close range photogrammetry to structural monitoring for building construction. Without the vision metrology surveys performed, the engineering team would have needed to rely solely on complex CAD models of the overhanging atrium section for the second-order elastic analysis. Instead, the photogrammetric measurements yielded comprehensive deformation data to independently assess the structural models and to subsequently refine these models.

3.6 Reinforced concrete bridge beams

3.6.1 T-beam measurement requirements

A further application of automated single-sensor vision metrology to the monitoring of deformation was an investigation of shear failure mechanisms in reinforced concrete bridge beams (Fraser and Brizzi, 2002). Concrete road bridges represent a category of civil engineering infrastructure that is subject to increasing structural demands. On the one hand, legal vehicle loads are increasing, and on the other hand, existing bridges are deteriorating with time. There is thus the prospect that many existing bridges will eventually fall below current design standards in relation to their capacity to carry increased vehicle loads. Given the ever-present constraints upon capital expenditure for new infrastructure, there is an increasing emphasis upon utilising existing bridges more effectively in an effort to extend their usable life. Two areas of current research interest in civil engineering are the determination of improved strength assessment models for existing bridges, and the strengthening of reinforced concrete bridges.

One of the requirements of bridge strength assessment is to ascertain bridge beam shear and bending strength, which can in turn require the measurement of beam displacement and shear crack development as the beam is incrementally loaded to failure. The subsequent non-linear finite element analysis, which is employed to better model plasticity, requires an extensive array of surface displacement data gathered over a number of load conditions. Photogrammetric techniques are ideally suited to such a 3D displacement measurement task, with notable attributes of the technology for this application being full measurement automation coupled with fast data acquisition, high accuracy and the ability to accommodate dense object point fields of 1000 or more monitoring points.

Vision metrology was employed in a project to measure the displacement of four bridge T-beams strengthened with carbon fibre reinforced plastics as they were subjected to incremental loading. Figure 3.14 shows one of the four monitored beams positioned for load testing, the loads being progressively increased until shear failure occurred, generally at around a 400 kN load. A first requirement in the measuring of beam displacement to quantify shear crack development

Figure 3.14   *Carbon fibre strengthened T-beam under load.*

Figure 3.15   *Target array in shear failure zone, with carbon fibre stirrups shown.*

and support finite element analysis is a dense array of beam surface points, albeit on only one lateral surface of the reinforced T-beam. Such a requirement poses no problem for vision metrology, and for each of the beams tested between 1000 and 1500 targets were employed.

Figure 3.15 shows a number of surface targets on a beam that has failed in shear. It should be noted that although photogrammetric measurements were to be correlated with data from displacement transducers, a total of only ten or so transducers per beam was feasible. A second fundamental requirement for beam displacement measurement to support finite element analysis is high accuracy. For the particular 6 m T-beams considered, the bending displacement between the states of load free to fully loaded reached a maximum of about 3 cm before shear failure occurred. Given that as many as 20 measuring epochs per beam needed to be employed, an accuracy of 0.1 mm was sought for the epoch-to-epoch point displacement measurements.

3.6.2 Network design

In order to achieve the necessary spatial triangulation precision, a network of 20 camera stations was adopted, with a single image per station being recorded with a GSI INCA 4.2 digital camera from a set-back distance of about 7 m. Figure 3.16 shows a sketch of the convergent imaging geometry, which ensured that in spite of some obstructed target situations, all object points were recorded in at least nine images.

Application of the empirical design formula (Equation 3.1) yielded an anticipated mean object point coordinate standard error of $\bar{\sigma}_c = 0.05$ mm for an imaging scale of 1:390 ($S = 390$), an image coordinate measurement standard error of $\sigma = 0.3$ μm (0.03 pixel), one exposure per station, an empirical design value of $q = 0.7$ and a k value of 3. Moreover, a quite homogeneous distribution of accuracy would be expected for all object points.

An important design requirement for the vision metrology process in this case was rapid data acquisition. As each static design load on the beam was reached, a steady state was held for the duration of the image recording. To help ensure that the beam remained shape invariant during this time, especially at higher loads, it was highly desirable to minimise the time needed

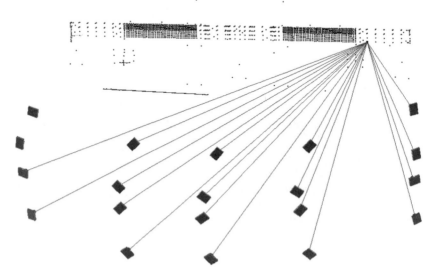

Figure 3.16   *20-station network geometry for beam survey.*

Figure 3.17   *Recording of imagery for T-beam deformation survey.*

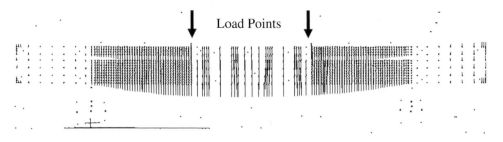

Figure 3.18 *Displacement characteristics for loaded concrete T-beam.*

for image capture. By using a small stepladder, it proved possible to occupy all 20 stations and capture the images in less than one minute (e.g. Figure 3.17). Subsequent analysis confirmed the validity of the assumption of object shape invariance to the desired tolerance during this short period.

Although the experimental testing did not call for "real-time" photogrammetric measurement results, full automation of the vision metrology process ensured that all computations related to a single measurement epoch would be completed within 3 minutes of the imagery being recorded. The measurement of 1500 T-beam surface points over 20 load increments, which involved the determination of (X, Y, Z) coordinates of 300,000 target positions, was thus completed some 3 minutes after the recording of the last image at the final epoch (i.e. when shear failure occurred) for each beam.

3.6.3 Displacement measurement results

Figure 3.17 shows one of the targeted T-beams being photographed, the dense array of monitoring points in the regions of anticipated shear cracking, and the less dense distribution of targets away from the expected failure zone should be noted. In most respects, the conduct of the photogrammetric measurement for each epoch followed that envisaged in the design procedure, and for all beams surveyed the desired accuracy of 0.06 mm in point positioning was surpassed, the realised point positioning standard error being close to the 0.05 mm level predicted via Equation 3.1.

The dense array of monitoring points provided through photogrammetry, all referenced within a uniform (X, Y, Z) Cartesian coordinate system, provided a number of analysis possibilities that were not feasible with displacement transducers alone. For example, it is possible to identify and quantify various rigid-body movements that contributed to shear strength. Figure 3.18 shows the basic displacement characteristics of a loaded T-beam prior to shear failure. Following failure, block shifts and rotations could also be analysed.

The photogrammetric measurement of the deformation of the reinforced concrete bridge beams as they were loaded to failure provided valuable new dimensional information for investigations of shear failure mechanisms and flexural capacity in the strengthened T-beams.

3.7 Ore concentrator/crusher

3.7.1 The deformation problem

A further example of structural monitoring employing automated vision metrology is the multi-epoch deformation survey conducted to quantify the relative displacement between the rotor and stator of the ore concentrator/crusher shown in Figure 3.19 (see Fraser, 1999).

Figure 3.19   *Electric ring motor showing targets on the rotor and stator.*

This large electric ring motor, located at a gold mine in Australia, is reputed to be one of the world's largest. The required structural monitoring centred on a determination of variations in the rotor–stator separation which accompanied changes in power. It had been observed that as amperage to the motor increased to a certain point, the stator appeared to deform. The rotor itself is 15 m in diameter, and when the stator (including housing) is included the overall diameter increases to 23 m. Yet this large, complex structure was required to be measured to an accuracy which would allow determination of point displacements to 0.5 mm precision, or about 1:50,000 of the size of the object.

The requirements of the vision metrology surveys at different power loadings centred upon the following:

- a determination of any shape deformation of the stator (as indicated by point displacements on the housing);
- a mapping of displacements occurring between adjacent points on the rotor and stator, and hence a mapping of rotor–stator separation for different power loadings;
- a determination of the impact of a "block shift" movement of the stator with respect to the axis of the motor, which is achieved by shimming the stator; and,
- a determination of any deformation of the drum of the ore concentrator unit.

In order to fully acquire the necessary dimensional data from which deformation parameters could be quantified, several epochs of photogrammetric measurement were required. Structural instability and the presence of significant vibration precluded the use of alternative 3D measurement technologies such as laser trackers.

Measurements of six states of the stator and drum were undertaken:
(1) The stator was measured to determine its shape in a "hot" state, while the ore crusher was running (i.e. the drum was rotating). Measurement of the rotating drum of the motor in the same survey was not possible with only one camera.

(2) The rotor (drum) and stator were measured at "rest" (no power).

(3) and (4) The rotor (drum) and stator were measured at two power loads, referred to simply as amperage **A** and amperage **B**.

(5) The amperage **A** state was again measured, this time with the stator translated (shifted), through shimming, in order to minimise the rotor–stator separation.

(6) Measurement (1) was repeated.

The six epochs of measurement of the front area contained more than 260 monitoring points. From the measured data at these six epochs, critical dimensions, point displacements and deformation patterns were quantified and analysed. Figures 3.19 and 3.20 show the photogrammetric targets which constituted deformation monitoring points for the "front" surveys. These comprised retro-reflective "dots", as well as coded targets for automatic exterior orientation.

On three occasions the back section of the motor, behind the 2.5 m wide stator, was measured. Since the surveys conducted of this section were very similar to those of the front section, they will not be further considered in this overview.

3.7.2 Network geometry

Figure 3.20 shows target point locations and a limited number of camera station positions for one of the six measurement epochs. Due to the complex nature of the object, the main criterion in the network design was to ensure that every object point would be recorded in a sufficient number of highly convergent images. In all, between 90 and 100 images were taken in each network, with each object point appearing in between 6 and 30 images. The average image scale was 1:600, which corresponds to a mean set-back distance of 10 m.

With this imaging scale, along with an image measurement accuracy of 0.03 pixel (approx. 0.3 μm) and the highly convergent multi-ray geometry, the anticipated mean object point coordinate accuracy obtained from the empirical formula (Equation 3.1) was 0.20 mm. This level of accuracy well exceeded that required that for deformation measurement, which was set at 0.5 mm. Upon performing the data reduction for each measurement epoch, it was

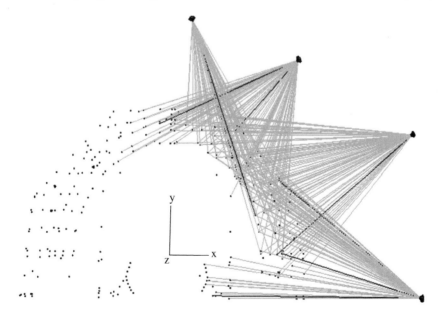

Figure 3.20   *Object point array, along with locations of four of the camera stations.*

found that the design expectations had been realised and the required triangulation accuracies achieved. As regards the time required for each survey, it took 25 minutes to record the 100 or so images at each measurement epoch (due to man-lift manoeuvring times etc.), yet less than 5 minutes was needed to automatically measure all images and complete the bundle triangulation and (X, Y, Z) coordinate determination.

3.7.3 Results of deformation analysis

Based on the photogrammetric measurements conducted at each epoch it was possible to perform a multi-epoch deformation analysis. Of principal interest in this analysis was the determination of both shape changes in the stator and variations in the rotor–stator separation, in accordance with the previously stated project requirements. It was verified at an early stage that the rotor/drum section of the electric motor was not subject to any shape change during the power loading tests.

The two most significant findings from the deformation analysis arose from the nature of the point displacements between measuring epochs, which are indicated in Figures 3.21 and 3.22.

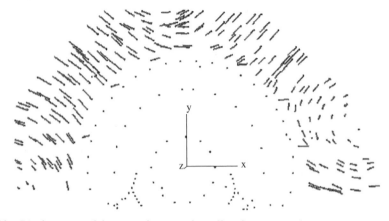

Figure 3.21 *Displacement of the stator between the null and amperage A.*

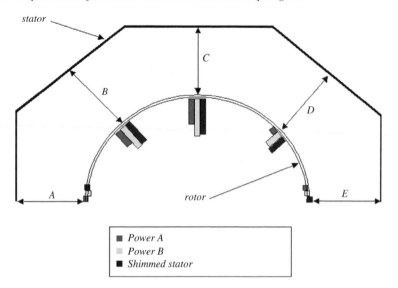

Figure 3.22 *Variations in separation distances between the rotor and stator.*

Figure 3.21 presents a plot of the point displacements on the stator which accompanied the increase in power loading from zero amperage (null state) to amperage **A**. A similar pattern, with further displacement, was seen at amperage **B**. What is particularly interesting in the figure is the systematic nature of the displacement vectors and the very compelling case for the deformation being induced through magnetic field effects. One can appreciate the strength of this field when one considers the massive steel structure of the stator. Figure 3.22 shows a plot of the variations in separation between the rotor and stator at five locations around the semi-circumference of the rotor. Two of these correspond to the amperage **A** and amperage **B** power loadings, whereas the third shows the change from the null position which accompanies the shimming of the stator. It is noteworthy that changes in separation distance reached several millimetres.

The multi-epoch deformation analysis of the ore concentrator offered a further illustration of the high accuracy, productivity and flexibility of off-line vision metrology applied to very large objects. Although the electric motor measurement would by any standard constitute a complex task, it was nevertheless a reasonably straightforward process to carry out this deformation monitoring project via a close range photogrammetric approach.

3.8 Super-hot steel beams

3.8.1 Monitoring thermal deformation

So far, all the structural monitoring examples considered have utilised single-sensor vision metrology. The basic assumption in these cases has been that the object is shape invariant during the period in which the imagery for a particular survey is recorded. In situations where the object is undergoing dynamic changes in shape, all photogrammetric imagery must be recorded at a single instant in time, which necessitates the use of multiple synchronised digital cameras. The last case study to be considered is an application of digital close range photogrammetry to the deformation monitoring of a series of seven super-hot steel beams. Measurements for each beam were conducted at 70–80 epochs over 2 hours as the steel cooled from 1100°C to near room temperature. An on-line, three-camera configuration was established to measure both stable reference points and targets subject to temperature induced displacement to an accuracy of 0.5 mm. Special targeting was required to accommodate the changing colour of the beams as they cooled.

Figure 3.23 shows a 9 m steel beam positioned within a thermal testing facility at the Technical University of Braunschweig, Germany. As part of a material testing process, there was a requirement to measure the thermal deformation of seven such beams, each of a different cross-section and different steel composition. Finite element models had been formulated covering the deformation of each beam type over a temperature range of 1200–0°C, and it was desired to evaluate the integrity of these models. This could only be achieved through a direct measurement of surface point deflections across a wide temperature range.

At normal temperatures this task would have been straightforward, but it became potentially very difficult due to the extremely high temperatures involved. Contact measurements were out of the question as were non-contact optically based 3D coordinate determination methods that could not sustain a measurement rate of 1–2 points per second. This was because of an expressed desire to monitor the beams over an approximately 2-hour period as they cooled, initially with an epoch of measurement occurring every 15 seconds. After a review of candidate 3D measurement technologies it became clear that a photogrammetric approach would be the most viable, though it would not be free of practical difficulties. Point positions for each of the targets indicated in Figure 3.23 were to be measured, from which point deflection data covering all

Figure 3.23   *Targeted steel beam, as viewed from the centre camera station.*

monitoring epochs could be compiled. Given the volume of data to be recorded it was necessary to design a highly automated, on-line data processing system, which had to incorporate special features to overcome problems associated with the high-temperature environment. A significant problem was that once the beam was uncovered and commenced cooling, its colour changed from a bright yellow (so-called white hot) through orange and red, and finally to brown.

3.8.2 Physical constraints, imaging configuration and targeting

The photogrammetric network needed to accommodate two categories of target, the first comprising the 10–15 monitoring points on each beam, and the second comprising about 30 targets upon the stable structure behind the thermal testing pit (see Fraser and Riedel, 2000). The stable points served two purposes. The first was to provide a fixed reference system with respect to which epoch-to-epoch point displacements would be determined, and the second was to provide the necessary broad spread of monitoring points to support photogrammetric bundle adjustment.

A three-camera configuration was employed for the photogrammetric network, which for any epoch comprised all targets on the steel beam along with 25 or so of the stable, background targets. Figure 3.24 illustrates the basic network geometry. The progressive scan CCD cameras had imaging arrays of 1300 × 1030 pixels and a pixel size of 6.7 μm square. The two outer cameras were fitted with 8 mm lenses, which provided a horizontal field of view of 57°, whereas the centre camera had an 85° field of view from its 4.8 mm lens. The wider field of view at the centre camera was necessary due to limitations in the available set-back distance for this imaging station. The average camera-to-object distance for the outer two cameras was 9.6 m (imaging scale of 1:1200) whereas that for the centre camera was 6.7 m (1:1400 scale).

In accordance with Equation 3.1, coupled with an image coordinate measurement accuracy of 0.1 pixel, the adopted three-station network indicated that it would yield a triangulation accuracy (rms, one standard deviation) of close to 0.5 mm for points along the beam. At each measuring epoch, the three images were acquired sequentially, and then processed via a single frame-grabber

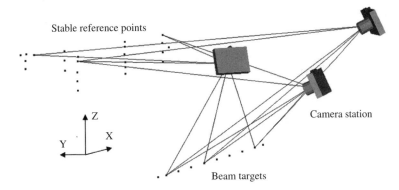

Figure 3.24 *Network of three camera stations and 40 object points.*

board and stored within the host PC in a total time of 0.2 seconds. It was intended that the processing of the triangulation of the targets would be on-line.

In contrast to the standard targeting used for the stable reference points, the targets which constituted deformation monitoring points on each beam had to both survive temperatures exceeding 1100°C and maintain high contrast against a background that was dramatically changing in colour over time. The targeting material selected was a highly heat resistant ceramic which maintained its off-white colour over the full range of temperatures. Although the 6 cm circular ceramic targets displayed sufficiently high contrast to be measured automatically to 0.1 pixel accuracy or better when the steel was cooler than 800°C, above this temperature, when the beam was white hot, manual assistance was often required to ensure accurate image measurement.

3.8.3 Image recording and processing

For each steel beam, 70–80 epochs of imagery were recorded, with the full test sequence taking 90–120 minutes. The first 40 measured epochs were required at 15-second intervals, after which the frequency was reduced to a measurement every minute for 20 minutes. The final 10–25 epochs were recorded at 5-minute intervals. All three cameras had been pre-calibrated and the first stage of data processing was to establish a preliminary exterior orientation/triangulation for each beam. With the cameras in place, imagery was recorded of the background points, and also, when possible, of points on the beam prior to its covering for heating. The subsequent bundle adjustment established the exterior orientation for the network, along with initial triangulated coordinates for points on the beam.

One aim of the adopted on-line data processing was to perform photogrammetric triangulation in such a way that all images recorded up to and including the current epoch would be included in the bundle adjustment. Thus, if there were 30 stable background points and 10 targets on the beam for each epoch, the bundle adjustment at epoch 50 would comprise 150 images and 530 object points. The stable points would be common to all epochs whereas the targets on the beam would be re-labelled and thus constitute "new" object points at each epoch. The use of all available data within the bundle triangulation at each epoch was very desirable, since a reference coordinate system common to all epochs was implicitly defined via this approach. The on-line photogrammetric processing system was realised through a modification of the *Australis* system for off-line digital close range photogrammetry. Extra features were added to the software system to support the cumulative, multi-epoch processing, a critical component of which was the automatic measurement of image points. This was a relatively straightforward process since the images from the previous epoch could be employed to "drive" the image measuring process at the current epoch.

3.8.4 Beam measurement results

The primary initial outcome of the measurement of the seven beams was a determination of the pattern of beam surface deformation which accompanied the dramatic changes in temperature. Figures 3.25 (see colour section) and 3.26 (see colour section) show plots of position versus time for the two axes, X (longitudinal axis) and Z (vertical), for targets on one of the beams. The maximum point displacement in both cases is about 5 cm, whereas the deformation in the lateral direction (Y) was considerably less, at under 2 cm. The measured point displacement data was delivered to the structural engineers for further analysis.

In spite of the presence of minor and unavoidable systematic errors due to targeting and radiometric variations, acceptable results were obtained with the final representative rms accuracy of triangulated object points being close to 1.0 mm for the monitoring targets. Although observational residuals within the imagery displayed a systematic error component, the effect of this could be anticipated to be near constant over a series of epochs. It was thus expected that the relative accuracy of measured deformations would be superior to the precision indicated by the variances associated with the triangulation. There was no practical means to rigorously verify this assumption, though the smoothness of the resulting curves of point displacements versus time supported the conclusion that sub-millimetre accuracy was achieved.

The final rms value of image coordinate residuals in the 210-station, 800-point bundle adjustments for 70 epochs of data averaged 1.6 μm (0.2 pixels). In terms of relative accuracy, monitoring points on each beam were positioned to about 1:9000 of the size of the object, which is modest by modern vision metrology standards, though not unexpected for the small format CCD cameras and the physical conditions involved. The successful multi-epoch deformation monitoring of the seven steel beams offered a good example of how vision metrology can be applied in difficult environments where the application of alternative 3D coordinate measurement technologies is precluded.

3.9 References

Fraser, C.S., 1996. Network design. In Atkinson, K.B. (Ed.), *Close Range Photogrammetry and Machine Vision*, Whittles Publishing, Scotland, 256–281.

Fraser, C.S., 1999. Multi-epoch deformation analysis via vision metrology. *"Towards A Digital Age"*, Proceedings of the 3rd Turkish–German Geodetic Days Conference, Istanbul, Turkey, June 1–4, Vol. 1, 49–58.

Fraser, C.S. and Brizzi, D., 2002. Deformation monitoring of reinforced concrete bridge beams. *Proceedings, 2nd International Symposium on Geodesy for Geotechnical and Structural Engineering*, Berlin, Germany, 21–24 May, 338–343 (also on CD-ROM).

Fraser, C.S. and Riedel, B., 2000. Monitoring the thermal deformation of steel beams via vision metrology. *ISPRS Journal of Photogrammetry and Remote Sensing*, 55(4), 268–276.

Fraser, C.S., Brizzi, D. and Hira, A., 2003. Photogrammetric monitoring of structural deformation: the Federation Square Atrium project. In Grün A. and Kahmen H. (Eds.), *Optical 3D Measurement Techniques VI*, Wichmann Verlag, Heidelberg, Germany. II, 89–95.

Fraser, C.S., Woods, A. and Brizzi, D., 2005. Hyper redundancy for accuracy enhancement in automated close range photogrammetry. *Photogrammetric Record*, 20 (11), 205–217.

Geodetic Systems Incorporated, 2005. See http://www.geodetic.com/ (accessed 23 August 2005).

Johnson, G., 2004. *Deformation determination and comparison of the trestle bridges on Puffing Billy using digital close-range photogrammetry*. Internal student project report, Department of Geomatics, University of Melbourne, Victoria, Australia.

Photometrix, 2005. See http://www.photometrix.com.au (accessed 23 August 2005).

4 Engineering and manufacturing

Stuart Robson and Mark Shortis

4.1 Introduction

4.1.1 General outline of the application

Close range photogrammetry has been in use for the measurement of engineering structures and manufactured objects for very many years (Atkinson, 1976), initially using imaging systems based on glass plates (Brown, 1971; Moore, 1973; Cooper, 1979), then imaging onto photographic film (Brown, 1984; Dold, 1990) and now almost universally employing digital imaging systems (Beyer, 1995; Fraser and Shortis, 1995). Photogrammetric techniques are typified by demands for sensor calibration, precise photo-coordinate measurement and rigorous least squares based network adjustment in order to deliver specified 2D, 3D and 4D measurements of engineering structures and manufactured artefacts constructed from a wide variety of different materials.

Typical scenarios involve objects with sizes in the range 1–10 m but characterisation of objects as large as 100 m or more is possible. Measuring accuracy is typically in the range of a few tens of micrometres to tenths of a millimetre. For example, relative precisions well in excess of 1 part in 100,000 (Peipe, 1997) can be achieved with cameras having an image resolution of the order of 6 million pixels. These levels of precision require high contrast targeted points in the object space and image measurement of their centroids, template matching or ellipse fitting (Luhmann et al., 2006). Applications are extremely diverse and include a wide variety of uses in the aerospace, automotive, shipbuilding, petrochemical and general manufacturing industries, as well as more general application to production line monitoring, static and dynamic structural monitoring and testing, and dimensional quality control. The significant advantage of photogrammetry as a measurement tool is that it is a reliable, rapid non-contact technique, able to provide an instantaneous measurement record that can readily be adapted to a range of object sizes.

Engineering and manufacturing are reliant on accurate, precise and reliable data; accordingly any measurements made must include traceability, typically to the level associated with a measurement standard (e.g. VDI/VDE, 2002). It is therefore appropriate to consider photogrammetry as one of a range of optical coordinate metrology systems that can be applied to this field. Other such optical systems in active use include laser trackers (Lau et al., 1986) and laser scanning systems. An outline of the capabilities of laser scanning systems is given in Chapter 10.

As with most photogrammetric applications, development in the field has paralleled advances in imaging and computing technologies. These advances have culminated in current camera systems which combine both light sensing and computing capabilities in order not only to record images but also to make 2D image measurements within the camera unit itself. Further development in computer graphics technologies have allowed photogrammetric systems to evolve in their output from tables of coordinates and simple line drawn graphics, through integration with CAD/CAM systems (Patias and Peipe, 2000) to real-time 3D visualisation of pertinent information and its relationship to the observed structure (Woodhouse *et al.*, 1999). The major effects on photogrammetry of these technological changes have been: more automation and less reliance on skilled operators; increased choice of camera systems; measurement techniques and data-processing options applied to many more application areas; very high accuracy (up to 1/1,000,000 of the size of the object (Fraser, 1992; Burch and Forno, 1983; Gustafson, 1997); and, opportunities for "real-time" (1/25 second and faster) image sequence acquisition and measurement.

The term vision metrology is often used to describe the technique (Fraser, 1997). The term conveniently embraces the combination of photogrammetric and machine vision processes that have been combined as image capture, measurement and data processing systems have become almost exclusively digital.

Photogrammetric systems for vision metrology include the use of, or rely upon the following:

- high contrast information on the object, for example retro-reflective targets;
- LEDs, projected light patterns and in some cases optimally illuminated edge and texture information;
- a network of images taken with a single mobile digital camera or a set of pre- or post-calibrated digital cameras;
- rapid computation of 3D spatial information; and,
- statistically rigorous computation of 3D spatial information to determine and maintain data quality.

4.1.2 Application requirements

For the purposes of this chapter manufactured objects are taken to be those which are repetitively made through the use of production line approaches such as aircraft and automotive parts whilst engineering measurement is applied to one-off constructions including structures such as oil rigs and bridges.

Manufacturing

Manufacturing is critically dependent on precision machinery, process knowledge, as well as computer integration and automation technology. In recent years a general goal has been to increase both productivity and quality through the process of "first part correct" (FPC) manufacturing (U.S. National Institute of Standards and Technology, 2006). FPC was defined as "the ability to transition from design concept to a finished product with absolute certainty of a correctly produced part or product". Emphasis is on the ability to produce a high quality first product and a transition from one to many without interruption. Accordingly there has been an emphasis on integrating quality control within the manufacturing process to gain the immediate feedback benefits of close integration during production of the "as built" component and its computer-based design. This contrasts with the use of traditional spot checking quality control processes which, in the worst case, required the manufacturing process to stop whilst the object was removed from the production environment for checking. Quality control was thus seen as a disruptive, but necessary, process that interfered with efficient production. Under current manufacture where components may be sourced from all over the world, the emphasis is not only on in-house metrology but also standardisation of metrology practice across the complete supply chain.

Advances in the manufacturing of multi-part objects such as aircraft wings and cars have seen a move from static mechanical reference systems (jigs) that physically describe the shape of the object to computer-based numerical reference systems that are directly integrated with the design of the manufactured article. Provided that the numerical reference can be physically established in the production environment, numerical systems provide great flexibility because, when the design is modified, the numerical reference will also be updated without the need to mechanically modify the reference jig. Numerical systems also allow the use of computer simulation in the design of the measurement network and remove the need for the expensive and bulky master gauges that were necessary to mechanically verify each jig.

Engineering

Engineering measurement, whilst embracing similar measurement technologies, is generally applied in a more *ad hoc* manner since engineered structures are typically individual and are built on a project-by-project basis. Typical requirements are knowledge of where the elements of a structure must be placed in the initial setting out process and monitoring of how the structure performs both during construction and subsequent use. Experimental engineering structures may also be built specifically for the purposes of ascertaining how different material types and construction methods can be expected to perform prior to their acceptance as a standard method. In some cases experimental structures are tested to destruction. Chapter 3 provided several examples of the process of monitoring engineering structures.

A list of possible applications in engineering and manufacturing would exhaust the space available in this chapter, but a representative list is as follows:

- analysis and validation of 3D shape at the manufactured component level and of complex structures;
- on-line measurement and verification whilst objects such as tooling jigs are built to a 3D design;
- measurement of large objects such as radio antennas, ships, aircraft, and oil and gas platforms;
- rapid or remote measurement in dangerous environments such as chemical or nuclear plants; and,
- measurement of dynamic situations such as crash testing, robot performance and experimental validation of structural components.

In meeting these requirements systems often have to:

- be flexible to deal with difficult and demanding tasks;
- be capable of measurement on-site away from a specialist laboratory;
- have the ability to deploy multi-station measurement solutions in order to measure the complete object or structure;
- be able to measure large objects, say greater than 30 m;
- accommodate a wide variety of users with diverse methods for the collection of data that are often constrained by the requirements and nature of the working environment;
- integrate with the CAD specification of the object to be measured to allow comparisons to be made; and,
- enable the rapid adoption of changes to the design or predicted performance of the object being measured.

Systems therefore need to be compact, mobile and have flexible configurations. The measurement system also needs to reference not only the physical object but also its numerical CAD definition, whether this is a series of identifiable surface locations such as bolt holes in a metal plate, a geometric shape such as a parabolic antennae or a complex surface definition such as an aircraft fairing.

4.1.3 Quality control: are the measurements fit for purpose?

An important aspect of industrial measurement is the need not only to achieve high accuracies but to be able to quantify in some way the quality of the spatial data that have been obtained. Furthermore, a specified tolerance is often set so it is important to design the photogrammetric measurement process to achieve that accuracy at reasonable financial cost. Accordingly, results such as 3D coordinates from photogrammetric computations are only of practical value if it is possible to describe fully their quality. Without this description it is difficult, except perhaps from past practical experience, to know whether or not the coordinates are fit for purpose.

Assessment of quality

Three measures of quality are important: uncertainty, reliability and accuracy (see Section 2.4.2). Definitions which are more familiar to those working in industrial metrology are that uncertainty is a measure of the repeatability or precision of the coordinate data and reliability is the sensitivity to gross errors (also known as blunders or outliers) in the data (UCL, 2001). Internal reliability refers to the ability to detect gross errors whilst external reliability is a measure of the effect of gross errors on the coordinate data.

Relative precision, the ratio of the mean target coordinate precision, as derived from the photogrammetric network solution, to the largest span of the target array, is often used to quantify uncertainty and compare the performance of various systems.

Accuracy is commonly defined as the closeness to the truth, and in the case of industrial measurement requires comparison with independent measurements. Due to commercial sensitivities and the difficulty and additional cost of providing independent measurement, accuracy is rarely reported on in the public domain. Relative accuracy may be reported if an independent assessment is available. Relative accuracy is the ratio of the target coordinate rms error to the largest span of the target array, where the rms error is computed from coordinate differences with respect to the independent determination of target coordinates after a three-dimensional similarity transformation between the two coordinate sets.

Potential sources of error

For photogrammetry, as with all metrology systems, the sources of error can be grouped into those associated with the following:

- Operator errors such as misidentification of targets, inappropriate set-up of the workpiece to be measured, poor network design or poor selection of dimensions to be measured. Automation of systems tends to minimise operator input and thereby reduce the possibility of some errors.
- Calibration or other systematic or gross errors associated with the instrumentation, minimised by the use of self-calibration.
- The manufactured object or structure to be measured may introduce errors. For example, the workpiece may have a number of components and the interface between, or stability of, these components may not be accommodated.
- Instabilities in the measurement environment such as temperature variation and vibration may introduce systematic or gross errors. Most practitioners in the field are familiar with

the complexities introduced by differential thermal expansion within the workpiece or of the problems associated with precise measurement in the presence of vibration. In many cases operators report that, in the absence of any systematic error or calibration artefacts, environmental effects become the dominant source of measurement error.

It is vital to establish operational procedures which ensure that the potential for these major errors are minimised or that sufficient redundant measurements are taken to ensure that such blunders can be readily detected.

Network geometry

The basic photogrammetric computational unit is the pencil of rays that originates at the object, passes through the exposure station and is produced as an image point. The technique involves simultaneous least squares adjustment of all rays from all exposure stations to all measured image points. The estimation procedure utilised in the bundle adjustment incorporates both a functional and stochastic model. Mathematical models can be envisaged as simplified descriptions of reality, in this case, the functional or deterministic part being an approximation of the image geometry. The relationship between the models is complex and any unmodelled physical effects tend to propagate into the stochastic model, leading to underestimation of the precision of the measurements. For self-calibration (see Section 2.4.1 for a brief introduction to the concept), the unknown parameters from the extended functional model may be simultaneously estimated during this procedure (Cooper and Robson, 1996). Analysis of the results of the least squares estimation solution for the bundle adjustment is vital to refining the extended functional model and detecting gross and systematic errors.

Given the extended collinearity equations as a functional model, the major limiting factors are the number and location of camera positions and the precision with which the photo-coordinates are measured. For a strong estimate of the camera calibration parameters (as required for this work) increased redundancy in the form of additional photographs is desirable, as is a convergent geometric configuration between the photographs and the object. Section 3.2 presented a mathematical formula to assist the description of these design considerations, but readers desirous of further explanation may wish to consult Granshaw (1980).

Figure 4.1 presents a simple example where 3D ellipsoids describe the uncertainty of intersected target points viewed from two camera stations. Ellipsoids have been generated by simulation and are exaggerated to demonstrate the effect of varying intersection geometry. The ellipsoids at the edges would meet a given 3D tolerance, but those toward the centre of Figure 4.1 would fail. In the extreme case (not illustrated) where a target on the workpiece and the two cameras are collinear it is obvious that no solution is possible (the target could lie anywhere along the line connecting the two instruments). The reader is referred to Chapter 4 in Luhmann *et al.* (2006) for a detailed explanation and graphical examples.

In order to describe a general case, Grafarend (1974) and Fraser (1996) have summarised the problem of geometric network design in terms of four interconnected stages:

- Zero-order design (datum) involves the choice of the datum definition such that specific criteria associated with the project are met. For example, the use of minimum constraints such that the datum introduces no shape distortion of the estimated bundles of rays. In general terms, the concern is to produce an optimum co-factor matrix of the estimated parameters.
- First-order design (configuration) concerns the creation of an optimised network geometry, as described in Section 3.2. In the case of calibration, a multi-station convergent imaging geometry has been shown to be desirable.

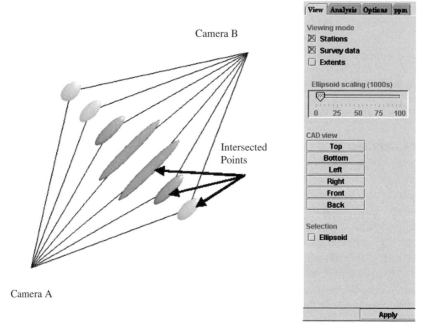

Figure 4.1 *Computed 3D ellipsoids demonstrating the uncertainty of intersected from two camera stations. Reproduced with permission of UCL (2001).*

- Second-order design (measurement) is aimed at maximising the precision of image coordinate measurement, such that the elements of the weight matrix approach the identity matrix. Potential for optimisation, apart from the obvious sensor resolution issues are imaged target size and measurement algorithm (Section 4.2.2) and the taking of multiple exposures.
- Third-order design (densification) concerned with the enhancement of precision by the addition of observations. Object point precision is largely independent of target array density for multi-convergent imaging geometries, so third-order design is not as important for optimisation of self-calibration. Densification is however very easily achieved with today's digital photogrammetric systems which can measure many thousands of image locations per second.

Automating the design of photogrammetric networks in order to account for complex object geometry typified by lines of sight of limited length and occlusions is work still in progress. Notable results have been reported by Mason (1994) using expert systems and more recently by Olague (2002) who employed genetic algorithms and heuristic computer simulations. An example of a semi-automated procedure including placement of the camera by a robot is given in Section 4.4.2.

Standards

An important driver in the industrial sector is to set internationally accepted standards that facilitate metrology which is transferable from engineering project to project and manufacturing supplier to central manufacturing facility. In this aspect coordinate measurement machines (CMMs) have led the way in conformance to recognised standards (Flack, 2001), whilst standards for measurement of targets have been established for laser trackers (ASME, 2005) and for photogrammetry VDI/VDE Part 1 in Germany (VDI/VDE, 2002). VDI/VDE Part 2 covers free-form measurement systems and Part 3, which is due for publication in spring 2008, deals with surface measurement systems that use some sort of orientation or navigation tool in order to register point clouds from different stations (Luhmann, 2006).

Good practice

On a more general note there are six guiding principles to good measurement practice that have been defined by the UK National Physics Laboratory (NPL) as part of the Computing Precisely Initiative. They are:

- The correct measurement: measurements should only be made to satisfy agreed and well-specified requirements.
- The correct tools: measurements should be made using equipment and methods that have been defined to be fit-for-purpose.
- The correct people: measurement staff should be competent, properly qualified and well informed.
- Regular review: there should be both internal and independent assessment of the technical performance of all measurement facilities and procedures.
- Demonstrable consistency: measurements made in one location should be consistent with those made elsewhere.
- The correct procedures: well-defined procedures with national or international standards should be in place for all measurements.

The industrial measurement specialist should seek to apply these principles through identification of the geometric, environmental and operational factors that could undermine system integrity and thus compromise conformity with these guiding principles.

4.2 Practical application of the technique

This section aims to provide the reader with an overview of currently accepted photogrammetric measurement capabilities and the key developments that have been necessary in order to convert photogrammetric techniques from their specialist, research-orientated beginnings into push-button measurement systems appropriate for integration into manufacturing and engineering applications.

Given the capabilities of self-calibrating bundle adjustment solutions with blunder detection, which have been available for the last two decades, the main advances in the technique necessary for its efficient deployment in an industrial environment are:

- digital camera technology;
- retro-reflective targets and their measurement in digital imagery;
- coded targets and coded objects;
- automated computation of camera positions and orientations; and,
- automated correspondence solutions which solve the issue of which target image emanates from which physical target.

4.2.1 Digital camera systems

Digital still cameras have gained wide acceptance for high precision vision metrology systems (Beyer, 1995). The acceptance of these cameras is based principally on the efficiency with which they can be used and accuracy levels which can be achieved. The cornerstone of this efficiency is the use of digital images, which can be rapidly acquired and processed (see Figure 4.2). In particular, advances in the automation of vision metrology systems (Fraser, 1997) would not be feasible without the use of cameras which capture digital still or video-rate images.

The key digital still image camera which started the widespread move to vision metrology is the Kodak DCS460 camera (Susstrunk and Holm, 1995; Peipe, 1997). The DCS460 is designed essentially for photo-journalism, but has many desirable features for a digital image photogrammetric camera. The high resolution of the charge coupled device (CCD) sensor combined with high image storage capacity and ready availability made the DCS the common choice for vision metrology applications. Some ten years on from the launch of the DCS series there are now many SLR other comparable and more sophisticated cameras available on the market. Table 4.1 gives the details of some current versions, the top four being representative of current high-end professional camera models and include the DCS460 for comparison. The principal differences between the older DCS and the newer cameras are those of cost, frame rate and battery life.

Digital cameras should be equipped with ring lights or flashes if they are to be used in conjunction with retro-reflective targets to acquire high contrast imagery of the signalised points of interest. The precise and accurate measurement of each imaged target location is a fundamental requirement if the stringent measurement tolerances demanded by industrial inspection applications are to be obtained.

Figure 4.2 *Examples of three digital camera systems used for industrial measurement purposes: (a) Nikon/Kodak digital SLR (Reproduced with permission of Jarle Aasland, NikonWeb.com); (b) iMetric ICAM 28; (c) GSI Inca 3.*

Table 4.1 Examples of digital camera systems employed within vision metrology systems. Note that this table does not include panoramic systems

Model name	Image dimension (mm)	Number of pixels (millions)
Kodak DCS 460	28 × 18	6.2
Nikon D100	24 × 16	10.0
Kodak DCS Pro 14	36 × 24	13.9
Canon D1 S	36 × 24	16.6
IMetric ICam 28	*86 × 49*	*29.4*
GSI INCA 6.3	*28 × 18*	*6.2*
GSI INCA 3	*31.5 × 21*	*8.2*

The last three rows in Table 4.1 (in italics) are purpose built for industrial measurement purposes and are representative of the state-of-the-art in the design of photogrammetric imaging systems. Their designs incorporate electronic ring flash units for even illumination of retro-reflective targets, on-board computer processing for effective retro-target image compression and measurement of target images at the camera. The latest model, the GSI INCA 3, includes a laser projection system to assist in pointing the camera system optimally at the object under poor lighting conditions.

A second type of camera, typified by the Redlake ES series, is able to deliver a continuous stream of digital images at rates of 30 frames per second and higher. Such systems have been synchronised and deployed in pairs and greater numbers for monitoring engineering structures undergoing dynamic change (Robson and Setan, 1996; Woodhouse and Robson, 1998). A key limitation to the use of such systems in delivering optimal photogrammetric networks is the length of the cable connecting each camera unit to its host computer which in general needs to be less than 10 m for error-free operation.

In common with any camera used for metric applications, calibration is necessary for digital still cameras as systematic errors in the collinearity condition must be modelled or eliminated. Both the off-line and on-line modes of operation typically deploy a camera (or cameras) with a fixed focus setting to minimise variation in principal distance and other focus dependent interior orientation parameters (Fryer, 1996).

Given due attention to network design criteria (Section 4.1.3), the measurement performance of vision metrology systems based on digital still cameras is dependent on the image resolution, image scale, image measurement precision and camera stability. The principal influence on accuracy for most digital camera systems is the resolution of the CCD sensor and the internal stability of the camera as it is moved from location to location in order to acquire the images that constitute the photogrammetric network (Fraser and Shortis, 1995).

4.2.2 Targets and target image measurement

Measurement with retro-targets

Retro-targets are made of a thin retro-reflective material which returns light very efficiently to the light source. Retro-targets first appeared in the photogrammetric literature in the early 1980s (Brown, 1984b) and have been widely adopted and adapted to a multitude of measurement tasks. Their application onto the object to be measured is fundamental to the use of vision metrology for high accuracy coordination, giving at least an order of magnitude greater precision than is available from the use of natural features such as edges or corners, which can suffer badly from spurious illumination effects (Luhmann *et al.*, 2006). Small circular retro-targets, typically employed in industrial measurement, produce a Gaussian intensity distribution in the image which provides point-to-point consistency and greatly simplifies the image measurement process. Images of larger retro-target images exhibit significant view dependent geometric shape variations, the elliptical shape of which must be accounted for in determining the photogrammetric centre of the target (Zumbrun, 1996; Dold, 1996).

The key to retro-target image measurement is the ability to estimate, to sub-pixel accuracy, the centre of each imaged target. There are two tasks to be performed in this process: recognition and location. Recognition of the target images is required to unambiguously identify targets within a scene and provide a coarse location for a local window or boundary. The precise location of the target image is generally a second process that determines the target image position within that local window or boundary.

Coarse locations can be determined using prior geometric knowledge of exposure station and target positions (Haggren and Haajanen, 1990), scanning the entire image based on a global threshold (Shortis *et al.*, 1994) or searching for coded targets (van den Heuvel *et al.*, 1992). Once located, target image centres can be computed using intensity weighted centroids (Trinder, 1989), edge detection and ellipse fitting (Zhou, 1986), or least squares template matching (Baltsavias, 1991). Centroids and ellipse fitting require the elimination of background or noise by the definition of a local threshold. Whilst template matching may be independent of high contrast targets, the technique requires an initial template to be defined and can become computationally inefficient unless there are sufficient algorithmic constraints (Grün and Baltsavias, 1988).

The following workflow illustrates a common procedure (see also Figure 4.3):

- Threshold the image to isolate target information from the background.
- Search for the edges of the imaged target and store the pixel grey values within this boundary.
- Compute the centroid of the data to provide an image measurement with sub-pixel accuracy.

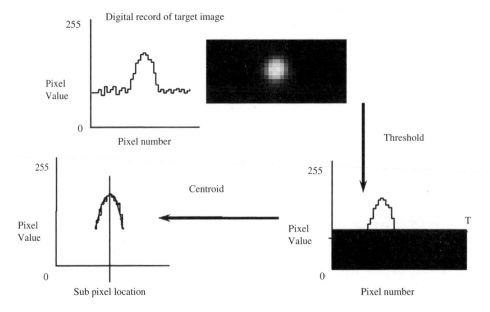

Figure 4.3 *Simplified sub-pixel location of a retro-reflective target image.*

A key principle of the selected image measurement method is that for repeated images with no physical variation in camera and object location, the coordinates of the computed centroid of each target in the image plane should remain constant at the level of functional resolution of the sensor. Analysis by simulation can predict the expected image space precision of target centring algorithms. Most studies are based on assumptions of grey-scale imagery, random noise and minimum target spans of a few pixels and predict image space precisions as fine as ±0.01 pixels (Stanton *et al.*, 1987).

There have been many practical tests of image space precision of target centring algorithms. Chosen test methods vary from single-target monitoring (Robson *et al.*, 1993) to multi-station convergent networks comprising many targets (Shortis *et al.*, 1995). The consensus, which has remained consistent for the past few years, is that in practice centroid-type algorithms can realise object space precisions equivalent to ±0.02–0.03 pixels. Few tests have included verification in the object space. Of those reported, object space results have generally been less favourable, indicating imaging accuracies of ±0.02–0.07 pixels (Fraser and Shortis, 1995). Whilst sub-pixel measurement techniques undoubtedly produce good results, it is prudent to consider what factors affect the geometry and radiometry of the digital image intensity distribution that is being measured.

The main practical drawback of the use of targets is the need to apply them to the object surface in such numbers that the presence of a target is assured wherever a measurement is required. Further, if the physical target location is common to a 3D point in the CAD model the target must be placed on that point. Several companies manufacture a range of metrology orientated retro-reflective targets attached to mounting studs, bosses and plates that ensure that they can be correctly located. Targets may also be accurately located onto particular objects so that the computed target locations can be used in conjunction with a CAD model to determine the respective position and orientation of individual components.

Such a scenario is greatly assisted by the use of coded targets. Coded targets are individually marked with, for example, a known constellation of surrounding targets, annuli or lines (Shortis *et al.*, 2003) in order to enable automatic decoding of their individual identification within the each image (see Figure 4.4).

If the target locations are known *a priori*, either from a prior survey or their accurate machining onto a stable 3D reference artefact such as the GSI Autobar (Figure 4.4), automatic identification also facilitates automation of the resection process (see Section 2.2.2 and Section 4.2.3). Further designs are possible in which a set of target codes fixed to an accurately machined right angle for example allow the measurement of edges by placing the right angle of the target in physical contact with the edge to be measured. Further examples of this type of technique are given in the first two case studies (Section 4.4).

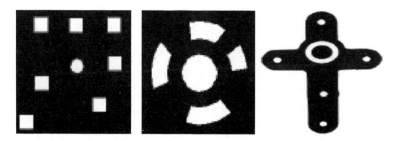

Figure 4.4   *Coded targets and a GSI Autobar orientation device.*

Other discrete target types

Passive, non-retro-targets may be used for a number of reasons, such as preventing damage to delicate surfaces or where thickness or weight are critical issues (Pappa *et al.*, 2001a; Shortis *et al.*, 2002). In these cases the lighting of the work piece or surface becomes an important consideration to ensure that high contrast target images are obtained.

Active LED targets have been employed in some specialist applications, such as aerospace models that have very high tolerances on the quality of flight surfaces (Burner *et al.*, 1997). A number of commercial, on-line systems also use active targets based on LEDs. Systems available from manufacturers such as Metronor, IKON, Metris and Optotrack are designed for the real-time tracking of the order of 50 active targets.

Measurement with projected targets

In cases where objects have surfaces on which it is either undesirable or not possible to place retro-targets, for example in a time constrained manufacturing environment, natural features, edges or projected targets must be used. There is a wealth of different solutions to such problems, two such examples being given in Section 4.4 of this chapter, namely the use of an electronic flash-based spot projector to measure solar sail structures and the use of imaged edges to determine the size and shape of steel plates in a shipbuilding environment.

In the case of the spot projector (Jones and Pappa, 2002), image measurement algorithms designed for retro-target based use can be applied directly. A key difference for projection systems when compared to physical targeting is that the projected target location is a combination of the object shape and its location with respect to the projection system. Accordingly the coordination and monitoring of discrete locations on the object is not generally possible with such techniques.

Measurement of natural features

Natural features and edge information are very attractive from a manufacturing standpoint to define points of interest without the use of targets. However, alternative image measurement algorithms are required which often have to be tailored to the individual circumstance. Further, the design of the system illuminating the object to be measured must be thought through with extreme care in order to maintain consistent image quality and to avoid systematic shadowing effects that are viewpoint dependent (Luhmann *et al.*, 2006). The use of edges also requires an accurate and reliable calibration and the use of epipolar lines to determine correspondences, or a paradigm shift from point-based photogrammetry to line-based photogrammetry (van den Heuvel, 1999).

4.2.3 Photogrammetric workflow

Photogrammetry requires lines of sight from two or more images from differing viewpoints in order to compute 3D coordinates. If the object to be measured can be regarded as being geometrically stable for the time required for image acquisition then the required images can be captured with a single camera that is moved from one image location, or camera station, to the next. It is also possible to keep the camera stationary and move the object to achieve the same end, provided that the shape of the object remains constant. The term "off-line" photogrammetry is often used to describe this mode of working. Image measurements from the resulting image network are then processed to yield the 3D location of each imaged target using the process outlined in Section 2.3. Basically this process involves: measuring image coordinates; the mathematical adjustment, which sometimes includes calibration of the cameras; confirming the reliability of the results; and preparing the results in a form suited to the client.

One technically interesting and demanding situation which can arise with automated processing for the coordinates of the targets on images involves solving what is known as the "correspondence problem", that is, which target or feature in any other image in the network corresponds to the selected target or feature in a given image (Maas, 1992b). One solution to this problem is illustrated in Figure 4.5 and involves the generation of an epipolar search space that defines the possible geometric relationship between the points in each image. The search space can be solved either on a piecewise basis or more efficiently using a tree structure (Chen, 1995).

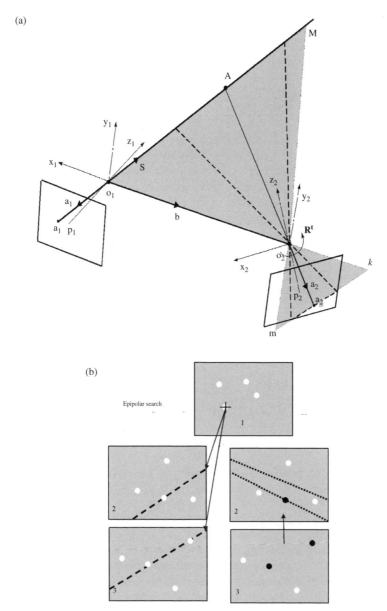

Figure 4.5   *Epipolar plane, epipolar lines and an epipolar search example.*

The example given in Figure 4.5 demonstrates three images in which the epipolar line from a candidate point in image 1 is projected into images 2 and 3. Image 2 contains no ambiguity, but image 3 has two target image points located on the epipolar line. The ambiguity between images 1 and 3 can be solved by projected epipolar lines from the two candidate points in image 3 back into image 2. The technique has been extended, using a variety of different approaches, to accommodate large numbers of images and imaged targets in most of the commercial photogrammetric software systems described in Section 4.2.5.

Off-line solutions

To attain the high accuracies required for industrial and manufacturing applications, networks of images are typically used. An example of such a process would be the confirmation of "as built" engineering structures to their design specifications, for example: wing sections, planarity and alignment of surfaces, full-scale building monitoring surveys.

Given a well-designed photogrammetric solution, accuracies of the order of 0.05 mm on a 5 m structure are typical. It is important to recognise that rigorous independent checks of object space accuracy, where available, can show degradations of the order of 100% when compared to internal measures of precision (Shortis *et al.*, 1995a; Fraser and Shortis, 1995). In other words, relative accuracy is typically poorer than relative precision. There are many potential sources of systematic error which may contribute to the disparity between relative accuracy and relative precision. The primary potential sources which have been identified are calibration stability (Shortis and Beyer, 1997), variation of distortion (Fraser and Shortis, 1992; Shortis *et al.*, 1996), unflatness of the CCD array (Fraser and Shortis, 1995), perspective distortion of targets (Dold, 1996) and optical aberrations (Robson and Shortis, 1998). Alternative calibration methods using finite element-based modelling have also been investigated with some success (Munji, 1986; Tecklenburg *et al.*, 2001). As previously noted, a standard for target-based off-line photogrammetry is available (VDI/VDE, 2002).

On-line solutions

An alternative method of image acquisition and data processing is used where the object to be measured is unstable. A network of synchronised cameras must be used to acquire the images that will contribute to the network. Such a process is often classified as "on-line" photogrammetry. Systems for this purpose typically employ two to four cameras and are designed for use with a targeted measurement probe which is brought into contact with the object to be measured. Often purpose-built on-line systems are deployed to measure structures that are either unstable or must be captured repeatedly over a period of time whilst undergoing deformation. Processes for this type of system are similar to those used for off-line systems except that the cameras are generally pre-calibrated and assumed to have a stable internal geometry for the duration of the image sequence. Coordination from pairs of cameras is governed by the geometry of two ray intersections and is typically limited to accuracies in the range of 0.1 mm on a 5 m object.

4.2.4 Vision metrology and machine vision

A traditional standpoint would regard vision metrology solutions as those that seek to measure objects for purposes such as conformance between a manufactured article and its original design, rather than machine vision orientated imaging systems that typically seek to automatically classify objects of different types from a database of possible types. Since both systems employ digital imaging technologies and computer-based data processing it is possible to carry out both in the same system. For a continually updated comparison the reader is advised to refer to the current International Society for Photogrammetry and Remote Sensing (ISPRS) Commission III and Commission V terms of reference (International Society for Photogrammetry and Remote Sensing, 2005) At the time of writing those for Commission V Working Group 1 are the most pertinent to industrial measurement. These terms of reference are summarised as:

- passive and active vision systems for industrial metrology, automation in vision metrology, image engineering;
- multi-sensor and hybrid systems;
- sensor and system calibration, accuracy assessment and verification, discussion of standards;
- free-form surface measurement techniques, illumination and projection methods (structured light, illumination algorithms);
- on-line and real-time systems, dynamic and high-speed processes;
- machine vision, industrial quality control and industrial robotics;
- CAD/CAM integration and modelling, use of domain knowledge for automation;
- new application fields;
- integration of system manufacturers and service providers;
- integration of CAD/CAM into the photogrammetric measurement process;
- digital photogrammetric systems for industrial mensuration; and
- transfer of photogrammetric technology to the industrial design, engineering and manufacturing sector.

Recent collaborative research has promoted the photogrammetric bundle adjustment into the machine vision domain as an optimal solution for processing data from networks of images (Triggs *et al.,* 2000) and is extending the technique, for example to provide a solution from very close spaced multiple views as might be delivered by a mobile video camcorder (Pollefeys *et al.,* 1999; Bartoli, 2002). Work is also ongoing to allow panoramic imagery to be optimally processed within a bundle adjustment (Grün and Parian, 2005).

4.2.5 Appropriate/available software

A variety of photogrammetric systems capable of industrial measurement is available. The list in Table 4.2 is by no means exhaustive but provides a sample of products in general use for photogrammetric measurement using retro-reflective targets.

Table 4.2 Products suitable for photogrammetic measuring

Product	Company	Website
Australis	Photometrix	www.photometrix.com.au
DPA-Pro	AICON 3D Systems GmbH	www.aicon.de
FotoG	Vexcel Corporation	www.vexcel.com
Hazmap	As-Built Solutions	www.absl.co.uk
RolleiMetric CDW	Rollei	www.rollei.de
PhotoModeler	Eos Systems Inc.	www.photomodeler.com
ShapeCapture	ShapeQuest Inc.	www.shapecapture.com
Tritop	GOM mbH	www.gom.com
VMS	Geometric Software P/L	www.geomsoft.com
V-Stars	Geodetic Services Inc.	www.geodetic.com

4.3 Alternative systems

The range of systems available for 3D measurement of either discrete points or surfaces to industrial specifications is enormous, demonstrating the fundamental demand for this type of measurement by a wide variety of end users. The following paragraphs describe the major types of system available.

CMMs rely on accurate knowledge of the 3D location of a touch probe which is brought into contact with the object to be measured. They are available in either gantry-based or arm-based designs. Gantry systems employ a three-axis motion system to move a contact probe into location. Capable of precisions of a few micrometres, they are static in their deployment and require location in a carefully controlled environment conferring the constraint that the object to be measured must be brought to the machine. Typically such systems are designed to measure objects approximately a metre in size, but purpose-built very large systems, for example on the A380 production line at Airbus UK, and specialist miniature systems, for example that developed by the UK NPL (Lewis *et al.*, 2001), are possible. Arm-based systems are of significantly lower cost and much more flexible in their deployment. Arm designs rely on a mechanically jointed arm in which angle encoders and known segment lengths are used to coordinate the location of a touch probe to the order of 0.1 mm or better. Measurement volumes are typically of the order of 1–2 m radius, but may be extended by the use of extension rails or pre-surveyed locations that allow relocation of the arm whilst maintaining a common coordinate datum. Both systems are capable of providing good solutions where only a few points need to be measured on a stable object. Gantry systems can be completely automated to deliver repetitive measurements in a production environment.

Alignment telescopes, manufactured by Brunson and Taylor-Hobson for example, allow precise lines of sight to be realised on the shop floor. These systems are closely associated with the alignment and maintenance of master gauges. Not being easy to integrate with the demands of a numerical reference system, they are being superseded by laser tracking systems.

Industrial theodolites that allow multi-station intersection to a single target have been available for many years. Whilst capable of measurement accuracies of a similar order as off-line photogrammetry they are largely manual in operation and cumbersome to use with operator fatigue being a key factor in their ability to deliver spatial data of consistent quality. In common with optical alignment systems, they are being superseded by laser trackers.

Laser tracker systems consisting of a laser-based range measurement system (either an interferometer or an absolute distance meter) combined with a two-axis rotation and angle measurement system have had a significant impact in the industrial measurement field. They are able to track an accurately manufactured retro-reflective target at rates of several thousand times per second. Tracker systems are heavily integrated into the numerical reference system through common interfaces such as New River Kinematic's Spatial Analyzer product. Manufacturers include Leica, API and Faro, who all manufacture instruments with spatial positioning accuracies of the order of 0.03 mm over ranges of up to 30 m. For larger structures, such as ships and bridges, total stations, consisting of an industrial theodolite equipped with an electromagnetic distance measurement unit, provide a less expensive polar measurement alternative, but with commensurately lower accuracy and precision. Examples of such systems are Sokkia Monmos and the Leica TDA 5000 series. These systems have had an impact on the extent to which photogrammetric solutions are used since they have distinct advantages where only a few tens of points are to be coordinated or a static surface needs to be measured. Photogrammetry maintains the advantage where instantaneous capture of many hundreds, and even thousands, of targets in a single instance is required.

Laser scanning systems (see Chapter 10) are also deployed in industrial and engineering applications, particularly on larger structures such as bridges and petrochemical facilities where the typical ±5 mm coordination ability of time of flight scanning systems is acceptable relative to typical design tolerances. There is an increasing number of scanning systems that are purpose built for the manufacturing environment and fall into two prominent categories. The Metris Laser Radar (formerly Metric Vision) is an example of a system that employs multiple lasers and very highly specified signal processing electronics to deliver polar scanning measurement at a precision of 0.03–0.5 mm over ranges in the tens of metres. The second group of systems are typified by the Leica T Scan and the Metris K Scan both of which utilise triangulation scanning systems that were originally designed to be mounted in place of the touch probe on a CMM. Both the T- and the K Scan use active tracking systems to compute the position and attitude of the laser scanner during the scan process so that the scanner can be hand-held rather than constrained by a mechanical CMM arm. These latter systems acquire data in swathes of the order of a few centimetres wide. The advantage of these systems is in their ability to capture dense surface information without recourse to any targets, edges or identifiable surface features. Currently they are limited in scanning rate such that the object to be scanned needs to be rigid (although not static).

A further system, "indoor GPS", manufactured by Arc Second (Arc Second, 2006) utilises a network of transmitters each of which contains a pair of rotating laser beams and a time reference strobe signalling device. The system is designed to be established on an infrastructure basis over an entire factory and facilitates the use of precision light-sensitive detection units located on the objects to be tracked that convert timing signals given out by the network of transmitters into 3D locations. The system is intended to have application in the final fitting of large aircraft and ship structures where it is necessary to continuously monitor the location and orientation of a variety of components. Measurement accuracies of better than 0.5 mm over ranges up to 30 m are claimed, with the possibility for average readings to deliver accuracies better than 0.2 mm over ranges of the order of 10 m. For outdoor structures the well-established GPS system can be used for coordination purposes of the order of a few tens of millimetres. In both cases, because measurement is at the receiver, multiple receivers can be simultaneously employed for different tasks whilst using the same set of transmitters.

Hybrid systems incorporating combinations of active sensors and photogrammetry include structured light (Maas, 1992a; Gartner *et al.*, 1995; Sablatnig and Menard, 1997) and coded light (Wahl, 1984). Such systems are limited in application to the range over which light may be projected and accordingly are generally designed to deliver measurements of smaller manufactured objects without recourse to using targets. The target point projection method delivered by the GSI ProSpot system could be regarded as example of a longer range system. A useful review of optical techniques for 3D shape measurement has been given by Chen *et al.* (2000).

4.4 Case studies

Given the extremely wide range of possible system options, the case studies selected for inclusion in this chapter focus on conventional photogrammetric methods and ignore hybrid systems. The first two studies can be regarded as exemplars of the use of customised retro-target-based techniques in manufacturing and engineering. The use of such targets is precluded for the third and fourth examples and these utilise projected targets and edges, respectively, in order to deliver data. The final examples are included to demonstrate the flexibility of the technique.

4.4.1 In-production measurement of the wing root of an Airbus aircraft using target-based photogrammetry

Airbus UK produces the wings for all of the Airbus variants at its Broughton, UK site. This example is drawn from in-production measurement for the single-aisle family of aircraft. Wings for these aircraft are manufactured to a fully equipped state in order that they can be attached directly to the fuselage at the final assembly lines in Hamburg (Germany) and Toulouse (France).

At Broughton, there is a requirement to measure the wing dimensions at the root ends prior to their dispatch. In the established process targets are placed on each of the stringers, on the top and bottom skins of the wing. Additional targets are placed on the front and rear spars and the "crown fittings". A reference frame and scale bars are placed around the targets (Figure 4.6). Coordination of the targets is conducted using a photogrammetric system. Due to the ageing tooling and the challenges of maintaining the equipment, the system has recently been replaced and upgraded using current state-of-the-art equipment.

The original system is showing its age for a variety of reasons. The camera used, a GSI INCA 2 (Figure 4.7), is considered large and heavy by those who use it on a regular basis and the ageing tooling is unsupported for breakdown. The process, as originally designed, required a large target consisting of three retro-target elements to facilitate the determination of both position and orientation at each location to be measured (Figure 4.6). The chosen photogrammetric methodology is an off-line process requiring one such target for each location, as opposed to the use of a touch probe and on-line measurement. Whilst the method has provided valuable quantitative data for a number of years, the targets are easily damaged in the production environment and their complexity makes it costly and difficult to replace them quickly. Accordingly any new method needs to demonstrate reduced target replacement and recertification costs. In addition to providing a modern replacement, the new system must have the capacity to accommodate the demands of build rates of the order of 30 wing sets a month. This requirement emphasises automation and the need to keep system set-up, measurement and subsequent data evaluation and validation times to a minimum whilst allowing non-specialist personnel to operate the equipment.

Several metrology systems were investigated to replace the existing process with on-site system demonstrations being conducted to determine suitability. The outcome was that the

(a)

(b)

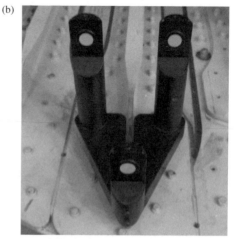

Figure 4.6 *(a) Airbus wing root fitted with targets for photogrammetric measurement; (b) detailed view showing target design. Reproduced with permission of Airbus UK.*

Figure 4.7   *The INCA 2 and INCA 3 cameras. Reproduced with permission of Airbus UK.*

photogrammetric process was the most suitable, but that Airbus UK and NTI Measure in association with GSI would work together to improve the original process with:

- a new more user friendly camera with a higher pixel count and a wider angle of view than the previous model to confer more accurate overall measurement;
- re-design of the targets and hard tooling to make the process more robust and easier to service and recertify; and,
- a new automated photogrammetric data processing software package that would allow a faster and more efficient way of delivering results.

The camera deployed in the updated system is a GSI INCA 3 (Table 4.1, Figure 4.7) with a fixed focal length of 17 mm and a 76° × 56° field of view. The lens focus is fixed such that the useful depth of field is 0.5–30 m. In comparison with the previous model, the INCA 3 has two million more pixels per image and is 3 kg lighter. More significant from a network design and optimisation perspective is the larger viewing angle which makes it easier to obtain images that include the complete wing root in each image in the network. To enhance ease of use on the shop floor, the camera is fitted with a self-calibrating ring flash unit that automatically adjusts its light output to deliver correctly exposed retro-reflective target images. A laser pointer is also included to allow the operator to more accurately align the camera according to the simulated design and generally to take images from more difficult angles.

A new target design which is significantly smaller and lighter thereby reducing the possibility of damage within the production environment completes the system. To enhance rapid placement in the correct position on the stringers (Figure 4.8 (see colour section)) targets are fitted into profiled rubber mats. This method also minimises damage to the target retro-reflective surfaces as they can be positioned without the need to touch the targets. To reduce replacement costs the target itself is a standard component that is used for several measurement purposes and can easily be replaced at relatively low cost.

Additional targets, shown in Figure 4.9, are a crown-fitting type and a vertical zero offset (VZO) hole-fitting design which employ target codes in order to facilitate automatic processing. The VZO target is not only coded, but is precisely machined so that it can use an additional two targets to provide the location of the hole centre into which it is mounted.

The new system is integrated into the production environment and is providing significant improvements:

- Software now processes data in 3 minutes, against the 50 minutes taken by the previous system.
- Complete automation of the photogrammetric process has removed a significant potential for user error.
- The system is suited to day-to-day repeat process conducted on the production line.
- Targets are pre-positioned by rubber matting which maintains repeatability.
- Tooling is smaller, more robust and less susceptible to damage.

4.4.2 Remote photogrammetric survey within the JET fusion reactor

JET, in Oxfordshire in the UK, is the largest experimental nuclear fusion device in the world and has been successfully operated by the United Kingdom Atomic Energy Authority (UKAEA) for EFDA (European Fusion Development Agreement) since 2000 and before that for 17 years by the JET Joint Undertaking (Figure 4.10 (see colour section)). Over the years, many thousands of different components have been installed within the doughnut-shaped vacuum vessel (Figure 4.11) in the search for the optimum operating environment for ITER, a new experimental facility which will advance the work in plasma physics already done at JET for electricity-producing fusion power plants. This new facility will be built in France. Because of a requirement to develop remote handling maintenance within fusion, all of the photogrammetry surveys are carried out remotely using the 13 m long articulated boom available in JET for in-vessel component installation and replacement.

The main challenge of a typical JET remote survey is that every aspect from target location to camera station and the subsequent geodetic network has to be pre-programmed with little opportunity for adjustment or improvisation on the day. Coupled to this is the fact that the survey results are on the critical time path of each installation, which normally requires three

(a)   (b)

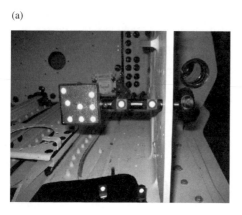

Figure 4.9 *(a) VZO Hole fitting target; (b) a clip on crown target. Reproduced with permission of Airbus UK.*

Figure 4.11 *A portion of the interior of the vacuum vessel. Reproduced with permission of EFDA JET.*

surveys. The first survey follows the welding of prime location lugs onto the vessel wall with a positioning tolerance of ±10 mm. The survey records the actual position of the lugs to an accuracy better than ±0.2 mm. This data is used to machine an adjustment interface block which is then attached to the welded lug and the second survey confirms the fitting of the adjustment block to the required accuracy. The final survey is to confirm the as-installed position of the main component onto the adjustment blocks as being within the ±0.5 mm tolerance necessary for in-vessel components in JET.

To achieve "fit first time" accuracy, it is necessary to simulate the survey in both remote handling and photogrammetry capability. The starting point is to select the component features that require measurement and build a target location within the CATIA CAD (Dassault Systems, 2006) model used for manufacturing. A remote target is then selected from a generic set or specifically designed to suit the component datum location feature. The target will consist of a "handle locking device" and a minimum of three coded retro-reflective targets. There are eight points per coded target, so three such targets introduce 24 points into the system to create volumetric information, including some redundancy, and are calibrated to give the position of the component datum location feature while automatically operating as templates within the bundle adjustment.

The CAD model, complete with targets attached to the components, is supplied to the remote handling software "Division MockUp version 2000i2", where robot pick-up, delivery and fitting techniques are defined and programmed for the targets.

The virtual robot positions cameras at each photo station at approximately 22° intervals from the survey epicentre, creating good triangulation locations. In reality only one (repositioned) camera is used but the simulation (Figure 4.12 (see colour section)) shows nine camera stations and 12 locations. The upper shots are in portrait, the lower are rotated 180° in portrait, and the mid-shots are taken twice in landscape at 90° rotations, which facilitates good self-calibration of the camera by locating the same targets in different regions of the lens. The 1 m minimum focal distance is accommodated by the coloured aiming bar extending from the virtual camera. The actual aiming point for the bar (Figure 4.12) is not determined at this time.

The camera stations, orientation and target positions are exported from MockUp to the GSI photogrammetry software "V-Stars" (Brown and Dold, 1995; GSI, 2006) where virtual photographs are taken with different aiming points to ensure that all targets are seen. In this simulation three aiming points were developed for each station making 36 photographs in total.

A least-squares bundle adjustment is run to predict measurement uncertainty of the network, to indicate potential weak points and thus to facilitate fine tuning of the camera stations (Figure 4.13 (see colour section)).

The real survey (Figure 4.14) gave better results than the simulation because greater camera separation (improved triangulation) was achieved owing to the flexibility of the MASCOT manipulator arms during the manually- driven aiming process. Also, each coded target was simulated as one point, whereas in reality the 66 coded targets actually contained 528 points.

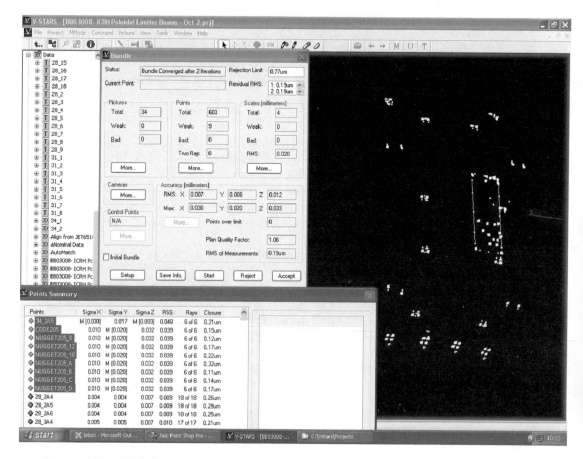

Figure 4.14 *GSI V-Stars network adjustment showing coordinated point locations and precisions. Reproduced with permission of EFDA JET.*

Photogrammetry, coupled to reverse engineering adjustment techniques, has been demonstrated over the last three major configuration changes in JET (and many dozens of surveys per shutdown) to be an effective solution to the "fit first time" requirement of remote maintenance and component upgrade, routinely achieving the required accuracy of 0.5 mm in this 10 m diameter machine. The adaptability of the GSI INCA2 camera being powered and fully remotely controlled via a computer network system, supported by rigorous simulation preparation, has resulted in quality and reliability of surveys equal to or better than that which can be achieved manually.

4.4.3 The measurement of solar sail surfaces using projected dots as targets at NASA Langley

New analytical and experimental methods for shape and dynamic characterization of future "Gossamer" space structures, such as large membrane reflectors and solar sails, have been developed at the NASA Langley Research Centre (LaRC) at Hampton, Virginia, USA. Accurate analytical methods are required for confident design of new or evolved membrane structures and for mission simulations (Jenkins, 2001). In order to support this requirement, photogrammetry has been used to determine the shape and dynamic characteristics of Gossamer research test articles and prototypes (Pappa *et al.*, 2001a, 2002; Shortis *et al.*, 2002). Such prototypes are typically scale models, in either air or vacuum environments.

The ability of on-line photogrammetry to "instantaneously" capture the shape of a complete structure is critical in this application since many Gossamer structures are so flimsy they can change shape from unintentional air currents. More problematic for the photogrammetric technique is that solar sails require highly reflective membranes for their operation in space (they are propelled by reflecting sunlight), and shiny membranes are difficult test objects because photogrammetry uses the diffuse component of reflected light, not the specular component. Accordingly, great care and attention to the use of retro-targets, projected dot targets and illumination is required in order to make reliable use of the technique.

In Figure 4.15a the large structure on the left is half of a four-quadrant 10 m sail concept where the length of each edge is 10 m. Those in Figure 4.15b are 2 m scale models of different sail designs. These research structures are in a 16 m diameter vacuum chamber, large enough to accommodate testing of a complete 10 m solar sail model in both horizontal and vertical

(a)   (b)

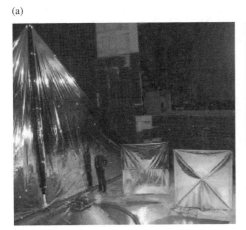

Figure 4.15 *(a) Three solar sail test structures located within a vacuum chamber; (b) structure undergoing vibration testing. Reproduced with permission of NASA LaRC.*

orientations. The image in Figure 4.15b shows a 2 m structure equipped with retro-reflective targets undergoing vibration testing. All these structures use aluminised Kapton membranes (25 μm thick) that are shiny, but with sufficient diffuse reflection for 3D photogrammetry. Useful space missions require sail sizes of at least 70 m with membrane thicknesses of less than 7 μm.

The two biggest technical concerns are proper deployment of the sail and controllability, followed by the shape of the deployed sail (Slade *et al.*, 2002). These aspects demand measurement of the operational shape of the deployed membrane (static shape) which relate directly to sail acceleration performance in space, and modes of vibration which relate to altitude control and deployment dynamics as the sail is mechanically manipulated from storage to its driving configuration. This section gives examples of some of the experiments that were carried out to establish specific measurement objectives and accuracy requirements in order to validate the photogrammetric technique and to determine methods for routine test methods. Alternative methods include laser vibrometry which is able to sequentially monitor a limited number of points on a sail structure (Dharamsi *et al.*, 2002).

Several different digital camera models were used for this research: four Olympus E-20 cameras, two Pulnix TM-1020-15 cameras and a Kodak D760m camera. The E-20 and D760m cameras were used for static-shape whilst the Pulnix cameras were used for dynamic measurement. Cameras were placed in convergent geometries to give intersection angles between 70° and 90°. In designing an appropriate network an additional limitation is that "hot spots" due to reflection will obliterate any projected target information. This case is compounded in the dynamic situation since the hot spots will move with the motion of the object rendering larger areas immeasurable.

Figure 4.16 shows the inherent difficulty in using target-based techniques with these structures. As well as exhibiting hot spots due to direct reflection from the illumination source, the use of a dense array of retro-reflective targets is inappropriate where detailed surface shape is required since the weight of the targets will significantly affect the response of the sail material. In cases where high-density is required, dot projection using a 35 mm slide projector and a slide containing some 1500 dots in a grid pattern has been used. Figure 4.16b is a measured image of a portion of the four-quadrant 2 m sail, which is shown in the lower-right corner of Figure 4.15a. It was illuminated with the slide projector in order to measure its static shape with high spatial resolution. The gaps in the target pattern are due to a hot spot and a natural gap in the material either side of a support member. In this case the E-20 camera system was used and required an exposure time of the order of 30 seconds to achieve good quality target images.

Whilst the 35 mm slide projection system is limited in its light output and hence only suited to small structures, its continuous nature potentially allows projection onto dynamic structures with continuous recording using the Pulnix camera system. However, the projected dots are not as useful as attached targets because they do not move with the structure, making analysis of motion time histories of specific points on the structure difficult. This means that with projected targets, it is only possible to measure 2D motion along the lines extending from the targets to the projector, it is not possible to measure in all three spatial dimensions.

An alternative solution suited to static and periodic, rather than continuous, recording is provided by an electronic flash-based spot projector, such as the GSI ProSpot. This system is designed to deliver a much higher level of light output and is therefore suited to providing an instantaneous record of larger structures. Figure 4.17a shows the ProSpot system being used to illuminate a diffuse sail test structure that combines conventional retro-reflective

(a)

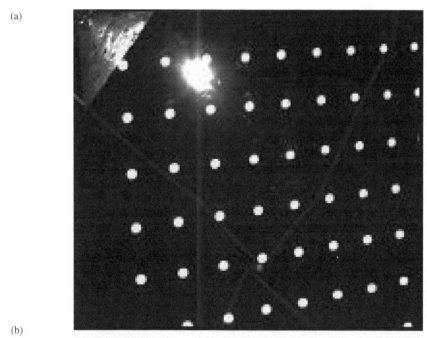

(b)

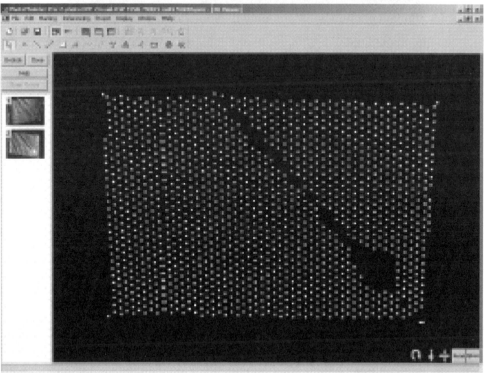

Figure 4.16 *(a) An enlarged view of a sail equipped with retro-reflective targets, the hot spot from the electronic flash illumination used is obvious; (b) shows an image, measured in PhotoModeler Pro, using targets projected with a carefully positioned data projector. Reproduced with permission of NASA LaRC.*

(a)

(b)

Figure 4.17   *(a) GSI ProSpot system being used to illuminate a diffuse white sail structure; (b) portion of a photogrammetric image of the illuminated structure. Reproduced with permission of NASA LaRC.*

targeting of its supporting structure with dot projection for measurement of surface shape. A corresponding image taken with one of the four synchronised E-20 cameras can be seen in Figure 4.17b.

Photogrammetric processing for many of the tests was carried out with PhotoModeler Pro software, but VMS and Australis are also used on site (Section 4.2.5). Since dot projection on shiny membrane surfaces yields images with a wide range of local illumination variations, the essential software feature for accurate and reliable measurement of these structures is target image location and subsequent sub-pixel measurement. A successful solution requires careful selection, preferably automated, of an appropriate threshold for each target (Section 4.2.2).

The example in Figure 4.18 (see colour section) is typical of the data produced and shows surface profiles across a sail structure computed by fitting cubic spline curves through the photogrammetrically computed 3D points. These cross-sectional slices make it easier to see the shape of the membrane. This plot uses different scales on the x and y axes, so the out-of-plane membrane shape (Z direction) is amplified approximately 20 times relative to the horizontal dimension (X direction). The data show a significant displacement of the upper region relative to the lower region by up to 6 cm. This warped shape, caused by the two upper rods of the sail bending under gravity, was the initial configuration of the structure.

Typical image space coordinate precisions are of the order of 0.25 pixels, with 3D coordination precisions for the smaller structures being of the order of 0.1 mm. In all situations correct scale is verified through the inclusion of known scale bar lengths.

Through a wide variety of tests, photogrammetry has demonstrated its ability as a flexible and robust approach with proven capability for measuring the static shape and dynamic characteristics of Gossamer-type spacecraft and components. It offers the simplicity of photography coupled with good to excellent measurement precision.

4.4.4 Measurement of flame-cut ship plates using multi-photo edge extraction

The shipbuilding industry provides a dynamic environment for a diverse range of measurement applications. This example focuses on the General Dynamics Marine Division at the Bath Iron Works Corporation at Bath, Maine, USA, who use a manufacturing process whereby large steel plates are cut to shape using a burning system (Figure 4.19a). This process is one in a series for which dimensional and accuracy control automation (DACA) systems are being developed to monitor the as-built dimensional quality of steel plates and other components as they are processed through the key fabrication, assembly, and ship completion stages of construction (Johnson, 1993, 2000).

The production requirements for DACA integration necessitated the development of a one-button type system for the automated measurement and analysis of the cut-outs in large flat steel panels to an accuracy of ±2 mm. Initial studies had used retro-reflective targets (Maas and Kersten, 1994; Johnson, 1996), but in this case no targets could be placed on the object to be measured to avoid interruption of the production line. The system had to be safe, user-friendly, cost-effective, reliable, maintainable, upgradeable and complement existing production throughput levels. An important note here is that the shop floor environment is very dirty, noisy and busy, with much vibration due the handling, transport and cutting of steel plates up to 4 m × 18 m in size.

Since 3D edge information was required, a photogrammetric solution was needed that could deploy accurate edge extraction algorithms across a network of images. An example of such a system is the optical tube measurement system developed by Bösemann (1996). The key factor in making edge extraction a possibility in the shipyard case is that each steel plate is relatively uniform in tone such that control of lighting to achieve consistent image quality and selection of an edge extraction algorithm capable of delivering sub-pixel information could achieve the task objectives. An example of an optimised photogrammetric image of a portion of a plate is given in Figure 4.19b. By using a static camera mounted above the plates on a gantry a sequence of such images could be built up simply by synchronising the camera shutter and electronic flash equipment with the motion of the automated table system that carries the steel plates around the factory floor.

(a)

(b)

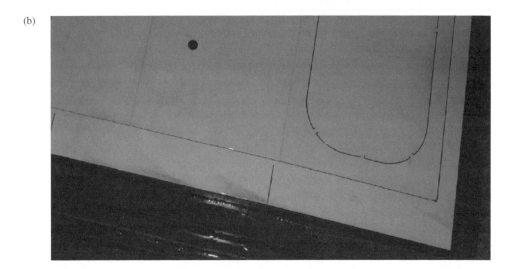

Figure 4.19 *(a) CM150 burning machine area; (b) a photogrammetric image of a portion of a cut plate. Reproduced with permission of G.W. Johnson (MidCoast Metrology) and S. Laskey (Bath Iron Works).*

A critical specification for all equipment was to maximise the use of consumer off-the-shelf components. Accordingly, the camera chosen was a Canon 1DS, an 11 megapixel, digital camera (CMOS sensor) which was connected to a host computer via an IEEE 1394a firewire interface (see Figure 4.20a (see colour section)). Part movement (and image acquisition rate) was determined by using three proximity switches set such that the system records one image for every metre of table travel. Images are downloaded from the camera to a host computer for photogrammetric data processing. Sequential orientation of the camera with respect to the cutting table was determined using coded targets, located on the cutting table edges which could be automatically found in the edge optimised images (Figure 4.20b).

Photogrammetric data processing for the DACA system is performed using VMS (Section 4.2.5). In this case VMS performs image measurement, coded target identification (Shortis *et al.*, 2003), initial orientation using the Zheng–Wang closed form resection (Section 4.2.3), least squares resection, edge image measurement and linking constrained by epipolar lines and, finally, the photogrammetric network computation. The camera is pre-calibrated due to the non-optimal geometry for self-calibration.

Testing with a variety of edge detectors (Gaussian-based wavelet, Canny and Sobel), demonstrated that the Sobel detector could provide satisfactory edge data. Whilst this may be surprising due to its simplicity and limitations when compared with the other methods, the quality of the edge images obtained was such that the Sobel method could provide edge information at the cut outs in significantly less processing time (Johnson, 2004).

Edge detection was followed by an edge linking procedure designed to filter out short unconnected edges such as those seen in Figure 4.21a (see colour section). A correspondence solution was then attained using epipolar geometry derived from each detected point on each edge into its neighbouring images. Intersection was used to compute the location of each edge point. Due to the close proximity of detected edges on each side of the cut lines, and the weakness in geometry of the single strip of images, a depth constraint which specified a maximum and minimum depth for the plate being measured was introduced to provide a robust solution to the epipolar search process.

The resulting edge point cloud data along with the associated statistical properties were exported into PolyWorks/Inspector as a point cloud for comparison between as-built and design plate layouts. The use of a scripting language enables the creation of a customised macro that can automatically import and align the CAD to the as-burned point cloud in order to conduct part to CAD comparisons (Figure 4.22a). Displayed attributes of interest include lengths, widths, squareness, straightness and flatness.

To verify system performance comparative measurements were obtained by sampling the burned plates on a daily basis using traditional manual techniques based on tape distances. The examples given in Figure 4.22b are based on shop floor measurements collected during the project set-up and initial integration period. Furthermore measurement quality has been enhanced through improved accuracy determined by scale bar validation, more complete data since many more samples are taken, and better reliability since three-dimensional coordinates with error statistics are available instead of simple two-dimensional linear distances.

(a)
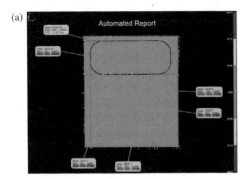

(b)

Control Accuracy (inches)			
Description	X	Y	Z
Baseline 1 (rms)	0.007	0.007	0.012
Baseline 2 (rms)	0.008	0.007	0.014
Deviations BL1 to BL2	0.001	0.001	0.002
Edge Measurement Accuracy (inches)			
Description	X	Y	Z
Integration + 0 Days	0.032	0.031	0.057
Integration + 30 Days	0.038	0.037	0.068
Best results	0.021	0.021	0.033
*Worst results (non fail)	0.195	0.193	0.320

Figure 4.22 *(a) Automated report output from Polyworks Inspector; (b) summary of results during initial implementation. Reproduced with permission of G.W. Johnson (MidCoast Metrology) and S. Laskey (Bath Iron Works).*

Due to the on-line integration approach, three-dimensional data can be measured, analysed and reported much more efficiently than the previous regime of manually taped two-dimensional linear distances, accuracy control check sheets, and statistical process control charting could provide. This cost effectiveness is complemented by increased data, higher accuracies, enhanced visualisation of results and increased sharing of reports. Simply stated, this provides an increasingly reliable picture of overall product and process quality to all appropriate levels of the company, within minutes of the measurement being performed. The targeted impact is that quality control type assessments are faster and more accurate while enabling all essential personnel to approach their jobs in a more informed manner.

4.4.5 Measurement of a dynamic object – NASA Langley solar collector Fresnel lens (stretched lens array)

This application describes photogrammetric measurements of the motion of a 1 m long Fresnel lens membrane used to concentrate light on solar collectors (Pappa *et al.*, 2002). The lens is composed of silicone rubber and is stretched between the end support arches (Figure 4.23). The lens is designed to focus solar energy into a thin line of light directed at the centre of a rectangular solar collector in order to improve the efficiency of the energy conversion and reduce the overall weight of the solar panels used in spacecraft.

The 250 mm high lens elements are designed to be assembled in banks of 35 on panels with a dimension of 3 m × 1 m. The panels support the flexible concentrator lenses and the solar cells, and also serve as heat radiators. The overall weight of the solar lens array is just 1.6 kg/m^2 and requires only 12% of the area of conventional solar cells for the same power output. The system is deployed by means of a structure composed of a series of hinged panels. Once deployed, the solar lens array will be capable of 360° of rotation to track the sun.

The experimental set-up to characterise the surface shape and vibration modes of the Fresnel lens is shown in Figure 4.24. The lens is mounted vertically and an exciter unit is used to simulate the vibrations from the shuttle reaction control system for in-orbit manoeuvres. Two small,

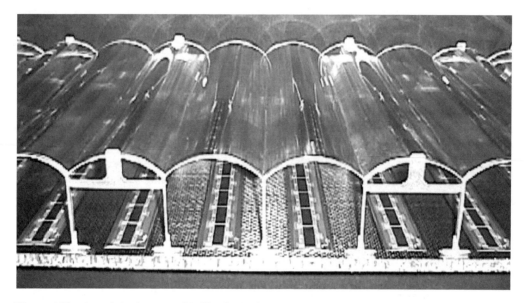

Figure 4.23   *A section of an assembled bank of solar cells and concentrator lenses.*

(a)   (b)

Figure 4.24   (a) Experimental set-up for the surface measurement and tracking of the Fresnel lens; (b) detail of the top camera and viewing slot.

circuit board CCD cameras were mounted on an independent fixture in order to image the inside surface of the lens through two viewing slots machined into the acrylic base of the lens element. This configuration was adopted to test the feasibility of in-flight monitoring of a lens array element that would be manufactured without the solar cells.

The two CCD cameras produced RS-170 monochrome, analogue video which was captured by a pair of Epix frame grabbers. The cameras were synchronised and VITC time code was injected into the 30 Hz video stream to allow each frame within the sequences of images generated for each test to be unequivocally identified. Similarly to the previous case study, retro-targets could not be used due to adverse reflections; therefore passive targets and careful control of illumination were used to avoid reflections off the lens surface into the photogrammetric images (Figure 4.25). The cameras were calibrated by rotating a small step-block target array within their overlapping fields of view to create a convergent multi-station network of ten exposures. Thirty-six targets were sufficient to calibrate the cameras and derive their relative orientation.

Image pairs of the static lens and a number of sequences of the lens under different modes of induced vibration were captured. As all the vibration periods were approximately 0.5 seconds or longer, target coordinates were computed using simple intersection techniques maintaining the relative orientation computed during the calibration process. Estimated target coordinate precisions were 40 μm.

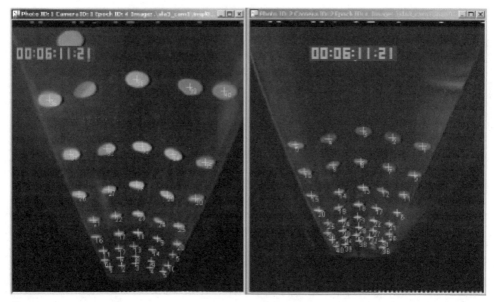

Figure 4.25   *Example top and bottom images of the Fresnel lens.*

In order to produce consistent data for visualisation of the motion of the lens array the unique identification of each target on its surface had to be tracked. Photogrammetric processing was carried out using VMS software with visualisation of the motion of the target surface (Figure 4.26 (see colour section)) being displayed using EngVis software (Woodhouse *et al.,* 1999).

Key photogrammetric issues associated with this application are in the very convergent geometry of the pair of cameras which was dictated by the desire to include the cameras in a location practicable for use with the lenses as a space deployable monitoring system. Variations in background intensity were also present, due to reflections off the membrane surfaces and ambient light sources. The effects of variations in target and background intensity were minimised by using a local threshold within the target image window for the centroid computation. The convergence of the cameras also caused a fall-off in the size and spacing of the targets across the objects, with the Fresnel lens showing a particularly extreme size variation. This effect was partially ameliorated by using a two-level adaptive window for the target image centroids.

4.5 Conclusions

Industrial photogrammetry is a flexible, accurate, non-contact measurement tool. It provides a precise and reliable method of determining surface shape, position and orientation for the monitoring of movement and change. The technique is optimal where several dozens to several thousands of 3D locations need to be recorded instantaneously.

Given the current state-of-the-art, the technique is a genuine alternative to applications in which single-target tracking solutions, at one end of the scale, and scanning technologies, at the other end of the scale, cannot be deployed. Whilst fully automated examples of vision metrology are confined to niche areas such as machine tooling and work-cell-based manufacture, many other measurement tasks are undertaken but generally rely on some level of user interaction in the image acquisition and data processing stages.

System robustness and geometric reliability of both imaging techniques and algorithms combined with advances in user friendly visualisation technologies are allowing non-specialists to directly use photogrammetric techniques. The role of the close range photogrammetrist is increasingly becoming that of a consultant, or a developer of automated, robust and reliable systems.

4.6 References

Abdel-Aziz, Y.I. and Karara, H.M., 1971. Direct linear transformation from computer coordinates into object space coordinates. In *ASP, Symposium on Close Range Photogrammetry*, American Society of Photogrammetry, Bethesda, Maryland, USA, 1–18.

Arc Second, 2006. White paper 063102 – Constellation 3Di Error Budget and Specifications, Arc Second Inc, Dulles, Virginia, USA. See http://www.indoorgps.com/PDFs/wp_Error_Budget.pdf (accessed 3 February 2006.

ASME, 2005. 89.4.19 Laser tracking performance evaluation. See http://www.mel.nist.gov/proj/dm.htm (accessed 2 February 2006).

Atkinson K.B., 1976. A review of close range engineering photogrammetry. *Photogrametric Engineering and Remote Sensing*, 42(1), 57–69.

Atkinson, K.B. (Ed.), 1996. *Close Range Photogrammetry and Machine Vision*, Whittles Publishing, Scotland.

Baltsavias, E., 1991. *Multiphoto geometrically constrained matching*. Ph.D. thesis, Institute for Geodesy and Photogrammetry, ETH Zurich, Switzerland.

Bartoli, A., 2002. A unified framework for quasi-linear bundle adjustment. *16th International Conference on Pattern Recognition*. Quebec City, Canada, Vol. 2, August 2002, 560–563.

Beyer, H.A., 1995. Digital photogrammetry in industrial applications. *International Archives of Photogrammetry and Remote Sensing*, 30(5W1), 373–378.

Bösemann, W., 1996. The optical tube measurement system OLM – Photogrammetric methods used for industrial automation and process control. *International Archives of Photogrammetry and Remote Sensing*, 31(B5), 55–58.

Bösemann, W. and Sinnreich, K., 1994. An optical 3D tube measurement system for quality control in industry. In *Automated 3D and 2D vision*, SPIE Proceedings 2249, 192–199.

Brown, D.C., 1971. Close-range camera calibration. *Photogrammetric Engineering*, 37(8), 855–866.

Brown, D.C., 1984a. A large format, microprocessor controlled film camera optimised for industrial photogrammetry. *XV International Congress of Photogrammetry and Remote Sensing*, Rio de Janeiro, Brazil.

Brown, D.C., 1984b. Tools of the trade. In *Close range photogrammetry and surveying – state of the art*. American Society of Photogrammetry, Bethesda, Maryland, USA, 83–252.

Brown, J. and Dold, J., 1995. V-STARS – a system for digital industrial photogrammetry. In *Optical 3D Measurement Techniques III*, Grün, A. and Kahmen, H. (Eds.), Wichmann Verlag, Heidelberg, Germany 12–21.

Burch, J.M. and Forno, C., 1983. The NPL centrax – a new lens for photogrammetry. *SPIE Proceedings* 399, 412–417.

Burner, A.W., Radeztsky, R.H. and Liu, T., 1997. Videometric applications in wind tunnels. In *Videometrics V*, El-Hakim, S.F. (Ed.) *SPIE Proceedings* 3174, 234–247.

Cooper, M.A.R., 1979. Analytical photogrammetry in engineering: three feasibility studies. *Photogrammetric Record*, 9(53), 601–619.

Cooper, M.A.R. and Robson, S., 1996. Theory of close-range photogrammetry. In Atkinson, K.B. (Ed.), *Close Range Photogrammetry and Machine Vision*, Whittles Publishing, Scotland, 9–51.

Chen, F., Brown, G.M. and Song, M., 2000. Overview of three-dimensional shape measurement using optical methods. *Optical Engineering*, 39(1), 10–22.

Chen, J., 1995. *The use of multiple cameras and geometric constraints for 3-D measurement*. Ph.D. thesis, City University, London, UK.

Dassault Systems, 2006. See http://www.3ds.com/products-solutions/plm-solutions/catia/overview/ (accessed 19 February 2006).

Dharamsi, U.K., Evanchik, D.M. and Blandino, J.R., 2002. Comparing photogrammetry with a conventional displacement measurement technique on a square kapton membrane. *AIAA Paper* 2002–1258, April 2002.

Dold, J., 1990. A large format film camera for high precision object recording. *International Archives of Photogrammetry and Remote Sensing*, 28(5), 252–255.

Dold, J., 1996. Influence of large targets on the results of photogrammetric bundle adjustment. *International Archives of Photogrammetry and Remote Sensing*, 31(B5), 119–123.

Flack, D., 2001, NPL Best Practice Guides: GPG (043): CMM Probing. See http://npl.co.uk/length/gpg. html (accessed 9 November 2006).

Fraser, C.S., 1992. Photogrammetric measurement to one part in a million. *Photogrammetric Engineering and Remote Sensing*, 58(3), 305–310.

Fraser, C.S., 1996. Network design. In Atkinson, K.B. (Ed.), *Close Range Photogrammetry and Machine Vision*, Whittles Publishing, Scotland, 256–281.

Fraser, C.S., 1997. Innovations in automation for vision metrology systems. *Photogrammetric Record*, 15(90), 901–911.

Fraser, C.S. and Shortis, M.R., 1992. Variation of distortion within the photographic field. *Photogrammetric Engineering and Remote Sensing*, 58(6), 851–855.

Fraser, C.S. and Shortis, M.R., 1995. Metric exploitation of still video imagery. *Photogrammetric Record*, 15(85), 107–122.

Fryer, J.G., 1996. Camera calibration. In Atkinson, K.B. (Ed.), *Close Range Photogrammetry and Machine Vision*, Whittles Publishing, Scotland, 156–180.

Gartner, H., Lehle, P. and Tiziani, H.J., 1995. New, high efficient, binary codes for structured light methods. In *Three-Dimensional and Unconventional Imaging for Industrial Inspection and Metrology* Descour, M.R., *et al.* (Eds.), *SPIE Proceedings* 2599, 4–13.

Grafarend, E., 1974. Optimization of geodetic networks. *Bollentimo di Geodesia a Science Affini*, 33(4), 351–406.

GSI, 2006. V-STARS Industrial Photogrammetry Systems, Geodetic Services, Inc. Melbourne, Florida, USA. See http://www.geodetic.com (accessed 20 October 2006).

Granshaw, S.I., 1980. Bundle adjustment methods in engineering photogrammetry. *Photogrammetric Record*, 10(56), 181–207.

Grün, A. and Parian, J., 2005. Close range photogrammetric network design for panoramic cameras by heuristic simulation. In *Proceedings of Optical 3-D Measurement Techniques VII*, Vienna, Austria, (Grün, A. and Kahmen, H. Eds.), Vol. I, 237–244.

Grün, A.W. and Baltsavias, E.P., 1988. Geometrically constrained multiphoto matching. *Photogrammetric Engineering and Remote Sensing*, 54(5), 633–641.

Gustafson, P.C., 1997. Robotic video photogrammetry system. In *Videometrics V*, El-Hakim, S.F. (Ed.), *SPIE Proceedings* 3174, 190–196.

Haggrén, H., 1994. Video image based photogrammetry and applications. *Videometrics III*, El-Hakim, S.F. (Ed.), *SPIE Proceedings* 2350, 22–30.

Haggrén, H. and Haajanen, L., 1990. Target search using template images. *International Archives of Photogrammetry and Remote Sensing*, 28(5/1), 572–578.

van den Heuvel, F.A., Kroon, R.J.G.A. and Le Poole, R.S., 1992. Digital close-range photogrammetry using artificial targets. *International Archives of Photogrammetry and Remote Sensing*, 29(5), 222–229.

van den Heuvel F.A, 1999. A line-photogrammetric mathematical model for the reconstruction of polyhedral objects. *Videometrics VI*, El-Hakim, S.F. (Ed.), *SPIE Proceedings* 3641, 60–71.

International Society for Photogrammetry and Remote Sensing, 2005. See http://www.isprs.org (accessed 14 March 2006).

Jenkins, C.H.M. (Ed.), 2001. *Gossamer Spacecraft: Membrane and Inflatable Structures Technology for Space Applications*, Progress in Astronautics and Aeronautics, Vol. 191, American Institute of Aeronautics and Astronautics, Reston, Virginia, USA.

Johnson, G.W., 1993. *Digital Photogrammetry a New Measurement Tool*. BIW Corporate Summary of 3D Measurement Technology Comparisons for Unit Erection. Internal Report, Bath Iron Works, Bath, Maine, USA.

Johnson, G.W., 1996. Practical integration of vision metrology and CAD in shipbuilding. *International Archives of Photogrammetry and Remote Sensing*, Vol. XXXI(B5), 264–273.

Johnson, G.W., 2000. Advanced application of digital close-range photogrammetry in shipbuilding. Presented at Commission V Working Group V/3 *ISPRS XIX Congress*, Amsterdam, The Netherlands.

Johnson, G.W., 2004 Dimensional and accuracy control automation for shipbuilding. *ShipTech* January 27–28, 2004. See http://www.nsrp.org/st2004/st2004. html (accessed 9 November 2006).

Johnson, G.W., Robson, S. and Shortis, M.R., 2004. Dimensional and accuracy control automation: An integration of advanced image interpretation, analysis, and visualization techniques. WG V/1, *XXth ISPRS Congress*, Istanbul, Turkey. See http://www.isprs.org/istanbul2004/comm5/papers/692.pdf (accessed 14 March 2006).

Jones, T.W. and Pappa, R.S., 2002. Dot projection photogrammetric technique for shape measurements of aerospace test articles. *AIAA Paper* 2002-0532, January 2002.

Krauss, K., 1996. *Photogrammetry, Volume 2. Advanced Methods and Applications*. Dümmler Verlag, Bonn, Germany.

Lau, K., Hocken, R. and Haight, W., 1986. Automatic laser tracking interferometer system for robot metrology. *Journal of Precision Engineering*, 8(1), 3–8.

Lewis, A., Oldfield, S. and Peggs, G.N., 2001. The NPL small CMM – 3D measurement of small features. In *Laser Metrology and Machine Performance V*, Peggs, G.N. (Ed.), National Physical Laboratory, Teddington, UK.

Luhmann, T., 2006. Personal communication on the availability and scope of VDI/VDE 2634.

Luhmann, T. and Wendt, K., 2000. Recommendations for an acceptance and verification test of optical 3-D measurement systems. *International Archives of Photogrammetry and Remote Sensing*, 33(5), 493–499.

Luhmann, T., Robson, S., Kyle, S. and Harley, I., 2006. *Close Range Photogrammetry: Principles, Methods and Applications*, Whittles Publishing, Scotland.

Maas, H.G., 1992a. Robust automatic surface reconstruction with structured light. *International Archives of Photogrammetry and Remote Sensing*, 24(B5), 102–107.

Maas, H.-G., 1992b. Complexity analysis for the determination of image correspondences in dense spatial target fields. *International Archives of Photogrammetry and Remote Sensing*, 29(B5), 102–107.

Maas, H.-G. and Kersten, T., 1994. Close range techniques and machine vision. *International Archives of Photogrammetry and Remote Sensing* 30(5), 250–255.

Mason, S., 1994. *Expert system based design of photogrammetric networks*. PhD thesis, Institute of Geodesy and Photogrammetry, ETH Zurich, Switzerland.

Moore, J.F.A., 1973. The photogrammetric measurement of constructional displacements of a rockfill dam. *Photogrammetric Record*, 7(42), 624–648.

Munji, R.A.H., 1986. Self-calibration using the finite element approach. *Photogrammetric Engineering and Remote Sensing*, 52(3), 411–418.

National Institute of Standards and Technology, 2006. See http://fpc.ncms.org/fpc_perspectives.htm (accessed 3 February 2006).

Olague, G., 2002. Automated photogrammetric network design using genetic algorithms. *Photogrammetric Engineering and Remote Sensing*, 68(5), 423–431.

Patias, P. and Peipe, J., 2000. Photogrammery and CAD/CAM in culture and industry – an ever changing paradigm. *International Archives of Photogrammetry and Remote Sensing*, 33(B5), 599–603.

Pappa, R.S., Giersch, L.R. and Quagliaroli, J.M., 2001a. Photogrammetry of a 5-m inflatable space antenna with consumer digital camera. *Experimental Techniques*, July/August. 2001, 21–29.

Pappa, R.S., Lassiter, J.O. and Ross, B.P., 2001b. Structural dynamics experimental activities in ultra-lightweight and inflatable space structures. *AIAA Paper* 2001-1263, April 2001.

Pappa, R.S., Woods-Vedeler, J.A. and Jones, T.W., 2002. In-space structural validation plan for a stretched-lens solar array flight experiment. *Proceedings of the 20th International Modal Analysis Conference*, Los Angeles, CA, USA, February 2002, 461–471.

Peipe, J., 1997. High-resolution CCD area array sensors in digital close range photogrammetry. In *Videometrics V*, El-Hakim, S.F. (Ed.), *SPIE Proceedings* 3174, 153–156.

Pollefeys, M., Koch, R. and Van Gool, L., 1999. Self-calibration and metric reconstruction in spite of varying and unknown internal camera parameters. *International Journal of Computer Vision*, 32(1), 7–25.

Robson, S., Clarke, T.A. and Chen, J., 1993. The suitability of the Pulnix TM6CN CCD camera for photogrammetric measurement. In *Videometrics II*, El-Hakim, S.F. (Ed.) *SPIE Proceedings* 2067, 66–77.

Robson, S. and Setan, H.B., 1996. The dynamic digital photogrammetric measurement and visualisation of a 21m wind turbine rotor blade undergoing structural analysis. *International Archives of Photogrammetry and Remote Sensing*, 31(B5), 493–498.

Robson, S. and Shortis, M.R., 1998. Practical influences of geometric and radiometric image quality provided by different digital camera systems. *Photogrammetric Record*, 16(92), 225–248.

Sablatnig, R. and Menard, C., 1997. 3D Reconstruction of archaeological pottery using profile primitives. In Sarris, N. and Strintzis, M.G. (Eds.), *Proceedings of the International Workshop on Synthetic-Natural Hybrid Coding and Three-Dimensional Imaging*, Rhodes, 5–9 September 1997, 93–96.

Shortis, M.R. and Beyer, H.A., 1996. Sensor technology for digital photogrammetry and machine vision. In Atkinson, K.B. (Ed.), *Close Range Photogrammetry and Machine Vision*, Whittles Publishing, Scotland, 106–155.

Shortis, M.R. and Beyer, H.A. 1997. Calibration stability of the Kodak DCS420 and 460 cameras. *Videometrics V*, El-Hakim, S.F. and Sabry, F. (Eds.), *SPIE Proceedings* 3174, 94–105.

Shortis, M.R., Clarke, T.A. and Robson, S., 1995a. Practical testing of the precision and accuracy of target image centering algorithms. *Videometrics IV*, El-Hakim, S.F. (Ed.), *Proceedings SPIE* 2598, 65–76.

Shortis, M.R., Clarke, T.A. and Short, T., 1994. A comparison of some techniques for the subpixel location of discrete target images. *Videometrics III*, El-Hakim, S.F. (Ed.), *Proceedings SPIE* 2350, 239–250.

Shortis, M.R., Robson, S. and Short, T., 1996. Multiple focus calibration of a still video camera. *International Archives of Photogrammetry and Remote Sensing*, 31(B5), 534–539.

Shortis, M.R., Robson, S. and Beyer, H.A., 1998. Principal point behaviour and calibration parameter models for Kodak DCS cameras. *Photogrammetric Record*, 16(92), 165–186.

Shortis, M.R., Robson, S., Pappa, R.S., Jones, T.W. and Goad, W.K., 2002. Characterisation and tracking of membrane surfaces at NASA Langley Research Center. *International Archives of Photogrammetry and Remote Sensing*, 34(5), 90–94.

Shortis, M.R., Seager, J.W., Robson, S. and Harvey, E.S., 2003. Automatic recognition of coded targets based on a Hough transform and segment matching. In *Videometrics VII*, El-Hakim, S.F., Grün A. and Walton, J.S. (Eds.), *SPIE Proceedings* 5013, 202–208.

Shortis, M.R., Snow, W.L. and Goad, W.K., 1995b. Comparative geometric tests of industrial and scientific CCD cameras using plumb line and test range calibrations. *International Archives of Photogrammetry and Remote Sensing*, 30(5W1), 53–59.

Slade, K.N., Belvin, W.K. and Behun, V., 2002. Solar sail loads, dynamics, and membrane studies, *AIAA Paper* 2002-1265, April 2002.

Stanton, R.H., Alexander, J.W., Dennison, E.W., Glavich, T.A. and Hovland, L.F., 1987. Optical tracking using charge-coupled devices. *Optical Engineering*, 26(9), 930–935.

Susstrunk, S. and Holm, J., 1995. Camera data sheet for pictorial electronic still cameras. In *Cameras and Systems for Electronic Photography and Scientific Imaging*, Anagnostoponlos, C.N. and Lesser, M.P. (Eds.) *SPIE Proceedings* 2416, 5–16.

Tecklenburg, W., Luhmann, T. and Hastedt, H., 2001. Camera modelling with image-variant parameters and finite elements. In *Optical 3-D Measurement Techniques V*, Vienna Univesity of Technology, Vienna, Austria.

Triggs, B., McLauchlan, P., Hartley, R., Fitzgibbon, A., 2000. Bundle adjustment – a modern synthesis. In Triggs, B., (Ed.), *Vision Algorithms: Theory and Practice*, Springer-Verlag, Berlin, Heidelberg, Germany, 298–372.

Trinder, J.C., 1989. Precision of digital target location. *Photogrammetric Engineering and Remote Sensing*, 55(6), 883–886.

UCL 2001. *Best Practice for Non-contacting CMMs*, Project 2.3.1/2/3 – Large Scale Metrology, National Measurement System Programme for Length Metrology, University College London, UK. See http://www.ge.ucl.ac.uk/research/projects/dti.html.

VDI/VDE, 2002. *Optical 3-D measuring systems imaging systems with point-by-point probing*. VDI/VDE-Guideline 2634 Part 1, Beuth Verlag, Berlin, Germany.

Wahl, F.M., 1984. *A Coded Light Approach for 3-Dimensional Vision*. IBM Research Report, RZ 1452. IBM Thomas J. Watson Research Center, Yorktown Heights, New York, USA.

Woodhouse, N.G. and Robson, S., 1998. Monitoring concrete columns using digital photogrammetric techniques. In Allison, I.M. (Ed.), *Experimental Mechanics, Advances in Design, Testing and Analysis. Proceedings 11th International Conference on Experimental Mechanics*, A.A. Balkema, Rotterdam, The Netherlands, 641–646.

Woodhouse, N.G., Robson, S. and Eyre, J., 1999. Vision metrology and three-dimensional visualisation in structural testing and monitoring. *Photogrammetric Record*, 16(94), 625–642.

Zheng, Z. and Wang, X., 1992: A general *solution* of a closed-form space resection. *Photogrammetric Engineering and Remote Sensing*, 58(3), 327–338.

Zhou, G., 1986. Accurate determination of ellipse centers in digital imagery. Technical Papers Volume 4, *ACSM-ASPRS Annual Convention* (Washington, DC, USA), 256–264.

Zumbrun, R., 1996, Systematic pointing errors with retro-reflective targets. *Optical Engineering*, 34(09), 2691–2695.

5 Forensic photogrammetry

John G. Fryer

5.1 General outline of the application

5.1.1 Introduction

The adjective forensic has been defined as "... pertaining to, connected with, or used in courts of law for public discussion and debate". Usually this adjective is extended to encompass all forms of scientific knowledge applied to legal problems.

Forensic photogrammetry has been growing in acceptance in courtroom situations since the 1950s. Its initial usage was largely restricted to government mapping agencies, including units attached to police forces, for the making of plans and sometimes models, for car, train and aeroplane accident sites. In almost all circumstances prior to ready access to computers from the late 1970s, traditional stereoscopic pairs of photographs and mechanical/analogue stereoplotters were used to construct these visual aids for the court.

The use of single images was restricted to graphical interpretation to determine the relative locations of details in those scenes. The difficulties associated with the reconstruction of the spatial relationships of objects without computing packages limited the application of single images to qualitative purposes.

The task of providing the court with a representation of a crime scene is still very much a major role for forensic photogrammetry today. However, with the advent of low-cost computer graphics and image processing, these forms of scene depiction may be extended to include virtual three-dimensional scene reconstructions on a computer screen and even animations of, for example, cars skidding, turning and crashing.

Before discussing the features of forensic photogrammetry, it is worth reflecting on the growing impact that video surveillance on closed circuit television networks (CCTV) is having on the lives of those who live in big cities. The number of surveillance cameras is on the increase and by 2001 in Britain alone, cameras in public places had passed one million (Bramble *et al.*, 2001). It has been estimated that on an average day in any big city, an individual could be captured by more than 300 cameras from 30 different CCTV networks. In London, a pedestrian is captured on a camera at least every five minutes and in central areas is likely to be watched by cameras at least half the time. The same sort of figures can now be expected in other large cities around the world. With this amount of video footage there is an increasing probability that criminal activity will be captured on video.

High rating television programmes in the last few years have focused public attention on the possibilities which image processing can bring to court. Fictional television programmes continually present white-coated technicians assisting forensic scientists with a bewildering array of high resolution computer-based tools which reveal otherwise hidden details in images. Unfortunately real life for forensic photogrammetrists is more mundane, and time consuming, than television would portray.

The examples in this chapter will illustrate that the consultant photogrammetrist may only be an "after-thought" in an investigation and must work with images which are sub-optimal, opportunistic and not taken with any prior knowledge of photogrammetric principles. An ability to work creatively with limited images and in conjunction with other surveying tools could aptly describe the job description. In brief, the forensic photogrammetrist's role can be very challenging.

5.1.2 General advantages of photogrammetry

The major advantage of imagery to forensic investigations, compared to other methods of recording a crime scene spatially, is that the totality of the information present at a scene is captured. All images may be archived and reused at a later date. In many cases, a plan is prepared for the investigating officers, but a vital piece of evidence may have been omitted. It may be a decade or more before it is discovered that an item at the scene of the investigation was overlooked. If the plan was prepared by ground survey, there may not be a chance to remeasure. Total recording of the scene is the great advantage of image-based techniques.

To illustrate this point, consider the circumstance of a high profile political bombing attempt outside the Hilton Hotel in Sydney, Australia in 1978. It was claimed that a bomb had been placed in a road-side rubbish bin. Plans of the entire area were produced from stereoscopic photographs taken by the police. A person was convicted for the incident but claims of his innocence persisted in the press for many years. Several years later, long after all such types of rubbish bins had been replaced throughout the city, a court appeal heard that the hole at the top of the bin was too small for that type of bomb to have been placed in it. The archival photographs were reviewed and an opening at the top of a nearby identical bin measured and, with other evidence, the person was acquitted. A ground-based survey may not have measured the scene to such detail.

"A picture is worth a thousand words" is a familiar old saying. This is exactly what images can provide to a court. A simple example involved the robbery of a taxi-driver where the crime scene (see Figure 5.1) was captured on an in-taxi security camera located just above the rear-vision mirror. The resolution of such devices, at the time, was extremely low with an imaging array of only about 300×200 pixels.

In this case, the prosecution believed the key to a conviction was to determine the length of razor blade held against the driver's throat. The method adopted was to try and replicate the crime scene and some photographs such as those shown in Figure 5.2 were taken in the car park at the police station with the kind cooperation of the taxi-driver. By superimposing Figures 5.1 and 5.2 the blade length was estimated at 143 mm. Police had found a razor with a blade of length 145 mm at the suspect's house. In preparing his report for the prosecutor, the photogrammetrist managed to obtain several digital images of the passenger from the taxi's security company and made some high quality prints at A4 size. Only poor quality facsimile prints from the taxi company had been seen by court personnel previously. The clever piece of photogrammetry to find the length of the blade was quickly overlooked, as upon seeing a high quality printout of the robber's face, the case for the defence collapsed.

Figure 5.1   *An image captured on in-taxi security camera during a robbery.*

Figure 5.2   *Photograph taken to replicate crime scene to measure length of blade.*

5.1.3 Some disadvantages of photogrammetry

It is unwise to believe that photogrammetric methods are some sort of panacea for forensic techniques. As the example illustrated by Figures 5.1 and 5.2 proves, the use of a high quality printer to produce one good clear image may be enough for a conviction.

The inference by E.H. Thompson (1962, see Chapter 1 for full details) that photogrammetry is only useful when all other simpler measurement methods are impractical, still has some resonance over 40 years later. Computer software has almost eliminated the problems associated with tilted photographs and uncalibrated cameras, but the disadvantage which remains is the low public knowledge or acceptance of photogrammetry as a measurement technique. Many otherwise highly qualified and technically proficient people, including most lawyers, have never heard of it. As a result they simply do not know when a photogrammetric investigation can help their case.

Another problem sometimes faced by photogrammetrists is that the photographs they are given to reconstruct a crime scene from are opportunistic images. This can lead to a problem in having the expert photogrammetrist's evidence accepted by the court. The photographs may not have been taken with any background knowledge of good intersection angles, adequate lighting or any regular form of control points. The photogrammetrists may well be given prints which have been cropped by unknown amounts taken from cameras which are not available for inspection. There may be no details on the camera, its focal length, lens distortion or other technical details one normally associates with a planned photogrammetric survey.

There can be claims that the images have been deliberately modified, or digitally altered. Such claims of tampering can be difficult to repudiate if it is not possible to establish a strong linkage (chain of evidence) between the original photographer, the images, those involved with handling the images and the forensic photogrammetrist. During the earlier and mid-parts of the twentieth century, in some European countries only photographic images taken on glass plates could be admitted as evidence in court to try to overcome the problem of claims of image tampering. Once video surveillance became commonplace later in that century, it was also subject to similar scrutiny, but now its acceptance by the courts is almost routine.

Scene reconstruction from a single opportunistic photograph is possible, but there must always be some known information about the photograph, camera or the scene. Such restrictions apply to a lesser extent with stereoscopic pairs of photographs. With single images, the required information could be the camera's focal length, lines or planes visible in the image being either parallel or perpendicular (such as walls, ceilings and floors) or some identifiable features being measured in position to act as control points. The older the photograph, the more difficult it becomes to find unaltered features at the scene to act as control.

The minimum requirements for solving the resection problem of locating where the photograph was taken depend on whether any knowledge of the camera's focal length exists (or can be reasonably assumed). Assuming the focal length is available, then the need is for three identifiable control points to be surveyed. Another control point must be added if the focal length is unknown and still another one if it is suspected that the principal point is not in the centre of the image. It is always advisable to have more than the bare minimum to produce a robust solution and have check measurements.

Competent photogrammetrists can usually produce a result for the spatial relationships of objects in a single photograph, but the big problem comes in convincing the court of the method used, especially if it is not a routine process. If the judge and the lawyers cannot readily understand the technique as described by the photogrammetrist, it is unlikely the matter will be allowed to go before a jury of laypeople. The judge may believe, perhaps rightly, that such a jury may not understand the computational complexities needed to obtain a result. However, if the method adopted has been published in a textbook or scientific journal and examples can be quoted of its acceptance, then even complex mathematical solutions may be accepted by the court.

If the photogrammetrists can make simple analogies and drawings to illustrate certain principles then the evidence may be acceptable. Figures 5.3–5.5 illustrate how the height of a bank robber can be estimated at the crime scene by using a piece of string to represent a grazing light ray from a known object across the robber's head to the camera. The mathematics to solve for the height of a robber with a tilted camera, lens distortions etc. may be complex, but if the principle can be shown with simple diagrams, the court may look more favourably on the evidence.

Figure 5.3 *A typical bank robbery scene. Figures 5.4 and 5.5 show that by noting the position of the suspect's feet on the patterned carpet, and his head in relation to the teller's grille, a determination of his height may be made.*

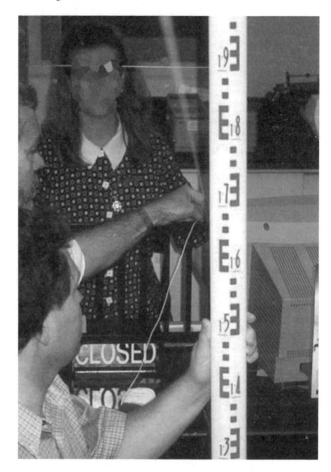

Figure 5.4   *A piece of string used to replicate a light ray from the security camera to the teller's grille.*

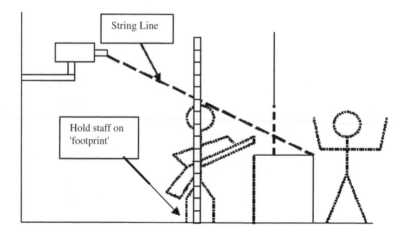

Figure 5.5   *Simple diagram to illustrate a method of height determination.*

In most surveillance photographs, bank robbers are not standing erect but are normally leaning forward, walking or otherwise concealing their true height by the wearing of a hooded coat or a hat. The forensic photogrammetrists can only measure the apparent height at the time of photography, but is usually asked to express an opinion as to the true height of the suspect. Photographs taken to show the reduction in height due to walking, stooping shoulders etc., such as that shown in Figure 5.6, can be used to enhance the credibility of numerical evidence, especially if the person chosen to partake in the exercise is of similar physical characteristics to the suspect.

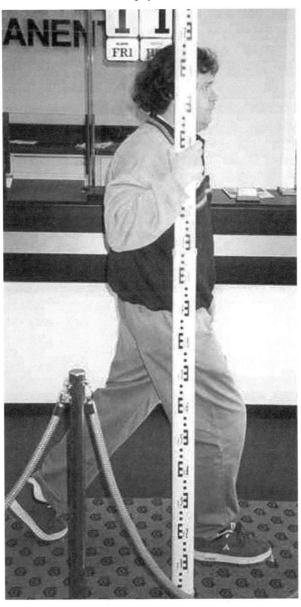

Figure 5.6 *The difference in height between a suspect walking and standing upright can be simply estimated by using a person with similar physical characteristics.*

5.1.4 Brief history of forensic photogrammetry

A full history of forensic photogrammetry is yet to be written. In the literature there are many examples of, usually, successful applications of photogrammetric principles to court cases. The judicial rules and conditions under which forensic photogrammetrists operate vary widely from continent to continent and then within countries inside those continents. Readers desirous of a summary of forensic photogrammetry in the USA and in Europe may find McGlone (2004, 1050–1062) interesting.

In some judicial systems, the forensic photogrammetrist's role is restricted to providing a report which others present to the court. In some countries, only government officials and the police present photogrammetric evidence, whilst in other jurisdictions private photogrammetric consultants can make a living from preparing and presenting evidence as an *expert witness*.

The concept and practice of being an expert witness in a courtroom situation also varies between countries. One thing in common with most forms of expert evidence, is that it is intended to benefit the court and the court's deliberations. It is not intended to be biased to try to further the view of one of the two parties involved. In this respect, it is not uncommon for an expert witness brought forward by the legal team of one of the parties to try to clearly explain to the judge and jury the true facts as uncovered by photogrammetric investigation.

In most countries, expert witnesses have full access to all notes, calculations and reports they have prepared and can refer and look at these whilst under examination. An ordinary witness can usually only rely on their memory during their presentation of evidence and under cross-examination. It is not unusual for the judge to ask questions directly to the expert witness for the clarification of all present in the courtroom.

In the last decade in several countries there have been changes in legislation to further assist the court to gain a better understanding of the events which have occurred at a crime scene. If two experts have been commissioned, one to represent each legal team, and they disagree in their reports, then before the court proceedings commence they are bound to meet. The experts must then write a common report which details what they agree upon and what they disagree upon and why they disagree in their findings, the different methods they each adopted and so on. They can still be subjected to examination and cross-examination, but much valuable court time is not wasted over the conundrum of "what to believe when experts disagree".

5.2 Distinctive aspects of forensic photogrammetry

There are some distinctive features of forensic photogrammetry. The provision of evidence to a court may involve several different types of images, including:

- an isolated snapshot taken by a passer-by, or even a newspaper reporter at the scene;
- conventional stereopairs taken deliberately by police or others with a knowledge of photogrammetry;
- aerial photographs, either in stereopairs usually used for map-making or single photographs, perhaps taken from a helicopter;
- multiple images, often on video surveillance cameras of low resolution, from a variety of directions which may require a complex bundle adjustment to reach a solution; and,
- the combination of historical imagery with that taken contemporaneously at the crime scene.

Another factor which distinguishes forensic applications is that the amount of known control may vary from ample to barely adequate to non-existent. For example, in the case where the police authorities are responsible for photographing a car crash scene, they will usually take

careful stereoscopic pairs and lay down lines of traffic cones (see Figure 5.12) which are measured in position by tapes or surveyed using the electronic distance measuring instruments.
When a passer-by or a newspaper reporter takes a single photograph at a crime scene, there is
often a delay of days, weeks, or months before it is recognised that there may be vital information in that photograph. The forensic photogrammetrist must then attend the scene to try and
find some details which can be used as control measurements. If the photograph is taken along
a street or in a room, then there are facts such as the floor or footpath being horizontal and perpendicular to walls which can be used to aid the scene reconstruction process.

  If the photograph is of an incident which occurred in a wholly natural environment such as
a beach then other facts must be gathered. Figure 5.7 depicts the instant of the fatal shooting
at Sydney's Bondi Beach in 1998. The exact time of the shooting was well documented so the
lengths of shadows cast by the sun from policemen (of known height) were available to provide
some control distances. The slope of the beach and the locations of the distant buildings can
be surveyed and the state of the tide can be ascertained. The horizon can be seen in this photograph, thus allowing camera tilts and rotation angles to be deduced. A key question faced by the
coronial inquest was "How far away were the police from the victim at the instant of shooting?"
(Hand and Fife-Yeomans, 2004, 185).

  A limitation which applies in many applications of forensic photogrammetry is the quality
of the images which must be analysed. Poor lighting is an obvious constraint. Many traffic
accidents and crimes happen at night and it is often imperative to clear away the debris as
quickly as possible. Artificial lighting is sometimes not powerful enough to clearly illuminate
the whole scene. By slowing the exposure time on the photographs, a grainy, low resolution
image may occur and some inaccuracies in the final plan may result.

Figure 5.7 *The instant of a fatal shooting at Bondi Beach, Sydney, Australia. A lack of apparent control
points is a feature of opportunistic photographs. Reproduced with permission of J.P. Firgoff-Bratanoff,
JP Imaging, Bondi, Australia.*

Where the imagery is provided to the photogrammetrist, who has had no control over how it was gathered, the quality of the images may be extremely poor. Consider Figure 5.8a which shows a scene extracted from some surveillance camera video in a liquor store shortly before a robbery was committed. Obviously the video recording equipment was in a poor state of repair. The relevant feature on which the police required information was the logo on the baseball hat, worn back-to-front on the young man's head. Given the poor resolution of the imagery, an interlacing problem with the recording video camera and the small size of the baseball hat, many hours of image processing led to the inconclusive result achieved in Figure 5.8b. Such is the lot of the forensic photogrammetrist!

(a)

(b)

Figure 5.8 *(a) A low resolution security image inside a liquor store. Note the baseball hat on the young man in the background; (b), the result of several image combination and enhancement attempts. The logo on the baseball hat could not be clearly discerned.*

5.3 Requirements for forensic photogrammetry

5.3.1 Accuracies

The accuracy required from a forensic photogrammetric exercise can vary from the sub-millimetre level (for example measuring a bite mark and comparing it to a set of dentures) to millimetres (length of a sawn-off gun barrel) to centimetres (the height of a suspect at a robbery) to decimetre or even metre level (the length of skid marks at a vehicle accident). These few examples represent a range of scales of accuracy in excess of 1000, yet the same photogrammetric procedures can usually be followed.

One feature with forensic photogrammetry is that the accuracy achieved will nearly always be different for every crime scene. Of course, photogrammetrists strive to achieve the best accuracy they can for every task and set in place checks on the level of accuracy they believe is possible given the circumstances. For example, when determining the height of a bank robber, it is always prudent to measure as many physical objects as possible in the near vicinity of the robber to provide checks. Another aid to the photogrammetrist is that surveillance cameras often take images at a rate of two per second, so that upwards of ten or more suitable images may be examined and a mean value calculated to give an estimate for the reliability of the result. The entire repeating of the measurements and subsequent calculations on an image, usually done on another day, is another way of gaining confidence that an answer is reliable.

In Figure 5.9, the height of the advertisement stand and other physical features could have been determined by photogrammetry at the time of calculating the suspect's height as a check

Figure 5.9   *Note the advertising signs which can be used as for accuracy checks when determining the height of the suspect.*

Figure 5.10 *Three months after the surveillance imagery in Figure 5.9. The carpet, tiles on floor and length of the serving counter had all changed, but the front door and overhead air-conditioning unit had remained unaltered.*

on gross errors. Unfortunately the interior of this bank had been extensively renovated in the three months between the robbery and the time the police found a suspect and contacted the photogrammetrist (see also Figure 5.10). Only three key features remained unaltered: the front door, the serving counter and the overhead air-conditioning unit. These were fundamental to determining the suspect's height.

It is extremely rare for the photogrammetrist to know beforehand the height of a suspect, the length of a gun, a shoe size or the length any other physical object that needs to be determined. A common question upon cross-examination in court is "Did you know, or were you told before-hand, how long the object was which you were asked to measure?" A positive reply will cast doubts on the evidence of even the most respected expert witness, so forensic photogrammetrists must not know whether their solution is correct or not until well after the court has finished its deliberations. This makes the requirement for check measurements all the more imperative and repeated measurements essential.

5.3.2 Software

Given the vastness of the possible combinations of imagery, it is difficult to suggest that one software package or tool could solve all forensic tasks. For aerial and most other stereoscopic pairs of photographs, analytical and digital photogrammetric software, and traditional photo-grammetric operators using analogue, analytical or digital stereoplotters, will provide a solution for a vital measurement, a plan or whatever is required.

Scenarios which involve opportunistic imagery require software which can accommodate single images. For those situations PhotoModeler, Shapequest and Rollei's Metric Single Rectification System are well-known examples. Computer packages designed to accept and adjust in bundle solutions the low oblique images typically taken around crime scenes are required when multiple images are available. Examples of such software include iWitness or RolleiMetric's Close Range Digital Workstation (CDW) in addition to those mentioned above. In some cases, only a more skilful consultant photogrammetrist with a good knowledge of applied geometry may be able to derive a solution to a single image which has peculiar geomet-ric constraints such as those in Figure 5.7.

On occasions, it is advantageous not to use software at all if an elegant solution using well-known measuring tools such as tape and plumb-bob will provide an equivalent result. A judge and jury can easily understand a simple diagram and the use of elementary surveying tools. Although the same result may be academically possible from a complex mathematical solution involving rotational matrices and lens modelling, it may have less impact in court on a judge and jury than being derived from a technique which laypersons believe they could have done themselves. The forensic photogrammetrist cannot rely solely on commercial software packages to do all the work.

5.3.3 Cameras

Almost every type of camera and sensor has appeared in court cases. The opportunistic nature of snapshots taken of a crime scene means that the forensic photogrammetrist has no control over whether the images arrive via film, digital or video camera. When the forensic photogram-metrist is in control of taking the images though, for example when working for the police, a government agency or as a consultant assisting an investigation, then it is possible to choose a camera specifically for that scene.

Most forensic photographers would have a range of cameras at their disposal. A moderately high resolution (say 8 megapixel) digital camera of the SLR type with a range of interchange-able lenses would be the first choice for photography inside buildings, close-ups and even of outdoor traffic accident scenes. Back-up photography on a medium-sized format, say 70 mm, using film would be prudent as there are usually time constraints at crime scenes to record everything before cars, bodies, broken glass and other key features suffer physical movement.

A small automatic focus and exposure digital camera with a zoom lens (say 5 megapixel resolution) would be ideal for general photography of the scene. Although it may not be the intention to make forensic measurements from such a pocket-sized camera, extra images should be taken from all angles, just in case.

A recreational-style video camera is often used to record the whole of a crime scene, not so much for the purpose of measurement, but for portraying to the court the total environment sur-rounding the incident. A digital video camera is of course preferable, as down-loading of still scenes to a computer is facilitated. There have been instances where some of the images taken on the video camera have been used for later measurement. One such case involved a domes-tic murder and although the room in which it occurred was thoroughly photographed, it was much later when viewing the general video imagery that a gardening tool shown in the laundry became a key item of evidence. That gardening tool had been imaged on the video from two or more different directions, so a strong photogrammetric solution was made to determine certain features of interest on its handle. Video cameras can often provide such unintentional stereo or multiple images. Put simply, you can't have too many images!

A tripod is a most important item for use with the primary forensic camera as light levels may not be optimal and timed exposures may be necessary. Supplementary lighting may be required, but usually this requires an electricity outlet in a building or a portable generator or large batteries at an outside location.

5.3.4 Control targets and object preparation

The control marks visible in a forensic scene vary from extensive to non-existent. This wide difference occurs as a consequence of whether it was possible for the forensic photogrammetrist to prepare for the photography or whether the imagery was captured by surveillance cameras or an opportunistic snapshot. In this latter case, the photogrammetrist must survey some permanent features at the scene to establish control points. This should be done as soon as possible after the imagery is received. As can be seen in Figures 5.9 and 5.10, in even a few months at a commercial premises, considerable renovations may make the task of providing control measurements difficult. It is a unique feature of forensic photogrammetry that a considerable time, even years, may elapse before an image is recognised as containing an important object, the size and shape of which must be determined.

In some forensic scenes where the photogrammetrist is called upon to take the imagery, there may be no physical restrictions to limit the amount of control points which can be placed to aid the accuracy of the final result. Small cones, numbered discs or cards with targets on them can be spread around the object to be photographed and surveyed or simply measured by taping to establish control point values. These targets are often referred to as evidence markers (see Figure 5.12).

One technique which is useful for some situations uses a grid projected onto the object which may otherwise not be amenable to the placing of control points on it. For example it may be required to photograph a human body part such as a face, arm or leg. The physical placing of "hard" control points on such surfaces may not be allowed, so the use of a slide or overhead projector can allow a graticule of crosses, lines or even a random pattern to be projected onto the body part of interest. This is especially useful if the object is smooth and shows little texture. An example of a pseudo-random pattern would include the projection of a map of the world onto the object. Photographs taken from a variety of angles will then allow the photogrammetric reconstruction of such a surface (see also Figure 5.11).

With the photography of inanimate objects, there may not be any restrictions on placing control markers, but practically there may be other constraints. It is not possible to unduly delay traffic on busy roadways, and when victims and their vehicles must be moved, the photogrammetrist must work quickly.

The size and shape of control targets is really a consequence of the scale of the scene and subsequent photography. At a road accident, traffic cones with numbers or letters placed on or near them are the usual form of control. These can be located in position with a surveying instrument or by taped measurements. These targets may be coordinated as a result of a photogrammetric bundle adjustment solution applied to the whole set of photographs.

Traffic accidents are identifiable as a clear sub-set of forensic photogrammetry and, being relatively common, are described in more detail in Section 5.5.3 in this chapter. It suffices here to state that like most forensic surveys they are often carried out under trying circumstances, but

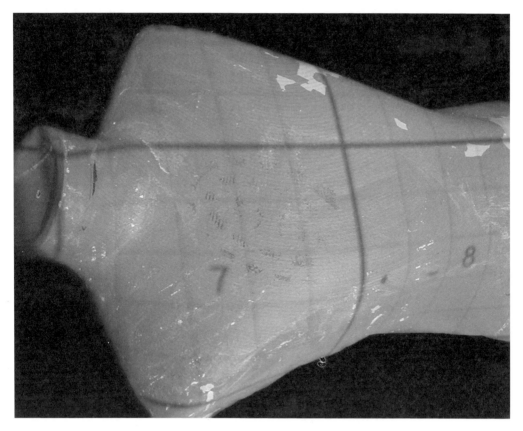

Figure 5.11 *A projected grid used to assist point identification on an otherwise untextured surface. Note also the imprint of a boot on the torso.*

in addition may occur at night, outdoors without access to conventional power for lighting, and in bad weather conditions.

For an indoor scene, mini-versions of these cones may be used along with other control points such as small tripods. These evidence markers may simply consist of randomly numbered black and white circular targets on flat pieces of metal distributed around the objects of interest and be surveyed for their position.

For objects under, say, one metre in length, the use of rulers as control lengths or set-squares can provide the control reference needed for a photogrammetric solution. Consider Figure 5.13 where the tread pattern of a boot was required. The tread was initially rubbed with charcoal and an imprint made on white paper, then the sole of the boot was rubbed with chalk to highlight the wear pattern. To directly compare these two patterns, the image of the actual boot was reversed.

As with all forms of photogrammetry, if a three-dimensional object is under investigation, it is important to try and have a good three-dimensional spread of the control points. It is usually not sufficient, if three-dimensional aspects of an object are required, simply to have all control points in one plane, lying on the floor for example. Such a situation can lead to a poor determination of the height of an object and it is always preferable to have an object surrounded in all

Figure 5.12   *Evidence markers and traffic cones used as control points at a traffic accident. Reproduced with permission of DeChant Consulting Services, USA.*

three dimensions by control points. The use of a surveying staff for scale is recommended in most scenes as this can provide an independent method to assess likely accuracies, if not needed as control for the actual photogrammetric set-up.

The author's experiences would suggest that a forensic photogrammetrist should always have access to a small step-ladder. Apart from the obvious benefit of getting an almost vertical angle to take photographs of objects prone on a floor, overall images of crime scenes can be seen in perspective and some occluding obstacles overcome.

Where images from surveillance cameras are involved (see Figure 5.14), on-site surveying measurements must be made to relate the position of the camera(s) to fixed objects in the scene. The height of the surveillance camera above the floor level is a crucial measurement as is its relationship to the wall(s) to which it is attached. The use of the step-ladder is again encouraged as viewing the crime scene from exactly the same perspective as the surveillance camera often provides clues as to which fixtures should be measured to act as control.

5.3.5 Forensic outputs

The output of a forensic photogrammetric exercise may be a plan of an accident site, an estimate of a suspect's height, a shoe size, a blade length or three-dimensional plots showing a comparison of dented panels on two vehicles which are the subject of a possible fraudulent insurance claim. Photogrammetrists are increasingly being required to provide the basic coordinates for all the objects in a scene, so that these measurements will allow the reconstruction of build-

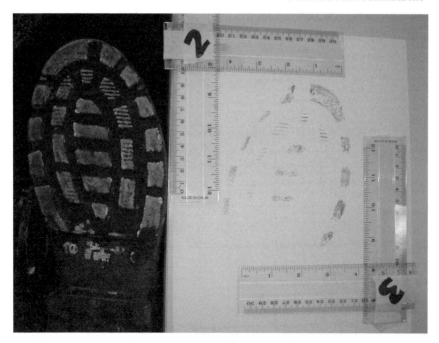

Figure 5.13 *The use of perpendicular rulers as evidence control markers for analysis of a boot and its tread pattern.*

Figure 5.14 *The accurate location of a surveillance camera is a crucial control measurement.*

ings, stairways, roads and paths in the form of computer models. Animations simulating a sequence of events leading to a crime or accident may then be crafted and shown to a court. Although the photogrammetrists may not necessarily construct the animation, it is their measurements which make such a reconstruction possible and a photogrammetrist will be the first in the witness box to testify to the accuracy of those measurements. So, the output is totally dependent on the nature of the investigation.

The forensic photogrammetrist's results must be written up in a clear, concise, yet full report. This report may be scrutinised by the police, the prosecutor, the defendant's legal defence team, possibly by another expert witness, the judge or magistrate and perhaps the jury. Clearly, the report has to be self-explanatory and written in a style which others can readily follow. All the assumptions made and procedures used to obtain results must be disclosed. Several months or even years may pass before the report and the results are examined, given the vagaries of some judicial systems.

The forensic photogrammetrist must not rely on memory to explain any result. Field notes and calculation sheets should always be kept in separate files for each investigation, and given that appeals against court sentences can occur a decade or more into the future, they should be kept in a very secure place.

5.4 Alternatives to forensic photogrammetry

Throughout this chapter, the emphasis is on "photogrammetry", but using the term to mean measurements derived from images in its widest sense. In many instances where photogrammetrists are called upon by police or other authorities to provide evidence from images of a crime scene, any competent surveyor or geomatic engineer could extract the relevant measurements. Very often there is really little or no photogrammetry, as described in Chapter 2, but a simple understanding of the principles of imaging (basically rays of light travel in straight lines). Traditional surveying techniques such as resection to find where the image was taken from and then intersection to find the relative location of two or more points of interest are often the only skills required. However, because a photograph or other type of image is involved, the law enforcement and legal system go to the forensic expert database in their jurisdiction and search under the key words of photographs or photogrammetry, and overlook surveying.

As an example of a situation where a problem was solved more by surveying techniques than by strict photogrammetric principles, consider Figures 5.15a and 5.15b and the following scenario. In a region where there were many old coal mines, a new urban housing development was approved. Upon doing the earthworks, an old mine entrance was uncovered and, fortunately, photographed from three different directions by the local authority responsible for old mine workings. Its exact location was not surveyed on that occasion and the earthworks contractor proceeded immediately to back-fill and consolidate the mine entrance. The housing estate was completed, houses were built and occupied.

Four years later, a resident complained to that mining authority of large and unusual cracks in the brickwork and rooms in his house. He claimed that it must be due to the old mine workings collapsing and wanted compensation from the mine authority, the earthworks contractor, the housing estate developer etc. The question arose as to whether the house in question was really built over the old mine entrance or not? Fortunately for all concerned, the house was at the extremity of the estate and the trees seen in the background of Figures 5.15a and 15.5b, including a fallen log, were still accessible. A surveyor located the relative positions

(a)  (b)

Figures 5.15 *(a) and (b): Two of three images taken of an abandoned mine entrance discovered during earthworks for a large housing estate. Four years later these provided the only clues as to the mine's location.*

of approximately eight distinctive trees and the house and plotted them on a detail plan of the estate.

Using the three photographs of the mine entrance, three clear plastic overlays were constructed which showed lines drawn from the equivalent of the focal point of the camera's lens to each of the trees and to the left and right sides of the mine entrance. The plastic overlays were superimposed on one another and placed over the surveyor's plan so that all the rays drawn to the respective features intersected (to within a metre) and the location of the mine entrance could be determined. It actually plotted more than 10 m away from the house in question. After more investigations by engineers from the mine authority, it was shown that the cause of the house problems was faulty construction. The house builder was subsequently found to be at fault.

Another example which used resection and then intersection techniques but at a larger scale than the mine investigation described above, concerned a request to determine the dimensions of a trunk of a tree shown, fortuitously, in three photographs taken in the front garden of a house. Eight years had elapsed since a man was seriously injured in a fall from the basket of a mobile crane while lopping a tree. There were claims of faulty equipment. The owner of the house had taken the three photographs soon after the incident. Figures 5.16a and 5.16b show two of those photographs.

The primary defendant was a large insurance company which believed that misuse of the equipment was involved in that a section of tree trunk was being lowered illegally in the basket of the mobile crane and this caused the accident. The dimensions of that section of tree trunk which can be seen at the base of the tree were required for evidence.

Upon visiting the site eight years after the incident, there were no above-ground signs of any trees. The lawn was green apart from a circular brownish patch. Prodding with a steel spike found the remains of the stump about 50 mm under that section of lawn. Using visual clues such

(a)
(b)

(c)

Figure 5.16 *(a) and (b) show two photographs taken soon after the accident involving the operator in the basket of the mobile crane; while (c) shows part of the reconstruction in the field eight years later to find the length of the section of tree trunk suspected of causing the mishap.*

as the partial occlusion of parts of the house and adjoining property, it was possible to relocate, in the field, where the original photographs must have been taken from. From these resected points, surveying tapes (see Figure 5.16c) were laid out across the lawn and gardens towards features on the house such as corners of window frames and doorways which were in line with the end of the tree trunk as shown in the photographs. The three tapes intersected within 100 mm, and using the stump's edge as the location of the other end of the piece of trunk shown in the photographs, its length, on site, could be determined. The diameter of the tree trunk was easily scaled from this length, a volume calculated assuming a cylindrical shape, and then a tree specialist contacted to determine the weight of such a volume. This civil case was settled out of court one year later.

The examples shown in Figures 5.15 and 5.16 illustrate that while photographs provided the information to solve the problem, it was more traditional surveying techniques which enabled a demonstrable solution. Of course, these examples may have been calculated using computer-based photogrammetric resection and then intersection, or even via a bundle adjustment. However, in the case of the mine, the simplicity of seeing plastic sheets overlain on the plan of the estate quickly convinced the laypeople involved as to the mine's location. In the case of the mobile crane collapse, the forensic photogrammetrist could obtain an answer directly on site and demonstrate the technique with photographs.

5.5 Case studies

The case studies shown in this chapter emphasise the range of scales that photogrammetric solutions can provide. The fundamental principles of the techniques employed are similar, but the size and shapes measured vary considerably from the length of a shoe or the barrel of a gun; to the height of a bank robbery suspect; to a traffic accident scene on a highway involving a photogrammetric survey over 100 m or more. The incidents involving racing cars were not examined in conventional courts but still demonstrate the application of photogrammetry to determine the relative positions of objects at a specific point in time. The chapter concludes with an example from the "hi-tech" end of the spectrum where medical imaging and photogrammetry are combined to produce a computerised cadaver to assist a forensic pathologist.

5.5.1 Case study 1: shoe size

The tasks given to forensic photogrammetrists are neither always high profile nor fit the glamorous image of television's forensic investigation teams. Figure 5.17 shows a photograph taken by a security camera during a bank robbery. The item of most interest to the police prosecutor was the shoe, of the cross-trainer style, seen on the suspect. The sole of that shoe had a distinctive tread and colour pattern and the police had established that very few pairs like it had ever been available for sale in the country.

The task in this case was simply to determine the shoe's length. This was a straightforward forensic task. The shoe was imaged in a near vertical position, adjacent to the bank's counter. The vertical edge of the counter was clearly visible and a distinctive white stripe can be seen. One taped field measurement was all that was required to establish the vertical overhang of the counter. The calculations consisted of a simple cross-ratio between the lengths of the shoe and the counter in the image and the taped length of the counter. The estimate obtained by this technique proved to be only 1 mm different from that measured by the police who had confiscated the shoes when the suspect was arrested.

Figure 5.17 *What is the shoe size? Note the distinctive and easy-to-measure vertical overhang on the counter.*

In this case the distance of the shoe from the camera was so similar to the camera–counter distance that no allowances for differential scale had to be made to the calculation. Applying such scale corrections is often necessary and can be a source of error, especially if the features to be physically compared are imaged on different parts of the photograph. Questions of lens distortion and camera tilts can influence the result if the objects under consideration occupy a large part of the image or are physically separated by a considerable distance.

5.5.2 Case study 2: gun sizes

The solution to all image-based forensic problems are not necessarily solved with routine photogrammetric procedures such as resection, intersection, bundle adjustment or the use of a computer package specifically designed to extract data from single images. Some problems require a good knowledge of perspective geometry and a reference manual such as Williamson and Brill (1990) is worthy of study.

In this case there had been a series of armed robberies and the surveillance video evidence showed the robbers, who were masked, using guns with shortened barrels and butts. Some time later police arrested some men outside a bank who were armed with similarly sawn-off shotguns and automatic rifles. Along with some other evidence the police believed they had arrested the gang responsible for the series of robberies. In court, part of the defence team's tactics was to claim that the guns seized at the time of the arrest were not the same as those in the videos. See Figure 5.18a and 5.18b for a photograph of the guns at the time of arrest and an example of a video frame from a robbery.

The forensic photogrammetrist was given access to the actual gun exhibits during the trial and allowed to make measurements of key lengths such as the barrel, stock and butt relative to distinctive features such as the trigger. These measurements were made using only a surveying tape in a room adjacent to the courtroom under the strict supervision of court officers. A good time to take redundant measurements and check your field notes!

The key to providing evidence for the court to consider in this case was to understand the implications of a formula used to rectify tilted aerial photographs in the era long before desk-top

(a)  (b)

Figure 5.18 *(a) and (b) Are these guns the same? Sawn-off guns at the scene of the arrest and a frame from a surveillance camera.*

scanners and computer-based solutions. The formula is rarely used these days, but a search of the earlier edition of a well-known photogrammetric textbook (Wolf, 1983) showed its derivation and described the principle shown in Figure 5.19.

A photographic image is formed as a perspective projection, and the anharmonic ratio is a formula which links distances on an object to its corresponding image. Irrespective of the orientation the object has relative to the camera, the formula shown in Figure 5.19 holds true. This formula allowed the comparison of the ratio using the butt, stock and barrel lengths of the guns in the video with those presented to the court. The suspects were convicted but it must be stressed that this was not the only evidence presented to the court. In fact there were a total of 72 witnesses, but the forensic photogrammetrist was one of only two expert witnesses, the others being police who were variously involved with the actual arrest and intelligence-gathering activities.

5.5.3 Case study 3: motor vehicles and traffic accidents

Photogrammetry has had a relatively long history as a preferred method for recording motor vehicle accident sites. In parts of Europe and Japan it has been used routinely for more than 30 years with specially designed vehicles equipped with stereopairs of cameras and dedicated police units ready to perform subsequent mapping tasks. Initially the cameras used were of the stable metric type, mounted on a rigid metal bar, and used glass plates. The mechanical/analogue stereoplotters were matched to these cameras and all aspects of the plan production followed routine procedures.

The 1970s saw the widespread introduction of electronic distance measuring (EDM) devices attached to theodolites and eventually the electronic total station instrument which incorporated EDM and theodolite functions with on-line digital storage facilities. These devices started to replace some of the special photogrammetric units as local police authorities were trained in the use of this equipment. Some traffic accident units were equipped with lower accuracy devices such as tripod-mounted or hand-held laser range finders with angle encoders. The role of the photogrammetric units became reserved for the more significant crash scenes, often those where fatalities were involved.

With the advent of laser scanning devices towards the end of the 1990s, a change in technology once again has affected the role of forensic photogrammetry. Once the ease of use of the laser scanners improved, for some large disaster scenes such as train and aircraft crashes, photogrammetry began to play a subsidiary role in the preparation of the plan. The cost of laser scanners and

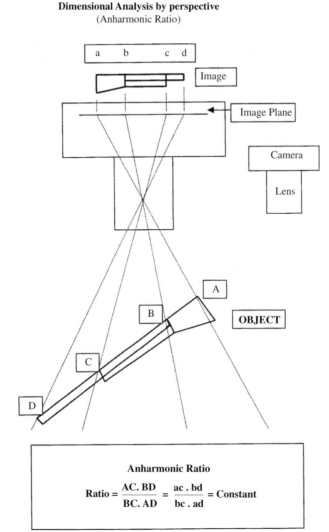

Figure 5.19   *The principle of the anharmonic ratio.*

related software is still of the order of US$100,000 (late 2006), so the widespread acceptance of this technology has not yet occurred. Given that expensive equipment purchases are difficult in many government authorities and that most local police units have total station technology, it is unlikely that laser scanners will have much impact on most traffic crash sites for some time yet. Only major accident investigation units are likely to have access to a terrestrial laser scanner in the near future and integrate its use with photogrammetric and other surveying techniques.

Despite the decline in traditional photogrammetric units dedicated to recording traffic accident sites, and the competing technologies of surveying equipment, there are some indications that photogrammetry may re-emerge as the preferred recording technology. Ease of use of digital cameras coupled with modern interactive software is causing this revival. It has been documented that at many traffic accidents, local police units have recorded the locations of vehicles and skid marks with total station instruments and then allowed the site to be cleared up for the resumption of normal traffic flow. This has led to key items of evidence not being recorded. For examples, minor dents on vehicle panels which later prove where pedestrians were first struck, patterns of oil and water spills under the vehicles and the location and orientation of items of street furniture are all features at a traffic accident which may not be recorded in the confusion of an accident scene. These features would be available for later examination if photogrammetry had been employed.

Figure 5.20 shows a vehicle which was involved in a fatal accident with a bicycle. An important feature, not immediately obvious, is a small dent on the roof of the vehicle, towards the rear. Using that mark and the obvious points of impact on the front bumper and the windscreen, the relative trajectories of the car and the cyclist were established. These were key pieces of evidence in this case.

Just as improvements in computer-based technologies made possible the introduction of total station instruments and laser scanners, so it has created a revolution in photogrammetric techniques. Digital cameras allow the photographer to immediately view the image, they are simple to use and they are sensitive and tolerant to a wider range of lighting conditions than their film-based counterparts. Most importantly, the images are available for direct download

Figure 5.20 *A small dent on the roof of this vehicle helped establish the trajectories of the vehicle and a bicycle immediately before collision.*

to portable computers on which software directly tailored to forensic investigation is available. The cost of an entire set-up of digital camera, portable computer and appropriate software is well under US$10,000, or about half the cost of a total station surveying instrument. The use of photogrammetric techniques at traffic accident sites seems certain for a revival, without the need for the dedicated units of 30 years ago. The benefits of reduced traffic disruption, collection of a permanent record and lower cost of equipment are being recognised worldwide.

A distinguishing feature of traffic accident sites is that they may be spread out over 100 m or more metres, usually requiring a long, narrow plan (consider the highway stretching out behind the vehicle in Figure 5.20). This means that a single pair of photographs will be insufficient and usually multiple pairs of photographs are taken along the highway every 5–10 m. Consequently control markers must be placed all along sections of road to be mapped, as well as specific evidence markers.

The natural geometry of long sections of roadway is not normally conducive to a good photogrammetric solution, as there is little opportunity for the control points to be placed anywhere but on a plane. It is difficult, unless a special tripod is available, to elevate the camera stations away from this plane of targets. The use of extra photographs taken from either side of the road to assist this geometric dilemma in any bundle adjustment is recommended.

The more complicated the scene of an accident or incident is, then obviously the greater the need for more photographs. In the case of a traffic accident, these should be taken as soon as possible after arrival since it is likely that vital pieces of evidence may be moved due to rescue operations or the actions of other agencies such as fire brigades, ambulance personnel, police etc. The following considerations should be made when photographing (and also taking video imagery) of a traffic accident site:

- Ensure photographs are taken of all skid and other marks from both directions along the road. Very often a skid mark can only been seen from one particular direction, and then with only the direction of lighting being favourable.
- Ensure key permanent features such as buildings, posts and signs are visible in these photographs so that future reference to them can identify their exact location on the roadway after rain has washed away their visible trace.
- Photograph the final resting positions of the vehicles from side-on as well as from the front and rear. These side-on photographs will show pieces dislodged from the vehicles, the skid marks and the vehicles' relationships to the sides of the road.
- Photograph the damage apparently caused by the accident to each vehicle from a number of directions for later reconstruction of those damaged panels.
- Photograph any other damaged parts of vehicles, sign posts, street furniture, fences etc. even though this damage may not seem to be part of the immediate crash site. At the accident, a lot of noise and confusion may be present and what may not seem a consequence of the accident at the time may become a very important piece of evidence later.
- After the vehicles, persons involved, and other items have been moved, those sections of the scene should be re-photographed as there may be water, oil or blood and small components not noticed previously.
- Very often the scene will have had to been photographed at night, so the forensic photogrammetrist should return the next day to review the scene and see if anything was overlooked.

The above checklist can be applied with modifications where necessary to almost any type of forensic scene. A key principle is to ensure that sufficient photographs are taken to record an incident: there really cannot be too many!

Figure 5.21   *Spray paint used as evidence markers to indicate recent liquid droplets.*

Stories abound about accident investigation surveys which went "wrong". In most cases, one or more of the points listed above had been overlooked. On one occasion a photogrammetric survey of a fatality along a 100 m section of highway was completed and all the points of detail, such as broken glass, blood splatter etc. recorded. What was not recorded was the location of the accident relative to adjoining fence-lines, light posts, traffic merging signs and other fixtures. This oversight was identified the following week during plan preparation, but when the photogrammetrists went back to the site, the highway authorities had already resealed that entire section and obliterated any paint or other markings on the road. During the court case, the judge requested a site inspection, but he could only be shown the approximate locality.

Although a two-dimensional plan is the most likely output of a traffic accident site, the third dimension sometimes is a most important consideration. For accidents occurring on high speed bends, a contributing factor is often the geometry and the cross-slope of the road. Questions can arise as to whether the super-elevation, or deliberate raising of the outer side of the road around a curve, has been properly engineered and constructed or whether after resurfacing the lane widths had been set out correctly. For major accident investigations, usually those involving fatalities, three-dimensional representations in the form of a still graphic, or even a animation, may be required. See Figure 5.22 for an example using CAD software.

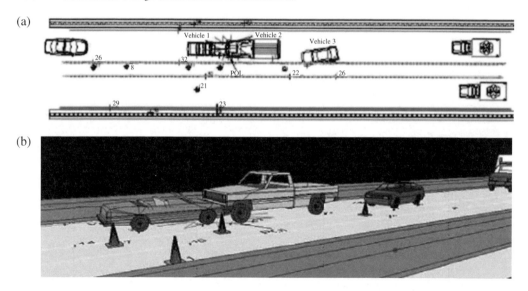

Figure 5.22 *CAD reconstruction of a traffic accident scene; (a) plan view; and (b) 3D reconstruction. Reproduced with permission of DeChant Consulting Services, USA.*

In instances where the accident is a major one spread out over a considerable distance, such as an aircraft crash, or other special circumstances are present, the photographs at the scene will become part of a larger plan utilising GPS (the global satellite-based positioning system), existing maps and aerial photographs from light planes or helicopters.

One example of special or unusual circumstances was a large number of cars in a crash on the M42 motorway in England in 1997. There was a fire at the scene which was so intense that the road surface melted. Police could not access the immediate vicinity for some time and the closely packed nature of the cars made any on-site survey, including the use of surveying total station instruments, very difficult. After all the wreckage had been cleared away, a collection of police helicopter and other press photographs as provided to the photogrammetrist who had the task of deciding which ones could be used to form a bundle adjustment. The resultant plan showing the positions of the vehicles achieved accuracies of the order of 10–20 cm. Forensic photogrammetrists must not be rigid in their approach to scene capture but be ready to adapt to local conditions.

5.5.4 Case study 4: racing cars

Photogrammetric experts have been used to assist in the resolution of several incidents in international motor racing events. Photogrammetrists have been integral to these investigations as they have the capability to derive and combine information about the cameras' lenses, physical measurements on the track, and the perspective views taken from both video and still cameras which can be confusing to the layperson. For instance, determining which driver was the cause of a high speed crash is considerably simplified if the vehicles are tracked over a period of time to produce a plan view of the event together with timing information. The investigators can clearly determine which vehicle altered its direction of travel and when.

Disputes have occurred as to the exact, or relative, positions of racing cars at the precise moment a warning light was shown. Warning lights are switched on when a dangerous situation occurs on one part of the circuit and rescue vehicles may have to come onto the track. During the period that such lights are on, all drivers must retain their relative positions in the race. If one car is overtaking another at that moment, a dispute may arise as to which vehicle was actually in front when the lights came on. The warning lights are displayed in prominent positions around the track and, for major races, a warning light is also activated at the centre of the driver's steering wheel.

An extreme situation occurred during the penultimate lap of the 2002 Indianapolis 500 race in the USA. The two leading cars were challenging each other when a crash occurred on another part of the track. Before the accident could be cleared away for the race to resume at full speed, the final lap was completed and the result was subsequently in dispute. Which car was ahead when the warning lights came on?

A combination of: video frames from fixed camera positions which depicted the leading cars; video sequences which showed when the warning lights came on; on-board telemetry data from one of the cars; and a survey of fixtures on the track, assisted in the recreation of the exact and relative positions of the two cars on the track. See Figure 5.23 for an image showing the two cars near the time of the incident. There were many other factors to consider in this case, not the least of which was whether the car on the inside of the corner was actually physically closer to the finish line even though it may have been overtaken by a small margin by the car on its outside.

In another similar situation shown in Figure 5.24, the camera was quite distant and at a poor intersection angle to the corner where one car was overtaking another. It is not obvious from the image of the curved track which car was in the lead. The question arises as to where a radial line should be drawn across the track to determine whether the car on the right had overtaken the car on the left before the end of the yellow-flagged section of the track (indicated by a marshal waving a green flag behind the safety screen on the right-hand side of the picture).

A survey of features that could be identified in the original racing video image (Figure 5.24a) was undertaken and the 3D locations of these features were used to resect the camera's position and orientation. Two locations (on either side of the track) of a radial line perpendicular to the track and intersecting the position of the green flag marshal were surveyed and captured on a second image (Figure 5.24b). These 3D locations were then projected onto the original racing image (Figure 5.24a) to allow the relative position of the two cars to be visually determined.

When a crash involving several racing cars occurs at high speed, it can be difficult to unravel the sequence of events leading up to the collision, even upon viewing the video of the event in slow motion several times. The changing perspective view of the cars as they race away from a camera, can cause the observer to reach false conclusions. Other factors such as the nearness to a corner, cars altering their line of progress, braking, and the fact that a sector of each driver's field of view is hidden due to the configuration of their safety helmet, neck brace and position of the rear vision mirror must be considered.

A crash shortly after the start of a Grand Prix event in 2003 was analysed to appeal against a penalty given to one of the drivers involved in this incident. An image taken immediately before the crash is shown in Figure 5.25a (see colour section). The positions of the three key cars involved were determined both across and along the track at a number of epochs. Across-track positioning was achieved by simple proportioning since the camera's view was along the line of the track.

Distances along the track required a more technical solution. Telemetry from sensors on one of the cars was used to establish the position of the car relative to CAD information regarding

Figure 5.23 *2002 Indianapolis 500. Which car was in front when the warning lights (not in view) came on? Reproduced with permission of Dr. T. Clarke, Optical Metrology Centre, UK.*

(a)

(b)

Figure 5.24 *Original racing image (a) with re-projected "equidistance" line derived from subsequent image (b), capturing radial lines. Reproduced with permission of Dr. T. Clarke, Optical Metrology Centre, UK.*

the track. The relative position of the other cars was determined by inspection of visual clues and a knowledge of the geometry of the cars. Where survey information existed it would also have been feasible to use the anharmonic ratio method depicted in Figure 5.19.

The resulting information allowed a plan view of the incident to be produced using the CAD plan drawing shown as Figure 5.25b. Such a plan allowed the investigation of the sequence of events leading to the crash to be studied without the misinterpretations possible from a simple viewing of the video of the race crash. A particular advantage was the incorporation into the diagram of the driver's blind spot which revealed that the driver depicted by the blue car had only a limited or no view of the car depicted in black.

These case studies on motor racing may have seemed somewhat incongruous to a chapter on forensic photogrammetry, where the word forensic usually conjures a mental picture of a crime such as a bank robbery and a traditional courtroom. These investigations were included because they show how the thorough analysis of images, combined with physical measurements, and the portrayal of the results by graphics, can lead to a clearer understanding of controversial incidents. The forensic photogrammetrist must be prepared to be innovative to provide an understanding to imaging problems in which non-experts have strongly vested interests in the outcome.

5.5.5 *Case study 5: virtopsy, a virtual autopsy*

Photogrammetric techniques applied to wounds on human bodies can only depict a surface model of those cuts, abrasions, and other disfigurements. It has been the role of medical doctors and forensic pathologists to examine such wounds more closely and to provide further medical evidence for the court. This is especially true if the incident caused death to the person. An autopsy may be performed to determine the extent of rupture or other damage to internal organs.

Virtopsy, meaning a virtual autopsy, is the term recently coined (e.g., Thali *et al.*, 2005) to describe the integration of photogrammetry with forms of medical imaging and scanning such as computerised tomography (CT) and magnetic resonance imaging (MRI). The surface shape of an injury is obtained by photogrammetry, and through the correspondence of some special reference marks which are visible on both the MRI scans and the photographs, a computerised composite three-dimensional model of the body part is constructed. As the following figures illustrate, it is possible to determine the effect of an impact on a body without the need for a normal autopsy.

Scenarios for the use of virtopsy which have been reported in the literature range from gunshot wounds to pedestrians being hit by motor vehicles. In the former case, the shotgun had been pressed against the chest of the victim and the photogrammetry of the immediate vicinity of the wound assisted in the identification of the type of gun. The medical imaging showed where ribs had been broken by shotgun pellets and CT sections through the body allowed the trajectory of the pellets to be tracked. In the case of the pedestrian, it was possible to relate the size and shape of a wound on the person's back to the rear-vision mirror of a vehicle. The depth of trauma inside the victim's body demonstrated the force involved at the time of the collision.

The successful matching of the photogrammetric surface model and the medical scans relies on the reference marks which must be visible on both sets of images. The radiological landmarks, also known as multi-modality markers, (which look remarkably like small metal washers), must be affixed around the wound area of interest before either the photogrammetry or the medical imaging takes place. These can be clearly seen in Figures 5.26 (see colour section), 5.27 and 5.28.

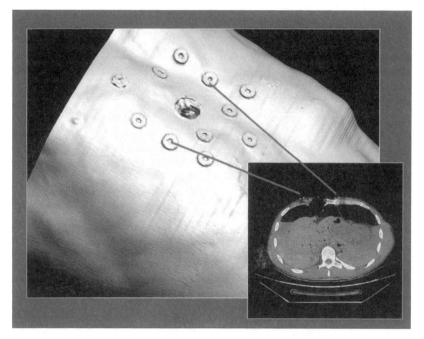

Figure 5.27 *A 3D reconstruction of a body with a gunshot wound from computerised tomography (CT) scans. After Thali et al. (2003b), reproduced with permission from ASTM International.*

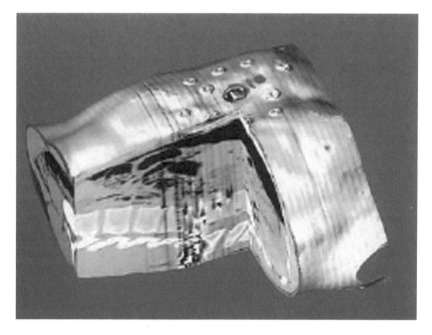

Figure 5.28 *Combining photogrammetric surface detail with CT slices to show the paths of shotgun pellets through the body. After Thali et al. (2003b), reproduced with permission from ASTM International.*

There are several advantages claimed for virtopsy. These include its non-invasive nature which could be extremely valuable in situations where for religious or other reasons it is not possible to perform a regular autopsy. Scanned internal medical images provide an observer-independent way of assessing the results of an autopsy which can remove the possible subjective nature of such procedures. Another benefit is that the entire set of imagery could be transmitted by the internet to another expert for a second opinion. The images can be stored on digital media for as long as desired, unlike the situation with cadavers. As our civilization moves down a pathway towards ever increasing degrees of specialisation, it is possible to predict centres of excellence will be established to cater for the analysis of such imagery.

5.6 Conclusions

Forensic photogrammetry is an interesting and diverse area of application. Many investigations, court cases and tribunal hearings can benefit from the clarity which the results of rigorous photo-grammetric investigations can produce. This chapter has highlighted a few areas where forensic photogrammetric techniques have been applied to the advantage of the legal process.

The distinctive features of forensic photogrammetry may be summarised as:

- Information on the sizes and shapes of objects and people may be extracted from single, stereo pairs or multiple images.
- The images may have been taken by a trained photogrammetrist in which circumstance there will be control points established prior to the photography.

The images may have been acquired fortuitously as snapshots by passers-by, as frames from surveillance cameras, or otherwise simply opportunistically. In these situations, it may be a long time before it is known that such images exist, there may be very little or no control information about the scene, there may be scant information about the camera, and the images may be damaged or cropped in some way.

In this latter scenario, there is always a need for the photogrammetrist to visit the scene and try to find objects and take physical measurements which can assist the recreation of the size and shape of certain objects as they were at the time of the incident.

The output provided by the forensic photogrammetrist to the court or legal authorities usually consists of a report which details the work undertaken. An integral part of that report should be a listing of the results for the size and shape of the key objects subject to investigation. The likely accuracy of the results must be made clear. The results may be displayed in the form of a plan, annotated image or even as a 3D model or animated sequence.

Most importantly, the forensic photogrammetrist must be aware of the technical limitations of many of those who seek advice, and try to present all results in terms which non-technical members of the legal profession, police and laypeople on a jury may comprehend.

Despite the undoubted benefits of forensic photogrammetry as a tool to assist investigation, its greatest drawback seems to be the lack of understanding about its potential, and knowledge of its very existence, by certain sections of the legal profession. The wider acceptance of forensic pho-togrammetry appears limited by these external factors, rather than any lack of the capabilities for photogrammetric investigation.

5.7 References

Bramble, S., Compton, D. and Klasen, L., 2001. Forensic Image Analysis, Version 2, *13th INTERPOL Forensic Science Symposium*, Lyon, France, 16–19 October 2001.
Hand, D.W. and Fife-Yeomans, J., 2004. *The Coroner*, ABC Books, Sydney, Australia.

Thali, M.J., Braun, M., Markwalder, T.H., Bruschweiler, W., Zollinger, U., Malik, N.J., Yen, K. and Dirnhofer, R., 2003a. Bite mark documentation and analysis: the forensic 3D/CAD supported photo-grammetric approach. *Forensic Science International*, 135, 115–121.

Thali, M.J., Braun, M., Wirth, J., Vock, P. and Dirnhofer, R., 2003b. 3D surface and body documentation in forensics: 3D/CAD photogrammetry merged with 3D radiological scanning. *Journal of Forensic Science*, 2003, 48(6), 1356–1365.

Williamson, J. and Brill, M., 1990. *Dimensional Analysis Through Perspective: A Reference Manual*, American Society for Photogrammetry and Remote Sensing, Bethesda, Maryland, USA.

Wolf, P.R., 1983. *Elements of Photogrammetry*, McGraw-Hill, New York, USA.

5.8 Further reading

Atkinson, K.B. (Ed.), 1996. *Close Range Photogrammetry and Machine Vision*, Whittles Publishing, Scotland.

Batterman, S.C. and Batterman, S.D., 2000. Motor vehicle, accident investigation. In *Encyclopedia of Forensic Sciences*, Academic Press, London, UK, pp. 33–42.

Braun, M., 2003. Use of Rollei d30 metric in forensics. *7th International Rolleimetric Police Workshop*, Neuss, Germany.

Bruschweiler, W., Braun, M., Dirnhofer, R.T. and Thali, M., 2003. Analysis of patterned injuries and injury-causing instruments with forensic 3D/CAD supported photogrammetry (FPHG): an instruction manual for the documentation process. *Forensic Science International*, 132, 130–138.

Buck, U., 2003. The use of photogrammetry with the Baden-Wurttemburg police. *7th International Rolleimetric Police Workshop*, Neuss, Germany.

Burg, H., 2003. The application of Rollei MSR by experts in accident reconstruction and integration into simulation programmes. *7th International Rolleimetric Police Workshop*, Neuss, Germany.

Carrier, B.D. and Spafford, E.H., 2004. Defining event reconstruction of digital crime scenes. *Journal of Forensic Science*, 2004, 49(6), 1291–1298.

Chainey, S., 2004. GIS and crime mapping – going beyond the pretty hotspot map. *Geomatics World*, 12(2): 24–25.

Chong, A.K., 2002. A rigorous technique for forensic measurement of surveillance video footage, *Photogrammetric Engineering and Remote Sensing*, 68(7), 753–761.

Ciccone, T., 1986. Seeing is believing. In *Photographs and Maps go to Court*, American Society of Photogrammetry and Remote Sensing, Washington, USA, 14–26.

Clarke, T.A., 2005. Personal communications and www.optical-metrology-centre.com.

Compton, D., Prance, C., Shears, M. and Champod, C., 2000. Systematic approach to height interpretation from images. *SPIE Proceedings*, 4232, 521–532.

Cooner, S.A. and Balke, K.N., 2000. *Use of photogrammetry for investigation of traffic incident sites*, Report TX-99/4907-2, Texas Transportation Institute, Texas A&M University System, College Station, Texas, USA.

Criminisi, A., Zisserman, A., Van-Gool, L. Bramble, S. and Compton, D., 1998. A new approach to obtain height measurements from video, *SPIE Proceedings*, 3576, 227–238.

DeChant Consulting Services – DCS, Inc. See http//:www.photomeasure.com (accessed 19 October 2006).

El-Hakim, S.F., Boulanger, P., Blais, F. and Beraldin, J.A., 1997. A system for indoor 3D mapping and virtual environments. *SPIE Proceedings*, 3174, 21–35.

Fraser, C.S. and Hanley, H.B., 2004. Developments in close-range photogrammetry for 3D modelling: the *iWitness* example. *International Workshop on Processing and Visualisation using High Resolution Imagery*, Pitsanulok, Thailand, November, 2004.

Fraser, C.S., Hanley, H.B. and Cronk, S., 2005. Close-range photogrammetry for accident reconstruction. In Grün, A. and Kahmen, H. (Eds.), *Proceedings of Conference on Optical 3D Measurement Techniques*, Vienna, Austria, 3–5 October 2005, ETH Zurich, Switzerland.

Fryer, J.G., 2000. An object space technique independent of lens distortion for forensic videogrammetry. *International Archives of Photogrammetry and Remote Sensing,* 33(A5), 246–252.

Fryer, J.G., 2000. Drawing the line on criminals. *Engineering Surveying Showcase 2000*, Issue 2, 33–35.

Fryer, J.G., 2004. Forensic geometry – not forensic geomatics. *Geomatics World*, 12(2), 22–23.

Harley, I., 2004. Forensic photogrammetry: a personal perspective. *Geomatics World*, 12(2), 19–21.

Hochrein, M.J., 2004. *A bibliography related to crime scene interpretation with emphases in geotaphonomic and forensic archaeological field techniques*. United States Department of Justice, Federal Bureau of Investigation, FBI Print Shop, Washington, D.C., USA.

Indy Racing League, Team Green, Inc. Appeal, Decision. Also for example of press coverage, See http://www.detnews.com/2002/motorsports/0206/05/sports-506086.htm.

Joseph, G.P., 2000. A simplified approach to computer-generated evidence and animations. *The Sedona Conference Journal*, 55–69.

Klasén, L. and Fahlander, O., 1997. Using videogrammetry and 3D image reconstruction to identify crime suspects. *SPIE Proceedings*, 2942, 161–169.

Kullgren A., Lie, A. and Tingvall, C., 1994. Photogrammetry for documentation of vehicle deformations–A tool in a system for advanced accident data collection. *Accident Analysis and Prevention*, 26(1), 99–106.

Maas, H.G., 1997. Concepts of real time photogrammetry. *Human Movement Science*, 16, 189–199.

Macquarie Encyclopedic Dictionary, 1990. Macquarie Library, Sydney Australia, 361.

McGlone, J.C., 2004. *Manual of Photogrammetry*, American Society for Photogrammetry and Remote Sensing, Bethesda, Maryland, USA, 1050–1062.

Photometrix. See http://www.photometrix.com.au (accessed 19 October 2006).

Photomodeler. See http://www.photomodeler.com (accessed 19 October 2006).

Regina v. Gallagher, 1999, District Court at Sydney, Australia, Kinchington QC DCJ.

Regina v. Hughes, 1999, District Court at Newcastle, Australia, Armitage QC DCJ.

Robertson, G., 1990. Instrumentation requirements for forensic analysis, In Grün, A. and Baltsavias E.P. (Eds.), *Close-range Photogrammetry meets Machine Vision, SPIE Proceedings*, 1395.

Robertson, G., 1990. Aircraft crash analysis utilizing a photogrammetric approach. In *Close-range Photogrammetry meets Machine Vision, SPIE Proceedings,* 1395.

Robertson, G., 2003. Forensic analysis of imprint marks on skin utilizing digital photogrammetic techniques. *88th International Educational Conference* sponsored by the International Association for Identification, 8 July, 2004, Ottawa, Ontario, Canada.

Subke, J., Wehner, H., Wehner, F. and Szczepaniak, S., 2000. Streifenlichttopometrie (SLT) A new method for the three dimensional photorealistic forensic documentation in colour. *Forensic Science International*, 113, 289–295.

Thali, M.J., Braun, M., Bruschweiler, W. and Dirnhofer, R., 2000. Matching tyre tracks on the head using forensic photogrammetry. *Forensic Science International*, 113, (1–3), 281–287.

Thali, M.J., Braun, M., Buck, U., Aghayev, E., Jackowski, C., Vock, P., Sonnenschein, M. and Dirnhofer, R., 2005. VIRTOPSY-Scientific documentation, reconstruction and animation in forensic: individual and real 3D data based geometric approach including optical body/object surface and radiological CT/MRI scanning. *Journal of Forensic Science*, 50(2), 428–442.

Thompson, E.H., 1962. Photogrammetry. *The Royal Engineers Journal*, 76(4), 432–444 and reprinted in *Photogrammetry and surveying: a selection of papers by E.H. Thompson, 1910–1976*. Photogrammetric Society, London, UK, 1977, 242–254.

van den Heuvel, H., 1998. Positioning of persons or skulls for photo comparisons using three-point analyses and one-shot 3D photographs. *SPIE Proceedings*, Vol. 3576, 203–215.

van den Heuval, F.A., 1998. 3D reconstruction from a single image using geometric constraints. *ISPRS Journal for Photogrammetry and Remote Sensing*, 53(6), 354–368.

Waldhaüsl, P. and Kager, H., 1984. Metric restitution of traffic accident scenes from non-metric photographs. *International Archives of Photogrammetry and Remote Sensing*, 25(A5), 732–739.

Walsh, J., 2004. Personal communications. Photarc Surveys Ltd, Harrogate, North Yorkshire, UK.

6 Quantifying landform change

Jim Chandler, with Stuart Lane and Jan Walstra

6.1 General outline of the application

6.1.1 Introduction

There is growing awareness that we live in an increasingly dynamic world. Global warming and associated environmental change is affecting the landscape and landforms around us in ever more dramatic ways. The wider media increasingly reports tragedies arising from: flash flooding, earthquakes, tsunamis and landslides, all of which wreak havoc on the supposedly stable and unchanging solid earth beneath our feet.

The desire to measure and quantify landform change has therefore grown in importance as scientists and society seek to grapple with the implications of an unstable world. It has not always been so; an investigation into older published literature reveals only a few instances where spatial landform change has been measured. This is somewhat surprising, particularly when it is realised that there is the distinct scientific discipline of "geomorphology" which studies the physical processes that "shape the solid surface of the earth and the landforms created by them", (Machatschek *et al.*, 1969). Why have geomorphologists not, until comparatively recently, fully embraced the mapping and spatial measurement technologies that have been available for many years? The unique capabilities of photogrammetry to extract spatial data in three dimensions and at various and flexible scales would suggest that photogrammetry is ideally placed to contribute to understanding landform evolution and associated processes. Why has photogrammetry not been used more frequently? One of the main problems is that usually there is insufficient time to observe how landscapes evolve because most geomorphological processes operate very slowly and often require thousands of years of time to elapse before there is a sufficient morphological expression that can be measured. Geomorphologists have tended to resolve such problems by using three main approaches and the case studies cited in this chapter have contributed to all three of these methodologies at different spatial scales.

The first approach is to model landscape evolution numerically by developing complex mathematical models to simulate the processes that shape landforms. Such approaches have gained increased popularity with the development of computers; particularly amongst fluvial geomorphologists and computational fluid dynamicists, examining the processes and impact of river flow. The general problem is the parameterisation of such mathematical models to ensure that output actually reflects change occurring in the real world. Use is often made of small-scale and laboratory-based physical models in which parameters such as discharge, gradient and streambed material are controlled, whilst key outputs such as stream velocity, turbulence and sediment transport are measured and compared with model outputs. The first case study (Section 6.7.2) demonstrates the way that photogrammetry has been used to accurately measure bedforms and sediment transport rates in flume experiments 2–12 m^2 in size.

The second approach used by geomorphologists has been to concentrate detailed field studies on processes and environments where the processes operate more quickly. Again, the second case study (Section 6.7.3) illustrates this approach in fluvial geomorphology which can apply to a very wide range of scales of river reaches, covering areas 100–10,000 m^2. Here photogrammetry has been used to quantify change occurring in dynamic natural river channels using a combination of oblique and vertical aerial imagery.

The third and final approach adopted by geomorphologists is the "space for time substitution", which is perhaps the most controversial and least scientific of all three. This assumes that in the modern landscape we see landforms at various stages of development and make inferences about changes through time based on the variety of forms seen at present (Paine, 1985). Such "ergodic transforms" have been widely used in geomorphology, particularly to provide a wider framework explaining change at the largest spatial scale. It relies on two main assumptions: that processes and systems structure identified today, also operated in the past; and that observed sequences reflect exclusively the passage of time and are attributable to no other cause. The final scale of case studies cited in this chapter (Section 6.7.4) attempts to obviate the need for ergodic transforms by exploiting one of the unique capabilities of photogrammetry: the extraction of spatial data from historical photography. This ability allows landform change to be quantified over 50, perhaps even 100, years of natural evolutionary processes. Although such time periods are lower than the thousands of years ideally required, it is considerably more than the timeframes available for the typical funded research project. Here, the case studies focus on large coastal and inland landslides occupying areas exceeding 3 km^2 (Section 6.7.4).

6.1.2 General advantages of photogrammetry

Although there is a range of measuring technologies capable of providing spatial data, photogrammetry is unique because of its flexibility of scale. As a single method it also offers many general advantages to geomorphologists interested in measuring landform change, many of these being associated with the recent advances in automated measurement.

Over the last 10 to 15 years the ability to extract dense digital elevation models (DEMs) has been the most significant reason why geomorphologists have begun to embrace photogrammetric methods. The automation of DEM extraction through image correlation has enabled data acquisition rates to exceed 500 points per second, allowing incredibly dense and detailed DEMs to be generated. These are exactly the data that geomorphologists require to interpret the landscape and to increase their understanding of earth processes. A parallel and supporting trend has been the development of capable terrain modelling packages using either grid or triangulated irregular network (TIN) based data structures. It is now comparatively easy to manipulate the (X,Y,Z) coordinates derived by photogrammetry and extract contours, cross-sections and volumes. Further processing allows the subtraction of surfaces representing multiple epochs so identifying areas of morphological change.

One interesting paradox is that automated DEM generation is regarded more favourably by geomorphologists than many photogrammetrists because the resultant surfaces are only of partial use to those involved in map production. The surface models derived by automated DEM extraction always represent the top of the "visible surface" and require further manual editing to remove unwanted features, typically trees and buildings. Substantive photogrammetric research has been undertaken to automatically derive the required ground surface, with mixed success. This problem is rarely of concern to the geomorphologist. The measured surfaces are normally natural and if actively eroding are generally devoid of vegetation. In addition, active surfaces are normally comprised of sand/gravel/rock particles which provide an image texture that is ideally suited to image correlation. The surfaces derived through image correlation are therefore generally exactly what the geomorphologists require and because no editing is required, they can be generated at the highest possible density, suiting automated DEM generation methods.

The ability to create orthophotos automatically using a DEM and the original imagery has also been highly significant. Retaining spectral information contained in the original image has allowed the geomorphologist to interpret the landscape directly without the simplification and abstraction that production of a conventional line-map inevitably introduces. Furthermore, the draping of the orthophoto on top of the terrain model allows impressive visualisations of landscape, which are very attractive and of particular benefit when trying to inform the expert and educate the layperson.

6.1.3 General disadvantages of photogrammetry

All measurement technologies present difficulties and photogrammetry has several significant disadvantages which account for its current niche status.

Chapter 2 identifies many of the traditional impediments that have prevented wider usage, particularly the need for specialised cameras and plotting instruments which are now unnecessary. Indeed, the case studies presented in this chapter have mainly used digital sensor technology that is becoming cheaper and is of increasing sensor resolution. The work has also been processed exclusively using a moderately equipped PC costing less than US$1000. Appropriate photogrammetric software is required and this remains expensive when compared to other common applications on the PC desktop. This is not surprising when the software development costs are compared to the potential market size. A single user licence for the Leica Photogrammetry System (Leica Geosystems, 2003) used for many of the applications in this chapter, currently costs US$15,000 (2006), a significant investment for most organisations and one that can only be justified if regular usage is envisaged.

Throughout this period of rapid evolution one disadvantage of a photogrammetric approach has remained: the need for expertise in photogrammetry. This disadvantage is always likely to be significant, for a simple reason: to the layperson the idea of extracting 3D spatial data from imagery remains a mysterious and perhaps even a "black art". The principles involved (Section 2.1) are not obvious and if any technique is not understood it is less likely to be accepted. New users always require convincing that photogrammetry can produce acceptable data. Once proven, users often gain sufficient understanding of the principles to be able to apply the approach in similar circumstances. With further experience and increased understanding they may even be able to apply the approach to solving new measurement problems. This learning cycle takes a considerable period of time and is therefore the most significant impediment to the wider usage and acceptance of a photogrammetric approach. Unfortunately, scientific papers and most textbooks rarely ease this process, being rarely written for the novice user. Indeed, this textbook aims to fill this particular gap in the market by encouraging the inexperienced user to begin to utilise photogrammetry.

A final disadvantage of photogrammetry is the varied quality of data that can be derived, although it can be argued that a technique that produces precision commensurate with scale is also a major strength of photogrammetry. Whether data is of an acceptable quality depends upon the needs of the user and their ability to quantify them. Expertise in designing an appropriate photogrammetric network to satisfy those needs is also crucial. A significant danger for novice users is the ability of photogrammetry to always obtain some form of output. It is much more difficult to ensure that derived data are of sufficient quality (normally accuracy) for the application. Unresolved systematic errors provide a fundamental constraint on the accuracy of photogrammetrically acquired data and for this reason it is essential to provide some form of independent check. The incorporation of independent checkpoints and comparison with photogrammetric data allows the accuracy of derived data to be quantified. However, application of this universal methodology does create certain difficulties in landform studies.

Geomorphologists are interested in the evaluation of process rate and so topographic derivatives are more relevant than the raw topography itself. In the presence of errors that are random and Gaussian, the uncertainty in slope (σ_s) can be expressed as a function of uncertainty in elevation (σ_e) and datapoint spacing (d) (Lane et al., 2003):

$$\sigma_s = \frac{\sqrt{2}\sigma_e}{d}$$

Although digital photogrammetry allows the collection of data at very high densities, tests have shown that elevation uncertainties in geomorphological applications are commonly degraded as compared with their theoretical estimates derived from considerations of image scale and pixel resolution. The result can be substantial slope uncertainty, which can cause serious problems when photogrammetric data are applied to geomorphological models. The situation is further compounded by residual systematic errors, which in a surface comprising many millions of automatically derived points, may be very difficult to isolate. Whilst the accuracy of derived data may be quantified, it can only be an estimate because the number of checkpoints available is commonly an order of magnitude smaller than the number of data points derived photogrammetrically (Lane, 2000).

6.1.4 Past and related work quantifying landform change

Although early photogrammetric methods were developed specifically for landform measurement (e.g. Finsterwalder, 1897), the applications cited in the older literature are comparatively limited. Interestingly, peaks in research output seem to coincide with the progression from analogue to analytical methods, and subsequently, the development of wholly digital photogrammetric techniques.

Important early work includes that of Petrie and Price (1966) who used a sequence of vertical aerial photographs to measure ice wastage on an Alaskan glacier. El-Ashry and Wanless (1967) similarly used a sequence to estimate sediment carried by longshore currents. Both studies used analogue photogrammetric methods which were time consuming and required significant investment in both hardware and expertise. These studies also used vertical aerial photography acquired with calibrated aerial cameras. A significant study by Lo and Wong (1973) used a 35 mm camera to examine the development of rills and gullies on a small section of weathered granite slope in Hong Kong. The trend to using such non-metric sensors was made possible through the development of analytical methods and the inherent ability to correct for a variety of systematic errors. Significant developments included: the direct linear transformation (DLT) by Abdel-Aziz and Karara (1971) which bypassed the need for fiducial marks; and polynomial-based models to compensate for lens distortion (Kenefick et al., 1972; Faig, 1975). Analytical methods also relaxed many of the traditional geometric limitations associated with image acquisition. Kalaugher et al. (1987) and Chandler et al. (1989) used oblique aerial imagery to monitor sea cliffs. At a larger scale, Collins and Moon (1979) were able to measure streambank erosion and Welch and Jordon (1983) used 35 mm imagery to derive digital terrain models (DTMs) to represent a small river-channel and computed volumes and change surfaces. In a series of studies Lane *et al.* (1994, 1995, 1996) used terrestrial/oblique imagery and analytical methods to measure change occurring in an active pro-glacial stream in the Alps, subsurface data being acquired using a total station. This study utilised a fully functional analytical plotter, an expensive device (US$300,000, 1990) which also required the measurement of points using wholly manual methods. A trend at the time was the development of cheaper PC-based systems, e.g. ADAM Technology's MPS2 (ADAM Technology, 2005) and Rollei's Rolleimetric (Photarc, 2005). Although these found some application (Heritage *et al.*, 1998), more widespread adoption was hampered by the time required for manual identification and digitisation of points.

The development of digital photogrammetric methods since the early 1990s provided the required automation through cheap PC-based hardware and reasonably priced software. Recognition of this potential provoked the British Geomorphological Research Group (BGRG) and Photogrammetric Society to organise symposia in the UK in 1995 and 2001 (see Lane *et al.*, 1998 and special issue of *Earth Surfaces Processes and Landforms*, Lane and Chandler, 2003). The work described in these publications and other significant international contributions will be reviewed briefly here, and structured according to the three case studies described later in this chapter.

Large-scale application: flume and micro-relief measurement

Although Lo and Wong (1973), Kirby (1991) and Merel and Farres (1998) had all used analytical photogrammetric methods to measure micro-relief features, it was the automation of DEM generation that provided the real impetus at this scale. The study by Stojic *et al.* (1998) demonstrated the massive increase in data density that could be created, these being necessary to represent the evolution of braiding processes in a flume 11.5 m × 2.9 m in size. Hancock and Willgoose (2001) also used scanned analogue imagery (Hasselblad film camera) to measure the evolution of a landscape surface in a flume 1.5 m × 1.5 m containing erodable fly ash, under the influence of a rainfall simulator. Chandler *et al.* (2001) and Lane *et al.* (2001) used a digital camera to measure the flume surfaces described in this chapter (Section 6.7.2). A similar digital sensor was adopted by Brasington and Smart (2003) also, to quantify the evolution of a drainage network in a flume 2.1 m × 1.0 m in size, with DEMs generated at a resolution of 5 mm.

Medium-scale application: rivers

Lane (2000) provides a comprehensive review of the role that photogrammetry has performed for the measurement of river-channel change. At the scale of application for the case study described in this chapter (Section 6.7.3), key work includes the study reported by Westaway *et al.* (2000, 2003) and Lane *et al.* (2003) which used digital photogrammetry to monitor a wide gravel-bed braided river in New Zealand. This included the development of an algorithm to correct for the effects of refraction in shallow and clear water (Westaway *et al.*, 2000) and the coupling of image analysis and digital photogrammetry to deal with cases where the bed is not visible. Although many studies have used vertical aerial photography, use has also been made of terrestrial (Lane *et al.*, 1994, 1995) and oblique terrestrial imagery (Pyle *et al.*, 1997; Chandler *et al.*, 2002) for medium-scale fluvial work.

Small-scale application: landslides and coastal change

Normal aerial photography has been used to monitor change occurring on landslides and users have been quick to take advantage of the digital revolution. Brunsden and Chandler (1996) used automated DEM generation methods to continue to monitor the Black Ven landslide on the south coast of England cited in this chapter (Section 6.7.4). The high cost of obtaining imagery using conventional aerial cameras and platforms has encouraged the use of small format sensors borne on light and even microlight aircraft (Graham and Mills, 2000). Henry *et al.* (2002) used the Ricoh

KR-10M camera to monitor the Super Sauze debris flow in the Alpes-de-Hautes in Provence, France; Mills *et al.* (2005) combined GPS and digital small format imagery to create DEMs to detect coastal change at Filey Bay in North Yorkshire, England. At medium scale, a study has been conducted recently on the coast of North Yorkshire by Lim *et al.* (2005). In this study, laser scanning is compared with automated digital photogrammetry to monitor cliff evolution.

One interesting development afforded by digital image processing is the potential to derive 3D movement vectors directly using image correlation procedures and multi-temporal imagery. Kaufman and Ladstädter (2002) used this method to derive velocity fields occurring in rock glaciers in the Alps.

6.2 Distinctive aspects of landform measurement

Studies concerned with measuring landforms and quantifying landform change, are by definition, in need of data to describe spatial morphology. The basic building blocks for datasets are (X,Y,Z) Cartesian coordinates and fundamentally, the more the better. Clearly if spatial change is to be detected then multi-epoch surveys must be instigated, which then require consistent datum definition to ensure that genuine morphological change is measured. The ability to create DEMs automatically using area-based image correlation, with rates exceeding 500 points per second, is one of the main attractions of photogrammetry. Geographers and geomorphologists are spatially competent and very capable of using these spatial coordinates. Basic descriptors of morphology such as slope, aspect and curvature are often derived using widely available software. More complex statistical descriptors of form such as semi-variograms and co-variograms can be obtained once basic DEM data can be provided. Although orthophoto creation and visualisation are useful by-products of the photogrammetric process, it is the simple *XYZ* coordinate that is the key to most landform change studies. The ability to derive these data using cheaper sensors at ever greater spatial and temporal resolution should ensure continued interest in photogrammetry by the geomorphological community.

6.3 Typical requirements

The wide range of scales presented later in this chapter (Section 6.7) and an even more diverse range of potential applications, prevents definition of "typical requirements". However, there are some similarities that are worth highlighting in this section.

6.4 Accuracies

Absolute accuracies are of course, scale dependent but it is possible to generalise using relative measures of accuracy. Although high precision, multi-station convergent imagery could be used for geomorphic applications, the basic need for high-resolution DEMs encourages conventional stereo configurations. Such an imaging design restricts accuracies to the 1 part in 1000 to 10,000 range, similar to mapping using normal aerial photography. This requires some attention to camera calibration issues, although these are not as stringent as those required for high precision industrial measurement (Chapter 4).

6.4.1 Appropriate/available software

The work discussed in this chapter was processed using a single manufacturer's solution that has evolved radically over the last 15 years; although it should be recognised that other proprietary software would be appropriate (Section 6.5.1). The digital photogrammetry package OrthoMAX was originally developed by Erdas/Autometric in 1991 and ran on Sun Workstations under the Unix operating system, the only hardware capable of manipulating high-resolution image data at that time. With the growing capabilities of PC platforms, Erdas subsequently developed OrthoBASE for Windows NT/2000. This subsequently evolved into the Leica Photogrammetry System (LPS), under the stewardship of Leica Geosystems and is currently available commercially.

6.4.2 Cameras/sensors

The case studies described in this chapter use the full spectrum of cameras currently available to conduct modern digital photogrammetry. At small scales, the traditional full format (230 mm × 230 mm) film-based aerial camera is still widely used and can provide imagery appropriate for a variety of landform change studies using routine photogrammetry. Small format cameras have shown potential at small scales (Chandler *et al.*, 1989; Mills *et al.*, 2005) but are not effective when working over large areas, typical for this scale of application. At the other scale extreme, digital small format cameras are particularly effective for large-scale evolutionary studies. Indeed, the latest generation of mass-produced consumer-grade digital cameras appears to be more than capable of filling this and related niches (Chandler *et al.*, 2005). Such sensors require some form of camera calibration but advances in software provision promise to make these steps routine also.

6.4.3 Object preparation

The desire to record the effects of natural geomorphological processes means object preparation has to be kept to a minimum. Typically, this involves installing photo-control points which, ideally, do not interfere with the very phenomena that are being recorded. Conventional photogrammetric practice encourages distribution of sufficient redundant photo-control points around the object, to avoid extrapolation errors. This should be achieved where possible, but a balance is often required between potential extrapolation errors and extracting unrepresentative data, so compromise is often necessary.

The scale of the measurement task has some impact on the type of control points used and their coordinate determination. For large- and medium-scale studies, pre-marked control points provide the user with full control over their number and location. Although a minimum of four points is required, six to eight should be installed per stereomodel, particularly if non-metric sensors which require *in situ* self-calibration are used (Section 6.7.2). Intersection methods using a total station have been found to provide a most effective means of determining coordinates of points, particularly if multiple survey stations are used and "3D variation of coordinates" least squares estimation software is available (Section 6.7.2). The new generation of total stations equipped with reflectorless EDM can also produce acceptable results, if a little less reliably.

For small-scale work the difficulty of predicting the date of survey flights means that natural photo-control points can be effective. Many geomorphological sites are surrounded by stable and visible features, (natural and man-made) which can be identified on the imagery and on the ground. In such situations differential GPS is perhaps the best survey tool for coordinate determination because the ground line of sight is not required. A conventional total station can also be effective but careful selection of survey stations is necessary to minimise the required fieldwork.

6.4.4 Required products/output

As stated in Section 6.2, the key photogrammetric product for landform change is the DEM, from which DEMs of difference can be derived. Orthophotos are valuable also, for visualisation

(Section 6.7.4, Black Ven) and interpretation/classification (Section 6.7.3). The ability to derive movement vectors (Section 6.7.4, Mam Tor) is also valuable for some studies and automated methods look to have potential (Kaufman and Ladstädter, 2002) for favourable image sequences.

6.5 Solutions options

6.5.1 Existing commercial systems or appropriate software

The applications described in this chapter (Section 6.7) use software produced through an evolutionary collaboration between three commercial organisations: Autometric, Erdas and Leica (Section 6.4.1), culminating in LPS (Leica Geosystems) currently used by the present authors. Other commercial systems provide the capability of automated DEM generation, which is of key significance for quantifying landform change. These packages include: VirtuoZo V-DEM (SupreSoft, 2005); SOCETSET (BAE Systems, 2005); Match-T (Inpho, 2005); Geomatica OrthoEngine (PCI Geomatics, 2005); ImageStation Automatic Elevations (Intergraph, 2005); and PI-3000 (Topcon, 2005). Many of these commercial systems have experienced a similar evolutionary pattern to LPS, with most software migrating from Unix to PC platforms and progressing from area-based to feature-based matching, as hardware and software developments have allowed.

6.6 Alternatives to photogrammetry for this application

Ground survey using either total stations or differential GPS has provided the traditional alternative to photogrammetry for landform change studies. Although these technologies have evolved, both require the physical placing of a reflector prism at desired locations, which restricts data acquisition rates to a maximum of 0.2 Hz, even using the latest motorised, reflector-tracking total station. This low sampling rate combined with potential disturbance of the very phenomenon being recorded has favoured a photogrammetric approach using automated DEM extraction at 500 Hz.

Laser scanning (Section 2.7) typically operates at 4000 Hz. Airborne and terrestrial versions of the technology are available and both can provide geomorphologists with incredibly dense data to represent surface morphology. The accuracy of airborne laser data is constrained mainly by the uncertainty in sensor position derived by differential GPS and inertial navigation systems. Original accuracy claims were over optimistic, but rms errors of better than 25 cm are achievable (Adams and Chandler, 2002; Lane et al., 2004), which is appropriate for many small-scale studies. At the medium scale, ground-based laser scanning offers potential and has been compared to terrestrial photogrammetry for cliff monitoring (Lim et al., 2005); this showed that beach-based photogrammetry had greater difficulty recording morphology on the higher cliff sections (50 m) than laser-based recording. Lasers also offer potential at large scale (Lane et al., 2001) though it is difficult to generalise.

Laser scanning methodology will undoubtedly have an increasing impact on landform change studies but an image/photogrammetric solution currently still has three key advantages: most scanners simply record three-dimensional spatial form and integrated systems which capture morphology and spectral data are rare, especially at close range; scanners remain expensive, even the cheaper ground-based sensors cost in excess of (US$ 80,000) and finally, and perhaps most significantly, scanner precision is suitable for one scale of measurement. As the case studies in this, and other, chapters forcibly demonstrate, the flexibility of scale provided by a photogrammetric solution remains unique.

6.7 Case studies

6.7.1 Introduction: case study selection

During initial discussions to decide upon the focus and philosophy of this book, it became apparent that flexibility of scale is the unique advantage of photogrammetry compared with other measurement methods. One way to demonstrate this forcibly was to seek chapters which illustrated this explicitly at a range of spatial scales. The case studies selected here range from measuring: bed surfaces of flumes just 0.4 m wide, to river reaches 200–1200 m in width and finally landslides of dimensions 2–3 km. A further distinction between the river and landslide case studies is the range of vertical scale, with the fluvial bedforms exhibiting a topographic variation of just 2 m compared with 150 m associated with the landslides.

Two of the case studies focus upon fluvial research which has, in particular, embraced photogrammetric methods more widely than other aspects of geomorphology. The linkage between fluvial processes and resultant morphological form is particularly direct and morphological change is often large and occurs in short time periods, all factors that have encouraged a photogrammetric monitoring approach. In addition, it has long been recognised that river channel topography operates at a number of different spatial scales (Leopold *et al.*, 1964) including: grain, bedform and channel scales which have all been identified explicitly. Although fluvial geomorphology has been important, it was felt that the third scale should be illustrated by another area of geomorphology, hence the application to landslide evolution. This choice also provided an opportunity to demonstrate another unique advantage of photogrammetry, the ability to extract spatial data from historical imagery.

6.7.2 Large-scale case studies: flume work

River flooding has become of topical interest in the media, particularly since the widespread flooding in the UK during the autumn of 2002. Although these unusual natural events helped to divert research funding in the UK, there were already various UK research projects that focused on understanding the interaction between river channel form, sediment transport and bedform development. The focus for much of this work was the Flood Channel Facility (FCF) supported by and located at Hydraulics Research Wallingford, UK. The FCF is a two-stage channel representing a section of a meandering river occupying a simulated flood plain. It is large by flume standards (30 m × 10 m) and contains a sinusoidal river channel of wavelength of 16 m and with a channel 1.6 m wide and 0.5 m deep (Figure 6.1). Water and sediment can flow down the main channel and optionally down the floodplain (hence the two-stage channel description) to simulate the fluvial processes occurring in a real river in a flooded condition. A funded research project provided access to the FCF and allowed the construction of a quarter-scale replica of the flume at Loughborough University, to allow for a more extensive range of experiments to be undertaken.

The photogrammetric challenge was to develop a method to measure the complex bedforms that developed in the sediment within the sinusoidal channel. Previous experience measuring the distribution of sediment in a braided flume in New Zealand (Stojic *et al.*, 1998) had demonstrated that area-based DEM generation methods could be applied successfully. However, accuracies had been limited because of the use of a film-based Pentax 645 camera, which was not equipped with stable fiducial marks and additional uncertainty was associated with scanning unstable film (Stojic *et al.*, 1998). Other funded research work (Lane *et al.*, 2001; Chandler *et al.*, 2001) had provided the finance to purchase the six mega-pixel Kodak DCS 460 digital camera. At that time (1998) the Kodak DCS 460 was the best digital camera available with proven photogrammetric

Figure 6.1   *Flood Channel Facility, Hydraulic Research Wallingford, UK.*

potential, particularly for industrial photogrammetry (Ganci and Shortis, 1996; Fraser, 1997). However, it remained a challenge to develop a recording and camera calibration methodology to extract high-resolution DEMs of both the FCF and the quarter-scale flume at Loughborough University (Chandler *et al.*, 2001).

Image acquisition

The geometry presented by the two meandering channels presented several difficulties and the configuration adopted for the larger FCF will be described here. A movable gantry platform was available and adapted to support the camera using a simple scaffold plank, with a 60 mm diameter drilled hole for the camera lens. The focal length of the lens of the DCS460 camera was 28 mm and with a camera-to-object distance of 4.2 m (photo-scale 1:150), each pixel represented an area of 1.4 mm^2 on the object. At this scale the footprint of each image was 4.1 m × 2.8 m, too small to provide full coverage of the meandering channel using one conventional strip. The difficult sinusoidal shape of the meander suggested that the most efficient means of providing the stereo-coverage necessary for DEM generation was to acquire a sequence of overlapping stereo-pairs. A design consisting of 10 overlapping pairs was selected, each consisting of two photos displaced laterally by 0.4 m (Figure 6.2). This design provided several advantages. The gantry crane could translate in the horizontal direction, which allowed the desired camera positions to be achieved easily, providing a base/distance ratio of 1:10 with an overlap of 90%. This highly redundant coverage combined with generous overlaps between pairs ensured that full coverage would be obtained, even if cameras were not placed in exactly the intended positions.

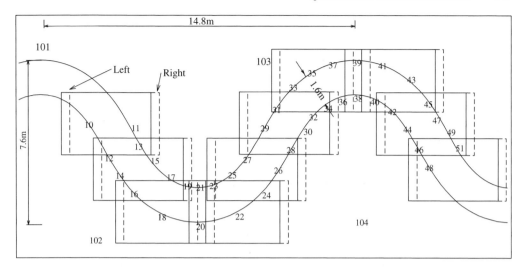

Figure 6.2 *Photo and control configuration Flood Channel Facility, Hydraulic Research Wallingford, UK.*

Although this configuration was effective, subsequent work and experience suggests minor modification. By acquiring slightly convergent stereo-imagery an improved base/distance ratio of 1:7 can be achieved whilst maintaining an endlap of 90–95%. This increases the precision of measured data and maximises stereo-coverage. However, it should be recognised that most commercial software fails if the camera axes cross before the object surface is reached.

It is important to always obtain adequately exposed images with the desired object represented using a wide range of pixel values (i.e. good contrast). The provision of adequate illumination and selection of appropriate camera exposure settings is therefore crucial. The FCF is located within a large building with no natural light. The existing lighting was poor and was supplemented by four portable arc lamps, each with an output of 5000 kW. The lamps were oriented to provide softer illumination by reflecting light from the ceiling, 6 m above the channel. One of the major advantages of using a digital camera for image acquisition is that exposure settings can be checked on-site by downloading images onto a portable PC. This was found to be invaluable and removed traditional concerns over the appropriateness of chosen exposure settings. The camera was operated in "aperture priority" mode, with an overexposure setting of +0.3, which ensured that illumination was adequate for the slightly darker sand grains occupying the flume bed. The typical exposure setting was an aperture of f/8 which, with the limited lighting, required a shutter speed of 1 second. "On-the-job" (Clarke and Fryer, 1998) or "*in situ*" self-calibration methods were used to calibrate the camera, and to prevent variation of the focal length during photo-acquisition, the auto focus ability of the camera was switched off and the lens taped at infinity focus. Image acquisition for the quarter-scale flume was similar, except the camera-to-object distance was reduced to 1.6 m with a photo-base of 0.15 m.

Photo-control

Forty control targets (60 mm × 40 mm in size) were distributed along the edge of the channel which yielded approximately 5–8 visible targets on each stereo-model (Figure 6.2). These points all lay upon the flat floodplain of the constructed channel and because on-the-job calibration methods were to be used, it was felt important to place markers at other lower elevations. It was undesirable to place markers on the sediment surface, but it was feasible to attach targets on the sloping sides of the concrete channel, just above the channel bed using silicon bathroom sealant. Horizontal and vertical angles were measured to these markers from four survey stations (Stations 101–104, Figure 6.2), each point being visible from at least three different stations. A Leica 1610 total station was used to measure all angles, with 570 measurements recorded on a data logger; these data were supplemented by distances measured using a steel band and height differences measured using a Leica NA3000 digital level. All measurements were combined in a "3D variation of coordinates" least squares estimation and error ellipses generated from this network suggested that the rms precision of these photo-control points was homogeneous and approximately 0.4 mm.

Photogrammetric processing and camera calibration

The imagery and photo-control was downloaded onto a Unix WorkStation and processed using Imagine OrthoMAX, the hardware and software available at that time (1999). If processed today (2006) a PC running LPS would provide an appropriate hardware and software combination. Initial measurement was routine with all control targets measured manually on all images supplemented by a series of tie points to create a stronger geometric network.

The Kodak DCS460 is constructed from standard professional camera components and is not designed for photogrammetric measurement, so some form of calibration procedure must be used before accurate spatial data can be extracted. Two principal camera calibration approaches are available, both of which are problematic. First, a convergent set of imagery of a retro-reflective test field may be acquired. Measurement and subsequent self-calibration using a free net adjustment (Granshaw, 1980) can yield precise estimates of "camera and lens calibration parameters" for that instant, but subsequent application of these may be inaccurate due to camera instability. Secondly, an "*in situ*" or "on-the-job" calibration approach may be adopted (Abdel-Aziz and Karara, 1971; Clarke and Fryer, 1998), in which imagery used for actual DEM extraction is measured and combined with object coordinates to calibrate the camera. Such an approach will yield less precise estimates but the parameters may prove more accurate, because they relate to internal camera geometry at that instant. The impact of selecting these differing approaches is discussed more fully by Chandler *et al.* (2001) but the study revealed that an *in situ* self-calibration approach was adequate and simple to implement.

At the time of processing, this required downloading the measured data from the OrthoMAX "triangulation adjustment" and using an off-line self-calibrating bundle adjustment to derive appropriate estimates for the inner orientation. This process was facilitated by a reformatting program that converted the Orthomax Triangulation report file into the appropriate files required for the self-calibrating bundle adjustment program GAP. This program, originally developed by Chandler and Clark (1992) was subsequently modified to include the camera model recommended by Patias and Streilein (1996) and Fraser (1997). Processing these data through GAP revealed the importance of recovering two parameters to model radial lens distortion and to estimate the focal length of the camera. Two additional parameters necessary to model the principal point offset proved to be of only marginal significance (Chandler *et al.*, 2001). These results confirmed the suggestion made by Fraser (1997) that, for medium accuracy work using digital cameras and stereo restitution, the important inner orientation parameters are simply an appropriate lens model and the focal length. Development of the more recent LPS software package includes the capability of performing a self-calibrating bundle adjustment, which has simplified the data processing required. Tests conducted by the author (Chandler *et al.*, 2005) have revealed that the self-calibrating capabilities of LPS now perform satisfactorily and access to an independent offline bundle adjustment is normally unnecessary for this type of work.

Generation of DEMs and orthophotos

Once an appropriate camera model had been established, the OrthoMax software was used to generate DEMs and orthophotos. For the FCF project, DEMs were generated from each stereo-pair at a resolution of 10 mm (Figure 6.3). Internal estimates provided by the software suggested that the automated DEM correlation procedure was successful, with between 80–85% of all points registering a successful match. Apparent failures were located in areas towards the edge of the overlap, and typically on the texture-less surface of the flat floodplain. The large overlaps provided by the conservative design of the photo-locations, allowed more inaccurate peripheral regions to be replaced with data from an adjacent stereopair. These eight DEMs were mosaiced together to create a composite DEM representing one full wavelength. A slope-shaded representation of this DEM (Figure 6.3) demonstrates how successfully the auto DEM extraction has been at qualitatively representing the streambed surface with all its macro-topographical features represented. Orthophotos were also generated and mosaiced together (Figure 6.4) which assisted in the identification and interpretation of features visible within the DEM.

Once the image acquisition and camera calibration methodology had been proven for the larger FCF flume it was necessary to adapt the approach to the quarter-scale flume at Loughborough University (Figure 6.5). This was a comparatively simple process and principally required reducing the flying height and camera base by a factor of four. This certainly demonstrates the flexibility of a photogrammetrically based measurement system. An existing and movable walkway was adapted to support the camera vertically over the flume, again utilising the scaffold plank to support the camera. A similar number of control targets were placed around the flume and positions were coordinated using the Leica total station. The target sizes were reduced, use being made of commercially available targets 10 mm in diameter. The first set of imagery was again used to derive a set of camera calibration parameters using *in situ* self-calibrating methods. Data was subsequently extracted in the form of DEMs (Figure 6.6) and orthophotos.

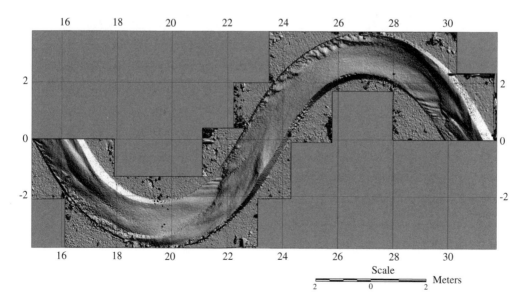

Figure 6.3   *Slope shaded DEM mosaic of Flood Channel Facility, Hydraulic Research Wallingford, UK (after Chandler et al., 2001).*

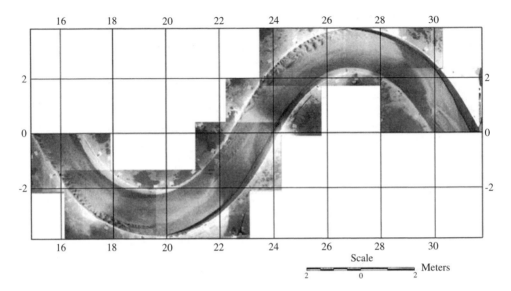

Figure 6.4 *Orthophoto mosaic Flood Channel Facility, Hydraulic Research Wallingford, UK (after Chandler et al., 2001).*

Figure 6.5 *Loughborough University flume.*

The real advantage of the quarter-scale flume was the degree of experimental control arising from the physical location of the flume. In excess of 30 experimental runs were carried out over a six-month period, during which extraction of DEMs became just one type of data recorded. Other key measured parameters included 3D turbulence vectors derived using laser-based methods. These revealed vortices arising from interacting "down-channel" and "down-floodplain" water flows that were shed at the flow boundaries. These became more prominent as the depth

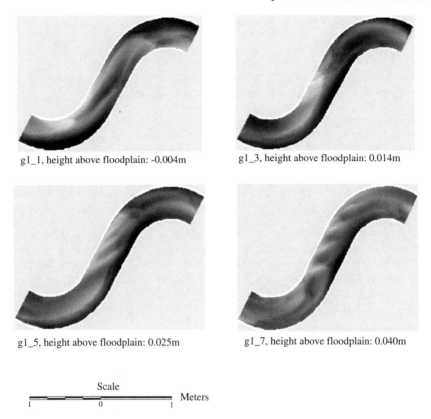

g1_1, height above floodplain: -0.004m   g1_3, height above floodplain: 0.014m

g1_5, height above floodplain: 0.025m   g1_7, height above floodplain: 0.040m

Scale

1   0   1   Meters

Figure 6.6   *Greyscale DEMs at different depths of flooding, Loughborough University flume.*

of flooding increased. The presence of these vortices was detectable in the morphology of the captured DEMs, where localised scour and deposition created the waves of sediment visible in Figure 6.6 (test g1_5 and particularly test g1_7).

One of the most satisfying aspects of this work was that image acquisition and consequent DEM generation became routine. The research assistant responsible was not a trained photo-grammetrist but learnt the required procedures. It was typical for images to be obtained during a 20 minute period at 10 a.m. in the morning and by 3 p.m. in the afternoon, the required number of DEMs had been generated and mosaiced together (Figure 6.6).

Accuracy assessments

As stated in Section 6.1.3, one of the drawbacks of a photogrammetric approach is that although data can be extracted from imagery comparatively easily, it remains far more difficult to ensure that the quality of data is appropriate for the task. An important stage following data acquisition is therefore data appraisal, which can only be achieved if accuracies are assessed by comparing measured data with accepted independent "truth" (Section 6.1.3).

Scientists at Hydraulic Research Wallingford had developed a "touch sensitive, incremental 2D bed profiling system" (Hydraulic Research Wallingford, 2005) which provided appropriate check data to assess the accuracy of the photogrammetry (Figure 6.7). The profiling system consists of a 10 mm diameter stainless steel tube with a machined rack and a probe, powered up

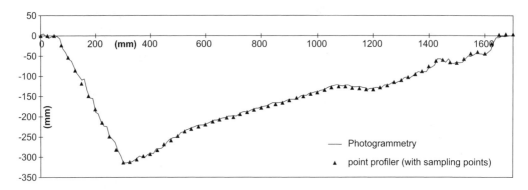

Figure 6.7 *Accuracy of DEM Flood Channel Facility, Hydraulic Research Wallingford (after Chandler et al., 2001).*

and down by a DC motor. The probe is mounted upon a carriage and the user is able to control the start point, total distance to be traversed and number of sampling points, and once started, all operation and measurement is fully automatic. This device was used to provide independent surface data along a series of cross-channel profiles, and allowed comparison with photogram-metrically derived data at these discrete locations (Figure 6.7). The horizontal sampling density is 25 mm for the profiler and 10 mm for the photogrammetry. The correspondence is good, with an overall rms error of 2.6 mm. Particularly close agreement can be identified within the steep channel sides, a region where the area-based correlation algorithm was expected to fail.

Conclusions

The "large-scale" case studies have demonstrated the efficacy of photogrammetry to measure DEMs representing flume beds. Although an expensive digital camera was used at the time, more recent work (Chandler *et al.*, 2005) has demonstrated that this type of measurement could be achieved using a cheap digital sensor costing less than US$400, provided appropriate camera calibration procedures are adopted (Chandler *et al.*, 2005).

Dense DEMs could also be derived using laser ranging technologies that have matured rapidly recently (Section 6.6). However, such sensors remain expensive and would not be so effective if multiple sensor locations were required, restricting application to much smaller areas than those conducted here.

6.7.3 Medium-scale case study: river channel change

Understanding morphological change in rivers is important in its own right, as well as part of broader studies where topographic data are needed to assess model predictions and to validate the topographic characteristics being reproduced in scaled laboratory models. This is particularly important in braided rivers that change rapidly, potentially over a large spatial extent, where the time required for ground-based survey (e.g. cross-section survey) can be significant. Up until the late 1990s, the prime focus of field measurement of braided river morphology was the cross-section survey. In general, this involved narrow, valley-confined braid bar systems (Ashworth and Ferguson, 1986; Goff and Ashmore, 1994), although there are examples of long-term survey of very wide braided rivers (greater than 1 km width) using cross-sections (Carson and Griffiths, 1989). These analyses were supported by quantification of the braided river planform measured using various forms of aerial survey (e.g. Laronne and Duncan, 1992). This situation began to

Figure 6.8   *The braided Lower Waimakariri River, South Island, New Zealand.*

change in the 1990s with application of oblique terrestrial analytical photogrammetry, coupled with ground survey of inundated areas, in order to obtain three-dimensional surface models (Lane *et al.*, 1994, 1995, 1996). The focus remained relatively small-scale, but was increasingly founded upon representation of the continuous variation of topography using DEMs. However, the most dramatic developments have come with: first, automated digital photogrammetry, primarily using airborne platforms (Westaway *et al.*, 2000, 2001, 2003; Lane *et al.*, 2003; Brasington *et al.*, 2003), but also using terrestrial oblique imagery (Chandler *et al.*, 2002; Bolla Pittaluga *et al.*, 2003); secondly, laser altimetry (Lane *et al.*, 2003; Charlton *et al.*, 2003); and finally, global positioning systems (GPS) (Brasington *et al.*, 2000). The result of these technologies is visualisation of braided river topography at a resolution and scale not seen before (Hicks *et al.*, 2002; Figure 2). In this section, the application of digital photogrammetry to the quantification of braided river morphological change, on a particularly wide river covering an area approaching 5,000,000 m^2, is reported.

Field site
The field site chosen for this study is a 3.3 km long reach of the lower Waimakariri River, South Island, New Zealand (Figure 6.8). The vertical relief in the study reach is small, generally less than 2 m, with a mixed gravel bed with a median grain size (D_{50}) of around 0.028 m (Carson and Griffiths, 1989). Monitoring of the lower Waimakariri River is currently undertaken by Environment Canterbury, the local regional council, who resurvey a number of cross-sections along the river, approximately every five years (Carson and Griffiths, 1989). In the 3.3 km study reach, there are four cross-section locations (17.81, 19.03, 19.93 (crossbank), and 21.13 km from the river mouth), which indicates the spatial density of survey (*ca.* 1 cross-section per km in this reach).

Data collection
Aerial photographs of the reach were taken in February and March 1999 and February 2000 by Air Logistics (NZ) Ltd, spaced so as to provide a 60% overlap for the photogrammetry. A feature of photogrammetric survey is the constant trade-off between the increase in coverage that comes from a reduction in image scale and the improvement in point precision that results

from an increase in image scale (Lane, 2000). This is particularly relevant with respect to large braided riverbeds, because of the low vertical relief (in the order of metres) compared to the spatial extent (in the order of kilometres). This imposes a minimum acceptable point precision for effective surface representation. In this study, a photographic scale of 1:5000 was chosen for the 1999 imagery, which given a scanning resolution of 14 μm, results in an object space pixel size of 0.07 m (Jensen, 1996). Based on this, the lowest DEM resolution that can be produced is 0.37 m, approximately five times greater than the object space pixel size (Lane, 2000). Analysis of the 1999 imagery had been completed in time for the redesign of the airborne survey to acquire imagery at a larger scale (1:4000) in order to increase texture in the image and so improve stereo-matching performance (see below). To achieve full coverage of the active bed, two flying lines were needed. Photo-control points were provided by 45 specially designed targets that were laid out on the riverbed prior to photography and positioned such that at least five (and typically 10–12) targets were visible on each photograph. Their positions were determined to within a few centimetres using real-time kinematic (RTK) GPS survey.

DEMs of the study reach were produced using the OrthoMAX module of ERDAS Imagine installed on a Silicon Graphics Unix workstation. DEMs were generated using the OrthoMAX default collection parameters with a 1 m horizontal spacing to give almost five million Cartesian coordinates representing the riverbed. An individual DEM was generated for each photograph overlap: two flying lines of eight photographs each meant that in total 14 individual DEMs were produced for the 1999 surveys. The larger scale of the 2000 surveys increased the number of DEMs to 20.

To aid in the assessment of DEM quality, and hence to identify means of improving data collection, the generated DEMs were assessed using checkpoint elevations measured by National Institute for Water and Atmosphere and Environment Canterbury field teams using a combination of total station and RTK GPS survey. An automated spatial correspondence algorithm was used to assign surveyed elevations to the nearest DEM elevation, provided the point was within a given search radius.

The challenges for digital photogrammetry in this application

Photogrammetric measurement of this kind of river represents a series of fundamental challenges for the digital photogrammetrist and these are summarised in Lane (2006), on which the following is based. First, the spatial extent means that remote survey remains the only feasible measurement technique. However, it involves a fundamental trade-off between the scale of survey and the resolution of data that can be achieved. Coverage of larger areas normally requires a higher flying height. A higher flying height will degrade possible data resolution and data quality. This is a particular problem in fluvial studies as the relief is low, such that flow and sediment transport processes may be driven by quite subtle changes in topography that can be readily swamped by noise in the data. Lowering flying height or repeating flight lines can be used to densify the surface and, in the case of photogrammetry, may be crucial for introducing sufficient texture into the imagery in order to assist the stereo-matching process, especially for low relief bar surfaces where the grain size is in the gravel or sand size range (Westaway et al., 2003; Lane, 2006). Lowering flying height raises the costs (money and time) of the survey because: greater levels of ground control may be required if the platform does not have on-board GPS/initial measurement unit (IMU); the time to acquire imagery is longer; and more data analysis and post-processing will be required. It follows that resolving this trade-off is not straightforward. Compromises have to be made.

The second problem is what to do with inundated areas (Lane, 2006). These are commonly the regions of most active change and hence most scientific interest. The problem can be reduced by constraining data collection to the lower flow periods, either seasonally or in relation to storm events. This does not necessarily yield data when it is required and most rivers retain some inundation even at low flows. Two solutions have been adopted. Both require image data, and so can make use of the particular advantage of photogrammetry as compared with other techniques, in that it is image based. The first solution applies to situations where the water is sufficiently clear for the bed to be visible. Provided the flow depth is less than the optical depth (which it commonly is in shallow rivers) and the flow is sub-critical (so the water surface is not "broken") then stereo-matching may successfully estimate subaqueous points (e.g. Westaway *et al.*, 2000) and these can be subject to a two-media refraction correction (Westaway *et al.*, 2001). If the water is turbid, the bed textural signature will commonly be lost. However, there may still be a depth signature. In theory, this could be theoretically derived from knowledge of the bed substrate colour and composition combined with sun angle and information on atmospheric characteristics. In practice, sufficient knowledge of the required boundary conditions is rarely available for physically based modelling and so semi-empirical approaches have been adopted (e.g. Westaway *et al.*, 2003, after Lyzenga, 1981, and Winterbottom and Gilvear, 1997). These require calibration data to establish a relationship between measured water depths and the spectral signature found in the imagery at the location of each depth at the time of data acquisition. These relationships are assumed to hold through space, despite possible variations in bottom reflectance. They will rarely hold through time due to variation in suspended sediment concentration, water depth and sun angle etc. Both the clear water and the turbidity based approaches involve a set of image processing with estimation of water surface elevations and water depths in order to estimate subaqueous point elevations. The methods are detailed in full for the clear water case in Westaway *et al.* (2000) and turbid water case in Westaway *et al.* (2003).

The third problem is how to identify and to correct possible errors (Lane, 2006). The DEM shown in Figure 6.9 consists of 4,200,000 data points. Individual point correction, using stereo-vision for instance, is simply not possible with this number of data points. Indeed, prior to using the DEMs derived using the digital photogrammetric methods, it was necessary to embark upon a lengthy process of data handling and management in order to determine and to improve the associated data quality (Lane *et al.*, 2004). First, it was necessary to identify possibly erroneous data points. Braided rivers are topographically complex but this complexity commonly involves relatively smooth surfaces separated by break-lines within which slope is locally much greater.

Figure 6.9   *DEM, Lower Waimakariri River, detrended for valley slope.*

This gives us basic *a priori* estimates of local elevation variance (essentially defined by an estimate of local grain size) and allows for development of locally intelligent point-based filters. Removal of those data points associated with these errors resulted in a significant improvement in the quality of surface representation (Lane *et al.*, 2004). The presence of vegetation in DEM data is also a form of error, and simple filtering routines based on image processing were used to identify and to remove vegetated locations (Westaway *et al.*, 2003). It was also necessary to replace removed data points. Fortunately, the richness of the associated data sets is such that removing many hundreds of data points does not reduce data quality significantly (Lane *et al.*, 2004), and this points to one of the key advantages of digital photogrammetry. However, in the vicinity of the locally more rapid topographic change (i.e. close to river banks), it was found that more intelligent interpolation of removed data resulted in better surface representation (Lane *et al.*, 2004). Even after identification and removal of possibly erroneous points, residual error remained. This imparts uncertainty into estimates of properties from the DEMs, including both slope and change between subsequent DEMs. Thus, Lane *et al.* (2003) used propagation of error techniques to develop a minimum level of detection (e.g. Brasington *et al.*, 2003) based upon a statistical analysis. In general, the theoretical data precision, as defined by sensor geometry, position and location, is significantly better than the actual precision and the dry bed precision is better than the wet bed precision (Lane, 2006). Degradation from theoretical precision was largely linked to the stereo-matching process in general and a significant loss of surface texture when 1:5000 scale imagery was used (in 1999). For February 2000, data collection was redesigned to use 1:4000 scale imagery and this resulted in a substantial improvement in matching success and hence in surface precision, approaching that of the laser altimetry survey. The magnitude of the wet bed precision emphasises that the spatial extent of wet bed inundation needs to be minimised in applications of this type. In this case study, this meant that data collection was restricted to either side of a flood event, and collection of within-flood data was not attempted.

Results

Figure 6.9 shows an example DEM, whilst Figures 6.10 and 6.11 portray DEM of differences showing fill and erosion, respectively. The first two photogrammetric surveys involved 1:5000 imagery, and preliminary analysis suggested a theoretical point precision of ±0.070 m. After automated error identification and correction, without recourse to the available check data, comparison with check data revealed actual accuracies of 0.261 m (February 1999) and

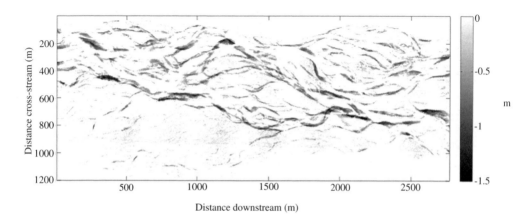

Figure 6.10   *Digital elevation models of difference, showing fill.*

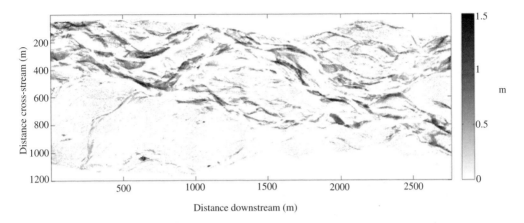

Figure 6.11 *Digital elevation models of difference, showing erosion.*

0.257 m (March 1999). By increasing the scale of data collection to 1:4500 in February 2000, the theoretical precision improved to ±0.056 but the post-correction comparison improved further to 0.131. Central to this was the improvement in image texture that came with the larger-scale imagery, and it was concluded that image texture was a crucial control on the success of this kind of application. The accuracy of the data derived from the wetted areas, based upon image analysis, was degraded as compared with the dry area values, ranging from 0.318 m in February 1999 to 0.219 m in February 2000. This improvement was largely because the image-based method relied upon estimation of water surface elevations at the margins of channel margins, and these were derived photogrammetrically. By improving the dry bed photogrammetry, the quality of the data from the wetted areas was also improved.

The obvious question is whether or not these resultant data could be used for geomorphological investigations. To assess this, classical error propagation techniques were used (Lane *et al.*, 2003) to determine the magnitude of change that the method could detect. This made use of the precision of each data type and each data epoch as the precision varied between type and epoch. This also allowed uncertainties in point elevation estimates to be propagated into volume of change estimates. The results were encouraging and confirmed that the DEMs could be used to estimate both patterns of erosion and deposition (e.g. Figure 6.10) as well as volumes of change.

Conclusions

The River Waimakariri case study demonstrates what can be achieved for a large braided river study using conventional aerial imagery. Key challenges involved deriving accurate elevation estimates where relief variation is minimal, and extracting height data in submerged areas. A semi-empirical approach which established a relationship between spectral signature and depth demonstrates the value of an image-based measurement solution.

6.7.4 Small-scale case study: landslides

Introduction

Two case studies will be used to describe the application of photogrammetry at small scale, both on active landslide systems in the UK. The Black Ven landslide is a coastal cliff system 1.5 km × 0.8 km in size, located between Lyme Regis and Charmouth on the Dorset coast. In contrast, the Mam Tor landslide is an inland system, 1.0 km × 0.8 km in size, located near Castleton in the

Figure 6.12 *Oblique photograph of Black Ven landslide, acquired in 1958 (Cambridge University Collection ©Crown Copyright/MOD).*

southern Pennine hills of England. Both sites are used here to demonstrate also the unique ability of photogrammetry to extract spatial data from archival imagery. This was initially achieved using wholly analytical methods (Chandler and Cooper, 1989; Chandler and Brunsden, 1995) on the coastal site. Subsequently, digital photogrammetry was used at both Black Ven (Brunsden and Chandler, 1996) and Mam Tor (Walstra *et al.*, 2004). In addition, the coastal site demonstrates the rapid evolution of photogrammetric methods and consequent increase in effectiveness of digital photogrammetry to quantify landform change.

Black Ven

The dynamic nature of the Black Ven landslide has attracted interest amongst geologists, botanists and geomorphologists over the last 200 years. Arber (1941) reviewed early geomorphic work but it was Brunsden (1969) who used a sequence of aerial imagery to demonstrate qualitative morphological change between 1946 and 1969. It was this sequence that was extended by Chandler and Cooper (1989) to include epochs dated 1948, 1958 (Figure 6.12), 1969, 1976 and 1988 (Figure 6.13). Both oblique and conventional vertical photography was utilised, but all imagery was typified by the general unavailability of camera calibration data and lack of conventional photogrammetric control on the ground.

   A series of analytical photogrammetric techniques were identified and developed (Chandler and Cooper, 1989) to extract quantitative data to represent landslide at those instances through time. These allowed the production of line maps, but it was the merging of 3D line strings representing key morphological boundaries with a regular grid DEM and the consequent processing

Figure 6.13 *Oblique photograph of Black Ven landslide, aquired in 1988 (personally acquired; after Chandler and Brunsden, 1995).*

through first-generation terrain modelling packages, which proved really productive (Chandler and Cooper, 1989). Contours, profiles, grid representations and statistical descriptions of slope angle could be created efficiently. The subtraction of DEMs at two epochs to create "DEMs of difference" was particularly powerful because this allowed the identification and quantification of areas exhibiting positive and negative morphological change. The data generated allowed geomorphologists to produce an "evolutionary model" for the landslide (Chandler and Brunsden, 1995). More significantly, these data proved the existence of "dynamic equilibrium", an important geomorphic model in which form and process are mutually adjusted, which had been only demonstrated in simple fluvial systems previously (Chandler and Brunsden, 1995).

At the time of original processing (1989) all points (10–11,000 per epoch) were measured manually using an expensive analytical plotter, each DEM requiring a week of measurement (Chandler and Cooper, 1989). By 1995, access to the automation provided by digital photogrammetry was available using Erdas OrthoMAX mounted on a comparatively cheap Unix WorkStation. This allowed DEMs to be generated at rates of 60–100 points per second, which compared to manual measurement on an expensive analytical plotter (5 seconds per point) was remarkable. A new epoch of imagery was commissioned and acquired by the UK Airborne Remote Sensing Facility (ARSF) in June 1995 at a scale of 1:4200. Two single frequency differential GPS receivers were also used on-site to establish photo-control and to link previous epochs into one unified system. Processing the 1995 imagery allowed the DEM of the same area to be represented by 800,000 points, all measured automatically overnight. Data derived from the 1995 epoch was compared with the evolutionary model that had been postulated prior.

This model was then extended through incorporation of landslide incidence and climatic data (Brunsden and Chandler, 1996).

A major landslide event occurred in January 2001 in which a 250 m section of cliff collapsed. This triggered acquisition of a new epoch of imagery, again flown by the ARSF at a scale of 1:7500 (Figure 6.14 (see colour section)). The massive improvement in PC power then available enabled the 2001 epoch to be processed entirely on a PC running OrthoBASE and also allowed the use of colour imagery, which triples data storage requirements.

The DEMs derived in 1995/2001 demonstrate the huge increase in data resolution when compared to the manual analytical approach (DVD; Chandler and Brunsden, 1995). DEMs at 1 m resolution enable the detailed representation of small morphological features without inter-pretation/identification and subsequent manual measurement. The production of colour ortho-photos (Figure 6.15 (see colour section), DVD) greatly assists the geomorphologists (and other scientists) because no interpretation has been performed during the measurement procedure. All features identifiable within the three colour bands of the visible spectrum are represented, but these now possess positional relevance.

The increase in data resolution has also allowed more detailed DEMs of difference to be created (Figure 6.16 (see colour section)). The use of grid-based DEM manipulation and full colour repre-sentation also allows an improved representation of "difference DEMs" to the contouring approach used in the past (Chandler and Brunsden, 1995). Differences are now represented at each sampling point and appropriately colour coded. If Figure 6.15 is related to Figure 6.16 the trained eye can now identify the spectral and morphological expression of a whole host of slope forming processes, including rotational failures, head-ward retreat, translational mudslide and basal erosion.

The enclosed DVD demonstrates the value of the dynamic image or video for many of the applications cited in this book. The DVD section relating to the Black Ven landslide demon-strates the value of the simple computerised animation for assessing landform change. One particular example is the "photo-map animation" comprising a sequence of orthophotos derived from the 1976, 1995, 2000, 2001 and 2005 imagery. The video sequence was generated as an "animated GIF" using readily available shareware software and could be rapidly generated from the orthophotos available. Not only did the animations demonstrate forcibly the huge changes that have occurred but also have identified those oases of apparent stability, often surrounded by regions experiencing cataclysmic change. The animation allowed landform changes to be viewed in a new and thought provoking way, which enabled these regions to be identified.

Similar animated sequences were generated using the difference DEMs available and the profiles measured using both analytical and digital photogrammetric methods. Again, such visualisations provide an important means to educate both the expert and layperson and forcibly demonstrate the value of photogrammetry for quantifying landform change.

Note that in January 2006, a further and significant landslide event occurred on Black Ven (14 January 2006) in which several individuals were trapped by mud and quicksand and had to be rescued. It is advisable not to venture onto active areas of the mudslides, particularly during the winter months.

Mam Tor

The value of combined digital photogrammetry and archival images for quantifying landform change is further demonstrated by work on the landslide of Mam Tor.

The landslide (Figure 6.17) is situated on the eastern flank of Mam Tor, at the head of Hope Valley near Castleton, Derbyshire, UK. The former main road between Sheffield and Manchester (A625) was constructed across the slide, but abandoned in 1979 as a consequence of continuous damage due to the moving ground mass. The initial rotational failure has been

Figure 6.17 *The Mam Tor landslide.*

Table 6.1   Characteristics of the acquired photographic epochs

Date	Source	Scale	Focal length	Scan resolution	Ground resolution	Media
1953	NMR	1/10,700	547 mm*	42 μm	0.45 m	Scanned contact prints
1971	NMR	1/6,400	304 mm*	42 μm	0.27 m	Scanned contact prints
1973	CUCAP	1/4,300	153 mm	15 μm	0.065 m	Scanned diapositives
1973	CUCAP	Oblique	207 mm	15 μm	–	Scanned diapositives
1984	ADAS	1/27,200	152 mm	15 μm	0.41 m	Scanned diapositives
1990	CUCAP	1/12,000	153 mm	15 μm	0.18 m	Scanned diapositives
1995	CUCAP	1/16,400	152 mm	15 μm	0.25 m	Scanned colour neg.
1990	Infoterra	1/12,200	153 mm	21 μm	0.26 m	Scanned colour neg.

NMR = National Monuments Record
CUCAP = Cambridge University Collection of Air Photos
ADAS = Agricultural Development and Advisory Service
* = estimated values from self-calibration

dated back to 1600 BC (Skempton *et al.*, 1989) and there is evidence that present movements are related to certain rainfall thresholds (Waltham and Dixon, 2000). During the 1990s, yearly monitoring schemes were set up by Nottingham Trent University (Waltham and Dixon, 2000) and the University of Manchester (Rutter *et al.*, 2003) providing detailed ground-based movement records.

The archival imagery and digital photogrammetry has provided an opportunity to extend these quantitative records back to the 1950s. A sequence of aerial photographs of varying scale and quality was acquired from the archive; key data relating to the different epochs is displayed in Table 6.1. Ground control was collected by means of differential GPS, with a ground precision of approximately 10 mm, and transformed to Ordnance Survey National Grid coordinates (OSGB36 datum). The main photogrammetric data processing was performed using LPS, though if camera calibration data was unavailable (1953 and 1971 images), the inner orientation parameters were estimated in an external self-calibrating bundle adjustment (as described in Section 6.6.2).

A measure of the accuracy of the photogrammetric models is provided by the rms errors of independent checkpoints (Table 6.2). As expected, there is a strong relationship between ground resolution and resulting accuracy. The accuracies of the 1953 and 1971 epochs are relatively low due to the rather poor quality of the scans, and geometric uncertainties arising from the lack of camera calibration data.

Table 6.2 Estimated accuracy of the photogrammetric models (rms error of checkpoints)

Epoch	X (m)	Y (m)	Z (m)
1953	0.56	1.31	4.62
1971	0.50	0.37	1.04
1973	0.13	0.16	0.35
1973-Obl.	0.29	0.10	0.21
1984	0.45	0.40	1.66
1990	0.26	0.40	0.43
1995	0.35	0.24	0.47
1999	0.26	0.30	0.73

High-resolution DEMs and orthophotos were automatically extracted from each photographic epoch and combined to produce perspective views of the landslide (Figure 6.18 (see colour section)).

The Mam Tor landslide is less dynamic than the coastal Black Ven system, which has implications for the viability of the different data analysis techniques. DEMs of difference proved to be less useful, since the vertical displacements between epochs were too small in comparison with the vertical accuracy achievable from the imagery. For Mam Tor, vertical displacements did not exceed 0.8 m between 1990 and 1998 (Waltham and Dixon, 2000), whilst the estimated vertical accuracy between any two combined epochs was greater than 1 m. However, the horizontal component of point displacement was typically much larger, and the horizontal accuracy far better than 1 m. Consequently, the orthophoto provides an excellent way of portraying and measuring horizontal displacements and orthophotos of the active central part of the landslide (Figure 6.19) show progressive surface change.

Because the surface deformations of the landslide are rather slow, clear surface features can be identified throughout the photo sequence and their horizontal displacements measured using manual methods. A total of 32 points were identified and measured, in all or at least many of the epochs. The difference in coordinates between successive epochs represented the horizontal displacement of a particular point during that period and visualised in vector maps (Figure 6.20 (see colour section)). Independent measures representing vector precision were provided by additional measurements of "stable points" outside the active landslide area. Any "displacements" arising at these points were used to generate error ellipses, exaggerated to the 95% confidence level. Movement vectors piercing their associated ellipse were considered to represent significant movement at the 95% confidence level.

Absolute displacements were subsequently converted to movement rates in metres per annum (m/a), to allow comparison with independent rates established by other authors using ground survey methods. Mean horizontal displacement of the landslide over the period 1953–1999 was found to be 0.21 m/a, varying from 0.09 m/a at the toe up to 0.74 m/a in the central part. These values are of comparable size to movement rates found by Rutter *et al.* (2003), 0.04–0.35 m/a during the last century and up to 0.50 m/a in recent years. Figure 6.21 represents

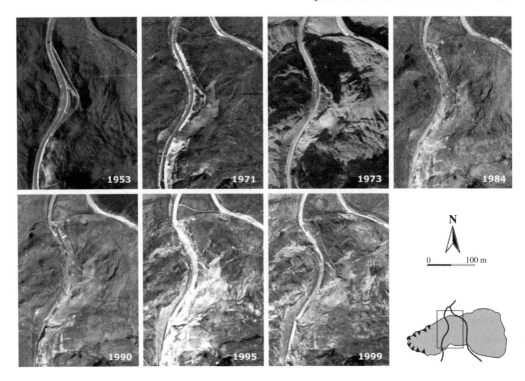

Figure 6.19 *Sequence of orthophotos of the central part of the Mam Tor landslide.*

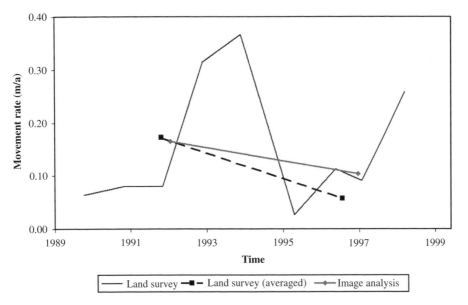

Figure 6.21 *Movement rates of a monitoring point, obtained by respectively land surveying (Waltham and Dixon, 2000; Rutter et al., 2003) and image analysis. The data of the land survey is yearly, while the averaged data present mean values in the periods 1990–1995 and 1995–1999, to allow direct comparison with the data from image analysis.*

the movement rates of a particular monitoring point obtained by land surveying, and the rate obtained by photogrammetry. To allow a good comparison, the yearly monitoring data were averaged over the same time interval, between the image epochs (1990–1995 and 1995–1999).

Image matching techniques have the potential for automated measurement of displacement vectors. Kääb and Vollmer (2000) used a correlation algorithm to map the velocity field of a rock glacier from a sequence of orthophotos. The high density and accuracy of the velocity data made it possible to extract meaningful strain-rate information. Kaufmann and Ladstädter (2002) developed this approach further by using the concept of pseudo-orthophotos, enabling automatic measurements in 3D. Unfortunately, attempts to apply these techniques in this particular case study failed, probably due to too much variation in image quality, unfavourable illumination conditions and large changes in vegetation cover between images from widely different epochs. The approach is currently being tested in a further case study where inter-epoch periods are lower.

6.7.5 Conclusions

Both the Black Ven and Mam Tor study have demonstrated the great value of multi-temporal imagery for mapping environmental changes occurring over wide areas. Where spatial changes are large, as in the case of coastal landslides like Black Ven, DEM-based methods are appropriate. Where exhibited spatial changes are smaller, orthophotograph-based methods of deriving movement vectors have proven effective. Although there is potential to use image processing methods to automate vector determination, the success of such algorithms is clearly image and case dependent. The specific ability to use historical photography to derive quantitative change data is unique and although it may not generate data of the same quality as ground-based survey, there is the potential of deriving movements through long time periods which can be invaluable when quantifying landform change.

6.8 References

Abdel-Aziz, Y.I. and Karara, H.M., 1971. Direct linear transformation from computer coordinates into object space coordinates. In: *ASP Symposium on Close Range Photogrammetry*, American Society of Photogrammetry, Bethesda, Maryland, USA, 1–18.

ADAM Technology, 2005. See http://www.adamtech.com.au/ (accessed 23 August 2005).

Adams, J. and Chandler, J.H., 2002. Evaluation of Lidar and medium scale photogrammetry for detecting soft-cliff coastal change. *Photogrammetric Record,* 17(99), 405–418.

Arber, M., 1941. The coastal landslips of West Dorset. *Proceedings of the Geologists' Association,* 52(3), 273–83.

Ashworth, P.J. and Ferguson, R.I., 1986. Interrelationships of channel processes, changes and sediments in a proglacial braided river. *Geografiska Annaler,* 68(4), 361–371.

BAE Systems, 2005. See http://www.socetgxp.com/ (accessed 17 December 2005).

Bolla Pittaluga, M., Repetto, R. and Tubino M., 2003. Channel bifurcation in braided rivers: Equilibrium configurations and stability. *Water Resources Research*, 39(3), 1046.

Brasington, J. and Smart, R.M.A., 2003. Close range digital photogrammetric analysis of experimental drainage basin evolution. *Earth Surface Processes and Landforms*, 28, 231–247.

Brasington, J., Langham, J. and Rumsby, B., 2003. Methodological sensitivity of morphometric estimates of coarse fluvial sediment transport. *Geomorphology*, 53, 299–316.

Brasington, J., Rumsby, B.T. and McVey, R.A., 2000. Monitoring and modelling morphological change in a braided gravel-bed river using high resolution GPS-based survey. *Earth Surface Processes and Landforms*, 25, 973–990.

Brunsden, D., 1969. The moving cliffs of Black Ven. *Geographical Magazine,* 41(5), 372–374.

Brunsden, D. and Chandler, J.H., 1996. Development of an episodic landform change model based upon the Black Ven mudslide 1946–1995. In Anderson, M.G. and Brooks, S.M. (Eds.), *Advances in Hillslope Processes*, Wiley, Chichester, UK, 869–896.

Carson, M.A. and Griffiths, G.A., 1989. Gravel transport in the braided Waimakariri River: mechanisms, measurements and predictions. *Journal of Hydrology*, 109, 201–220.

Chandler, J.H. and Brunsden, D., 1995. Steady state behaviour of the Black Ven mudslide: the application of archival analytical photogrammetry to studies of landform change. *Earth Surface Processes and Landforms*, 20(3), 255–275.

Chandler, J.H. and Clark, J.S., 1992. The archival photogrammetric technique: further application and development. *Photogrammetric Record*, 14(80), 241–247.

Chandler, J.H. and Cooper, M.A.R., 1989. The extraction of positional data from historical photographs and their application in geomorphology. *Photogrammetric Record,* 13(73), 69–78.

Chandler, J.H., Ashmore, P., Paolo, C., Gooch, M. and Varkaris, F., 2002. Monitoring river-channel change using terrestrial oblique digital imagery and automated digital photogrammetry. *Annals of the Association of American Geographers*, 92(4), 631–644.

Chandler, J.H, Cooper, M.A.R. and Robson, S., 1989. Analytical aspects of small format surveys using oblique aerial photographs. *Journal of Photographic Science*, 37(7), 235–240.

Chandler, J.H., Fryer, J.G. and Jack, A., 2005. Metric capabilities of low-cost digital cameras for close range surface measurement. *Photogrammetric Record*, 20(109), 12–26.

Chandler, J.H., Shiono, K., Rameshwaren, P. and Lane, S.N., 2001. Measuring flume surfaces for hydraulics research using a Kodak DCS460. *Photogrammetric Record*, 17(97), 39–61.

Charlton, M.E., Large, A.R.G. and Fuller, I.C., 2003. Application of airborne LiDAR in river environments: The River Coquet, Northumberland, UK. *Earth Surface Processes and Landforms*, 28, 299–306.

Clarke, T.A. and Fryer, J.G., 1998. The development of camera calibration methods and models. *Photogrammetric Record*, 16(91), 51–66.

Collins, S.H. and Moon, G.C., 1979. Stereometric measurement of streambank erosion. *Photogrammetric Engineering and Remote Sensing*, 45(2), 183–190.

El-Ashry, M.R. and Wanless, H.R., 1967. Shoreline features and their changes. *Photogrammetric Engineering,* 33(2), 184–189.

Faig, I.W., 1975. Calibration of close range photogrammetric systems. *Photogrammetric Engineering and Remote Sensing,* 41(12), 1479–1486.

Finsterwalder, S., 1897. Der vernagtferner. *Wissenschaftliche Erganzungschefte zur Zeitschrift der Deutschen und Osterreichen Alpenvereins*, 1(1), 1–96.

Fraser, C.S., 1997. Digital camera self-calibration. *ISPRS Journal of Photogrammetry and Remote Sensing,* 52, 149–159.

Ganci, G. and Shortis, M.R., 1996. A comparison of the utility and efficiency of digital photogrammetry and industrial theodolite systems. *International Archives of Photogrammetry and Remote Sensing*, 31(B5), 182–187.

Goff, J.R. and Ashmore, P.E., 1994. Gravel transport and morphological change in braided Sunwapta River, Alberta, Canada. *Earth Surface Processes and Landforms*, 19, 195–212.

Graham, R.W. and Mills, J.P., 2000. Small format digital cameras for aerial survey: where are we now? *Photogrammetric Record*, 16(96), 905–909.

Granshaw, S.I., 1980. Bundle adjustment methods in engineering photogrammetry. *Photogrammetric Record*, 10(56), 181–207.

Hancock, G. and Willgoose, G., 2001. The production of digital elevation models for experimental model landscapes. *Earth Surface Processes and Landforms,* 26(5), 475–490.

Henry, J.B., Malet, J.P., Maquaire, O. and Grussenmeyer, P., 2002. The use of small format and low altitude aerial photographs for the realization of high-resolution DEMs in mountainous areas: application to the Super-Sauze earthflow (Alpes-de-Haute-Provence, France). *Earth Surface Processes and Landforms*, 27(12), 1339–1350.

Heritage, G.L., Fuller, I.C., Charlton, M.E., Brewer, P.A. and Passmore, D.P., 1998. CDW photogramme-try of low relief fluvial features: accuracy and implications for reach scale sediment budgeting. *Earth Surface Processes and Landforms*, 23(13), 1219–1233.

Hicks, D.M., Duncan, M.J., Walsh, J.M., Westaway, R.M. and Lane, S.N., 2002. New views of the mor-phodynamics of large braided rivers from high-resolution topographic surveys and time-lapse video. In *The Structure, Function and Management Implications of Fluvial Sedimentary Systems*, International Association of Hydrological Sciences, *Wallingford, UK*.

Hydraulic Research Wallingford, 2005. *Touch sensitive, incremental 2-D bed profiling system for physical models*. See http://www.hrwallingford.co.uk/equipment/bed_profiling.html (accessed 10 August 2005).

Intergraph, 2005. See http://www.intergraph.com/isae/ (accessed 16 December 2005).

Inpho, 2005. See http://www.inpho.de/ (accessed 16 December 2005).

Jensen, J.R., 1996. Issues involving the creation of digital elevation models and terrain corrected ortho-imagery using soft-copy photogrammetry. In: Greve, C.W. (Ed.), *Digital Photogrammetry: an adden-dum to the Manual of Photogrammetry*, American Society of Photogrammetry and Remote Sensing, Bethesda, Maryland, USA, 167–179.

Kääb, A. and Vollmer, M., 2000. Surface geometry, thickness changes and flow fields on creeping moun-tain permafrost: automatic extraction by digital image analysis. *Permafrost and Periglacial Processes*, 11(4), 315–326.

Kalaugher, P.G., Grainger, P. and Hodgson, R.L.P., 1987. Cliff stability evaluation using geomorphological maps based upon oblique aerial photographs. In Culshaw, M.G., Bell, F.G., Cripps, I.C. and O'Hara, M. (Eds.), *Planning and Engineering Geology*, Geological Society, London, UK, 163–170.

Kaufman, V. and Ladstädter, R., 2002. Monitoring of active rock glaciers by means of digital photogram-metry. *International Archives of Photogrammetry and Remote Sensing*, 35(B7), 108–111.

Kenefick, J.F., Harp, B.F. and Gyer, M.S., 1972. Analytical self-calibration. *Photogrammetric Engineering*, 38(11), 1117–1126.

Kirby, R.P., 1991. Measurement of surface roughness in desert terrain by close range photogrammetry. *Photogrammetric Record*, 13(78), 855–875.

Lane, S.N., 2000. The measurement for the channel morphology using digital photogrammetry. *Photogrammetric Record*, 16(96), 937–961.

Lane, S.N., 2006. Approaching the system-scale understanding of braided river behaviour. In Sambrook-Smith, G.H., Best, J.L., Bristow, C.S. and Putts, C.E. (Eds.), *Braided Rivers*, IAS Special Publication 36, Blackwell, Oxford, UK.

Lane, S.N. and Chandler, J.H., 2003. The generation of high quality topographic data for hydrology and geomorphology: New data sources, new applications and new problems. *Earth Surface Processes and Landforms*, 28(3), 229–230.

Lane, S.N., Chandler, J.H. and Porfiri, K., 2001. Monitoring flume channel surfaces using automated digital photogrammetry. *ASCE Journal of Hydraulic Engineering*, 127(10), 871–877.

Lane, S.N., Chandler, J.H. and Richards, K.S., 1994. Developments in monitoring and modelling small scale river bed topography. *Earth Surface Processes and Landforms*, 19, 349–368.

Lane, S.N., Reid, S.C., Westaway, R.M. and Hicks, D.M., 2004. Remotely sensed topographic data for river channel research: the identification, explanation and management of error. In Kelly, R.E.J., Drake, N.A. and Barr, S.L. (Eds.), *Spatial Modelling of the Terrestrial Environment*, Wiley, Chichester, UK, 157–74.

Lane, S.N., Richards, K.S. and Chandler, J.H., 1995. Morphological estimation of the time-integrated bed-load transport rate. *Water Resources Research*, 31, 761–772.

Lane, S.N., Richards, K.S. and Chandler, J.H., 1996. Discharge and sediment supply controls on erosion and deposition in a dynamic alluvial channel. *Geomorphology*, 15, 1–15.

Lane, S.N., Richards, K.S. and Chandler, J.H. (Eds.), 1998. *Landform Monitoring, Modelling and Analysis*, Wiley, Chichester, UK.

Lane, S.N., Westaway, R.M. and Hicks, D.M., 2003. Estimation of erosion and deposition volumes in a large, gravel-bed, braided river using synoptic remote sensing. *Earth Surface Processes and Landforms*, 28, 249–271.

Laronne, J.B. and Duncan, M.J., 1992. Bedload transport paths and gravel bar formation. In Billi, P., Hey, R.D., Thorne, C.R. and Tacconni, P. (Eds.), *Dynamics of Gravel-bed Rivers*, Wiley, Chichester, UK, 177–200.

Leica Geosystems, 2003. *Leica Photogrammetry Suite OrthoBASE and OrthoBASE Pro users guide*. Leica Geosystems GIS and Mapping, LLC, Atlanta, Georgia, USA.

Leopold, L.B., Wolman, M.G. and Miller, J.P., 1964. *Fluvial Processes in Geomorphology*, Freeman, San Francisco, California, USA.

Lim, M., Petley, D.N., Rosser, N.J., Allison, R.J., Long, A.J. and Pybus, D., 2005. Combined digital photogrammetry and time-of-flight laser scanning for monitoring cliff evolution. *Photogrammetric Record*, 20(110), 109–129.

Lo, C.P. and Wong, F.Y., 1973. Micro scale geomorphology features. *Photogrammetric Engineering and Remote Sensing*, 39(12), 1289–1296.

Lyzenga, D.R., 1981. Remote sensing of bottom reflectance and water attenuation parameters in shallow water using aircraft and Landsat data. *International Journal of Remote Sensing*, 2, 71–82.

Machatschek, F., Graul, H., Rathjens, C., Clayton, K.M. and Davis, D., 1969. *Geormorphology*, Oliver and Boyd, Edinburgh, UK.

Merel, A.P. and Farres, P.J., 1998. The monitoring of soil surface development using analytical photogrammetry. *Photogrammetric Record*, 16(92), 331–345.

Mills, J.P., Buckley, S.J., Mitchell, H.L., Clarke, P.J. and Edwards, S.J., 2005. A geomatics data integration technique for coastal change monitoring. *Earth Surface Processes and Landforms*, 30(6), 651–664.

Paine, A.D.M., 1985. Ergodic reasoning in geomorphology: time for a review of the term? *Progress in Physical Geography*, 9(1), 1–15.

Patias, P. and Streilein, A., 1996. Contribution of videogrammetry to the architectural restitution results of the CIPA "O. Wagner Pavillion" test. *International Archives of Photogrammetry and Remote Sensing*, 31(B5), 457–462.

PCI Geomatics, 2005. See http://www.pcigeomatics.com/ (accessed 2 November 2005).

Petrie, G. and Price, R.J., 1966. Photogrammetric measurements of the ice wastage and morphological changes near the Casement glacier, Alaska. *Canadian Journal of Earth Sciences*, 3, 783–798.

Photarc, 2005. See http://www.photarc.co.uk/rollei.htm (accessed 2 November 2005).

Pyle, C.J., Richards, K.S. and Chandler, J.H., 1997. Digital photogrammetric monitoring of river bank erosion. *Photogrammetric Record*, 15(89), 753–763.

Rutter, E.H., Arkwright, J.C., Holloway, R.F. and Waghorn, D., 2003. Strains and displacements in the Mam Tor Landslip, Derbyshire, England. *Journal of the Geological Society*, 160(5), 735–744.

Skempton, A.W., Leadbeater, A.D. and Chandler, R.J. (1989). The Mam Tor Landslide, North Derbyshire. *Philosophical Transactions of the Royal Society of London, A. Mathematical and Physical Sciences*, 329(1607), 503–547.

Stojic, M., Chandler, J.H., Ashmore, P. and Luce, J., 1998. The assessment of sediment transport rates by automated digital photogrammetry. *Photogrammetric Engineering and Remote Sensing*, 64(5), 387–395.

Supresoft, 2005. See http://www.supresoft.com (accessed 1 November 2005).

Topcon, 2005. See http://www.topcon.com.sg/survey/ (accessed 1 November 2005).

Walstra, J., Chandler, J.H., Dixon, N. and Dijkstra, T.A., 2004. Time for change–quantifying landslide evolution using historical aerial photographs and modern photogrammetric methods. *International Archives of the Photogrammetry, Remote Sensing and Spatial Information Sciences*, Vol 34 (Part XXX), Commission 4, Istanbul, 2004, 475–481.

Waltham, A.C. and Dixon, N. (2000). Movement of the Mam Tor Landslide, Derbyshire, UK. *Quarterly Journal of Engineering Geology and Hydrogeology*, 33(2), 105–123.

Welch, R. and Jordan, T.R., 1983. Analytical non-metric close range photogrammetry for monitoring stream channel erosion. *Photogrammetric Engineering and Remote Sensing*, 49(3), 367–374.

Westaway, R.M., Lane, S.N. and Hicks, D.M., 2000. The development of an automated correction procedure for digital photogrammetry for the study of wide, shallow, gravel-bed rivers. *Earth Surface Processes and Landforms*, 25, 200–226.

Westaway, R.M., Lane, S.N. and Hicks, D.M., 2001. Airborne remote sensing of clear water, shallow, gravel-bed rivers using digital photogrammetry and image analysis. *Photogrammetric Engineering and Remote Sensing*, 67, 1271–1281.

Westaway, R.M., Lane, S.N. and Hicks, D.M., 2003. Remote survey of large-scale braided rivers using digital photogrammetry and image analysis. *International Journal of Remote Sensing*, 24, 795–816.

Winterbottom, S.J. and Gilvear, D.J. 1997. Quantification of channel bed morphology in gravel-bed rivers using airborne multispectral imagery and aerial photography. *Regulated Rivers: Research and Management*, 13, 489–499.

7 Medicine and sport: measurement of humans

Harvey Mitchell

7.1 General outline of the application

7.1.1 Introduction

The scope of medical photogrammetry is much wider than may at first be anticipated. Perhaps because medicine generally involves the internal functions of the body, it may be expected that external body measurement is of limited medical value. However, there have been attempts for over a century to use photography to assist in the measurement of the shape of a wide range of areas of the human body, including torsos, heads, faces, limbs, breasts, feet, skin, eyes, teeth, and the human body in motion. Measurement related to sport, especially via motion studies, is newer but is growing in value.

There are numerous situations for which measurements undertaken outside the body can be sought. Obviously there are disorders which can be examined externally, such as those which relate to the skin, the teeth and the eyes (including the retina). But there are many disorders which are made apparent through external conditions, especially those of the face and spine, and notably orthopaedic deformities. Assistance in treating the conditions which affect movement can be provided effectively by motion studies. Finally, because of natural concerns about physical appearance, it may be the external manifestations of a condition which are under treatment, rather than the underlying condition.

Medical and sports measurements often relate to anatomical studies of the mechanics, workings and other structural aspects of the human body, but there are other applications: in the detection of medical conditions by mass screening and the examination of individual patients, in the treatment of a disease or condition, and often in research into diseases and/or their treatment. Photogrammetric measurement has found applications in orthopaedics, ophthalmology, neurology, dentistry, occupational therapy, ergonomics, and many other areas related to human health. In recent years, the availability of digital video cameras has enabled photogrammetry to be used extensively for measurements of the human body in motion, for both medical and sporting purposes.

The external studies by photogrammetry (and other optical techniques) are relatively cheap and cause little distress to the patient. They can be more effective than current medical options, such as simple geometrical devices, including callipers, tapes and protractors, and can certainly be more valuable than observation by the human eye. Such external studies also need to be contrasted with internal, non-contact measuring techniques, such as radiography, ultrasound examination, computerised axial tomography (CAT) scanning and magnetic resonance imaging (MRI), which are all invariably intrusive, inconvenient, expensive or slow. They can even carry some risk, such as the ionising effect of X-rays.

Many hundreds of applications of medical photogrammetry have been published and these extend into many different fields and use many techniques. The applications which have seen the widest use and the most successful commercial developments have been in motion studies, especially for gait analysis. The existence of many other, non-commercial developments of photogrammetry for short-term usage indicates its strength in cheapness and flexibility and usability for the layperson. It is these lower level systems which are given emphasis here, as they can be implemented by the relative novice with cameras and computer. Less emphasis is given to expensive and/or specialised systems even though they are useful and productive.

7.1.2 Advantages of photogrammetry

An examination of the six requirements of human measurement, in Section 7.2, shows that photogrammetry is ideally suited to medical measurement. The first advantage of photogrammetry in medical applications arises from its basis in imaging, which gives it a surprising number of distinct benefits as a human measurement tool:

- Data collection can be quick enough to overcome people's movement.
- Touching people, with the risk of hurting or infecting them, can be avoided.
- The lack of contact avoids the risk of deforming the area of interest, or in motion studies, avoids the use of instruments that can interfere with the motion under study.
- Quick imaging is comfortable and convenient for humans.
- Image acquisition using small and familiar equipment is not daunting for medical patients (see Figure 7.1).
- Objects of various sizes can be tolerated.

Photogrammetry can be contrasted with other external measuring techniques, such as those involving electrogoniometers and accelerometers attached to the body for movement analysis, which are not only uncomfortable but can interfere with the free movement of the patient.

The second advantage of photogrammetry is its simplicity for spatial measurement, and allied to that, its low cost. It can often be implemented using two commercially available cameras, and software which is no longer expensive, operating on a common computer.

7.1.3 Disadvantages of photogrammetry

Photogrammetry's only significant disadvantage is the need for photography and data processing stages which can be laborious. However, the variety available in the photography and processing stages give photogrammetry its versatility. If a photogrammetric system is to be implemented for routine usage, procedures can be streamlined and customised for the given application.

7.1.4 History of medical and sports measurement

Throughout history, body form has been described using estimates made by eye or measurements taken with tapes and callipers. Egyptian artists of 40 centuries ago developed a system of proportions, using the width of the hand as a unit, for representing human body form. These efforts at quantification were frustrated by the difficulty of obtaining direct measurements on a living human body and to the lack of a convenient three-dimensional measuring system. The work of the American physician, Oliver Holmes in 1863, who used stereoscopic pairs of photographs to study gait, is often quoted as the earliest application of photogrammetry in medical measurement. The measurements led to an improvement in the design of artificial limbs for amputees maimed during the American Civil War. With the turn of the century, the number of applications began to increase and diversify into many areas of medicine, but progress was slow because of the interruption of two World Wars and also because the major effort was being directed into developing aerial photogrammetry which had a far greater potential. Since the 1950s there has been more extensive development and diversification. Medical measurement has been assisted by the introduction of other imaging systems such as moiré fringe interferometry and structured light. But, more recently, the increasing availability of computers has enabled the introduction of direct digital analysis of imagery, returning photogrammetric procedures to the cheaper end of the cost scale and offering a high level of automation.

7.2 Distinctive aspects of medical and sports measurement

Medicine and sport have common photogrammetric considerations because they both involve living humans. It is the consideration for humans that creates a crucial distinction for this chapter. This creates a class of close range photogrammetry with its own distinctive challenges and constraints which include:

- It is necessary to be concerned about the health, convenience, comfort and privacy of living humans. Cameras can generally be regarded as commonplace.
- Human beings, even when standing or sitting, continue to move continuously if almost imperceptibly, and so imaging needs to be both quick and finely synchronised.
- There are numerous photographic challenges with human objects, whether caused by the skin, with its low contrast and limited textural variation, or the eye, or by any body part, such as torso, limb or foot, which require all round coverage. Figure 7.2 shows differences occurring in imagery from different camera positions.
- The measurement also involves interaction with medical or other health practitioners. Users of the results may be involved in health in many diverse ways: as medical researchers, surgeons, clinicians, bio-mechanical engineers, occupational therapists, and so on. This can demand explicit communication about the constraints and requirements of the measurement, and the form of the output. One of the major challenges of medical photogrammetry is to provide information which is appropriate, usable and medically meaningful.
- Medical and sports measurement may mean that the imaging specialist will need to feel comfortable working with human patients and medical practitioners in unfamiliar surroundings.

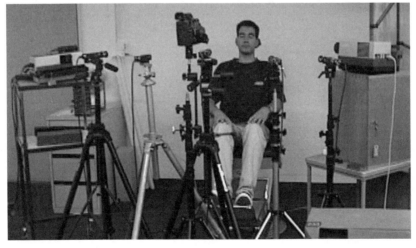

Figure 7.1 *Studio measurement of a human patient by photogrammetry. It is apparent that the equipment is generally not highly specialised, and it can be assumed that this equipment is not intimidating to the patient. Reproduced with permission of the Institute of Geodesy and Photogrammetry, ETH Zurich.*

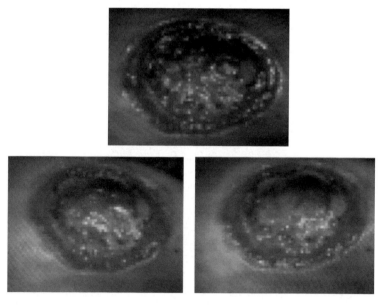

Figure 7.2 *Differences in positions of highlights are visible in this set of three images taken from three different camera locations, in a photogrammetric measurement for the purpose of studying the healing of bedsores. Reproduced from Malian et al. (2005) with permission of Blackwell Publishing.*

- Surprisingly perhaps, the cost of photogrammetric measurement usually needs to be kept low. The measurements are not normally crucial, and may often be dispensed with altogether, and replaced by a skilled surgeon with a good eye, or with simple callipers if it becomes too expensive. Measurements intended for mass screening or anthropometric data collection need to have a low cost per measurement to make widespread implementation viable. Despite the benefits of external measurements, they are virtually never indispensable to saving life. It is ironic that in the field of medicine, which is typified by expensive, high technology equipment, that a significant advantage of photogrammetry can be its low cost.

Measurement of humans can perhaps be extended to applications where the object of interest is not part of the live human, (even the case of a golf club, as measured by Smith *et al.* (1998), even though the issue was the dynamics of the golf club shaft during the swing!) because the challenge of photogrammetry involves an interaction with humans and a need to allow for them. On the other hand, measurement of inert items, such as prostheses or parts removed from bodies, is not discussed here, as this does not involve human beings, and the photogrammetric considerations are not characterised by those involved in the study of live patients, even though the client may be a health practitioner. Measurement of animals (Chapter 8) also has some crucial differences from measuring human patients.

7.3 Typical requirements

7.3.1 Accuracies

In comparison with many other uses of photogrammetry, the requirements of medical measurement for accuracy (expressed as a fraction of the total object size) are generally modest, compared with those expected when making industrial measurements. Typical accuracies range from about 1:500 to 1:5000. For example, a very precise 0.1 mm on a human back of 0.5 m dimension is equivalent to 1:5000.

7.3.2 Cameras

A major concern is for synchronised photography, because the state being measured is dynamic, even for a patient who is seemingly still. It is usually accepted that synchronisation to about 1/30 second is needed to ensure that the body does not move between images. There are few static situations when a single camera can be moved to different exposure stations to obtain coverage, so the need for synchronisation increases the number of cameras that are needed.

Synchronisation of cameras is not always easy. It may preclude the use of common off-the-shelf digital cameras. The acquisition of imaging by digital cameras which are connected to a computer is usually best achieved under software control. Electronic flash can provide a simple and precise means of synchronising the exposures of film cameras, if the photography is carried out in suitable conditions of low light levels. The shutters are opened, the flash fired and the shutters closed again. In some cases, however, a flash can startle patients enough to cause them to move by an unacceptable amount.

The importance of synchronisation is indicated by the procedure of Majid *et al.* (2005), who used imagery of a swinging object to assess their camera synchronisation for a facial photogrammetry study (see Figure 7.3). The simplicity of this arrangement is seen as contributing to the versatility of the photogrammetric technique.

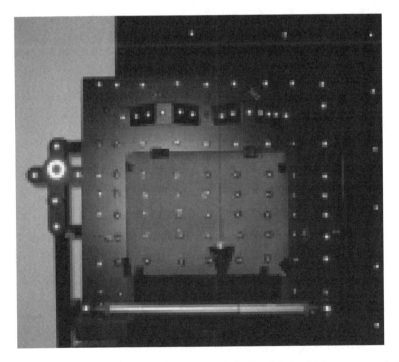

Figure 7.3 *Camera synchronisation test, using a surveyor's plumb-bob which was swung in front of the control frame. Reproduced from Majid et al. (2005) with permission of Blackwell Publishing.*

When the measurement relates to moving people, whether for medical or sports studies, this clearly creates a need for high-speed recording.

Medical imaging will generally be done at close range so that depth of field can be limited. And, at focal settings other than infinity, the principal distance of the lens may not be known during processing. Because accuracy requirements in medical imaging are not high by typical photogrammetric standards, the low resolution of some digital cameras is not a handicap. It is found that greater attention must be paid to various photographic matters, which depend on object size, the precision of measurement required, constraints on the camera and/or the object position.

7.3.3 Preparation for photography

The measurement environment can be unusual, perhaps a clinic or surgery or in open space, and rarely in a specialised photogrammetric laboratory, and this can create constraints for the design of the photographic procedure.

Medical and sports measurement may involve objects as small as teeth or retinas, or as large as sporting teams in action. Object coverage requires a consideration of camera field-of-view, adequate overlap, appropriate camera-to-object distance and suitable base-to-object distance ratio, which influence the achievable accuracy and precision. Camera configurations which are close to parallel allow for simplicity in later analysis, but those with a variable geometry are more versatile.

All-round coverage is required in some studies (such as for the entire torso or the head), and may require that images be obtained in several directions, using an array of synchronised cameras in coordinated positions. Coverage may be achieved in some cases through the use of mirrors, which has the additional advantage that it simplifies synchronisation.

As with all photogrammetry, it is crucial that the accuracy of the medical photogrammetric output be carefully assessed, so that accuracy levels can be quoted with confidence. Results provided to the client, even without a specific accuracy level, will have an implied accuracy, which needs to be realistic. It is also important to remember that the internal precision of results is not a true indicator of accuracy.

Object preparation is as necessary for medical photogrammetry as it is for other close range applications. There are two cases to consider: first, those applications for which targets are needed, when specific body points are to be located, and secondly, those applications involving general surface shape measurement based on a network of surface points.

Targets are often used in medical photogrammetry. For example, they may be wanted when anatomically relevant points are to be located, as in spinal shape detection. They are crucial for dynamic studies (see Section 7.6.3), whether to keep track of the same point on the body in motion or because of an interest in the motion of a certain point. Target placement on medically relevant sites can require the involvement of medically trained personnel, raising logistical and cost considerations. Both active (light emitting) and passive (reflective) targets find usage in medical studies. Highly reflective targets (usually retro-reflectors whose effect is illustrated in Figure 7.4) are especially popular since active targets, which require wiring, can constrain the poses adopted by the patient.

Targets may be used in surgical applications to locate probes, primarily in image guided surgery, as in Figure 7.5. The application can be regarded as medical, because the characteristics of the probe must take into account the needs of a patient, even though the patient may be anaesthetised when it is being used.

Skin often appears on an image with a uniform low contrast, especially the skin of younger children and babies, and hence it usually needs artificial texturisation. This is commonly

Figure 7.4   *Retro-flective targets seen in dynamic human studies. Reproduced with permission from Qualisys.*

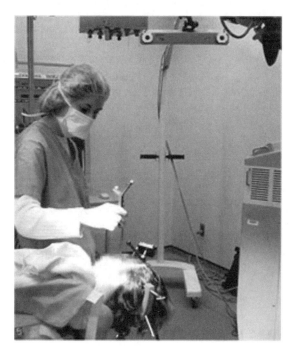

Figure 7.5   *Polaris camera system used to detect targets on a probe is mounted in an operating room. Reproduced with permission from Northern Digital Inc.*

achieved by the projection of patterns onto the skin, as in Figure 7.6 (e.g. Siebert and Marshall, 2000). On darker skin, laser light may be necessary. Powder has also been used to add texture. Waldhäusl *et al.* (1990) have used ultraviolet filters to obtain contrasting texture from areas of skin high in melanin, in conjunction with a double flash to avoid shadows. Placing an extensive array of marks on the skin can compensate for a lack of texture, but such a procedure is usually laborious. Colour photography is normally beneficial but orthochromatic emulsions and ultra-violet light can also be employed to good effect.

Control serves the same purpose in medical photogrammetry as in other close range applications, to assist with absolute orientation of the three-dimensional model. However, orienting the model to a particular coordinate system is often not needed in medical work and scale is usually the only absolute quantity needed. Accordingly, control points can be avoided in many applications.

On the other hand, there are some cases where a patient-based coordinate system is necessary, notably when comparative measurements over a period of time are needed, or when comparisons between a number of patients are desired. These arise, for example, when collecting anthropometric data on the distribution of shapes, when monitoring deteriorating conditions, and when recording the progress of a treatment on individual patients. When control is needed, there are often constraints, just as there are with targets. The placement of permanent control points, by tattoos or other skin markings, is normally unethical and otherwise unacceptable. Use can sometimes be made of blemishes on the skin, such as moles and warts, and other natural detail points, but the human can grow and simply change their shape through stance change, so that

Figure 7.6 *Human back texturised by casting a regular light pattern, from Mitchell (1994). The image indicates that the back shape itself need not be made apparent by the texturising.*

the entire point of having the control is lost. In just a few cases, some permanent features of the body, such as the eye-base, can be utilised to ensure consistent scaling for comparisons of a patient at different times. In other cases, a framework of control points around the patient has been resorted to, with a relationship between the patient and the frame then being provided by connections, such as rods to the ears.

7.3.4 Required products

The provision of medically meaningful output could be crucial issues for acceptance of photogrammetric procedures. Virtually all medical photogrammetric measurements require further interpretation and analysis to allow meaningful information to be given to the end-user. Only if this additional information is provided and understood will the photogrammetry be utilised in practice, and this should be recognised as an additional burden in the development of a medical measurement system. Medical information is rarely provided directly by raw spatial data, so simple measurement of the object is seldom sufficient to fulfil the requirements of the medical client. Extracting the medically relevant information can create the most difficult part of any application. For example, the human back is seen as an ideal surface for close range photogrammetric recording, but inferring the three-dimensional shape of the spine from the measurements is not straightforward. Similarly, gait observations on a moving person do not offer direct information about the medical condition that causes the movement problem.

Even the presentation of basic spatial information requires considerations of presentation to those not normally familiar with it, as recognised by Chadwick and Mitchell (1999). Diagrams, whether contoured plans or perspective wire-frame models, are rarely of direct use to the medical practitioner as they are not likely to present any extra quantitative information beyond that which can be provided by simple examination of the patient. They can be difficult to interpret for those unfamiliar with them. In some cases, other forms of output are needed such as volumes, areas and medical parameters. Measurement without appropriate presentation can result in a waste of the information that photogrammetry has extracted. Change in the patient's shape from one measurement epoch to the next can be the matter of primary interest in some treatments. Graphical depictions of movement of the human body can be very helpful, and have been achieved with graphics showing trajectories of motion, as in Figure 7.7.

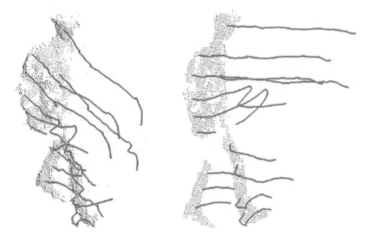

Figure 7.7 *Three-dimensional trajectories representing human body motion. Reproduced with permission from the Institute of Geodesy and Photogrammetry, ETH Zurich.*

7.3.5 Calibration and testing requirements

Low accuracy requirements in some medical cases can mean that users believe that calibration of some camera and sensor elements can be avoided. However, this omission is not recommended. In particular, sensor parameters including lens distortion and digital sensor array geometry, should be assessed. Modern digital cameras often have zoom lenses, which can leave the principal distances almost unknown.

It is also important for accuracies to be confirmed with test objects to ensure that the neglected errors are insignificant. Calibration is normally carried out by photographing objects with pre-surveyed targets, an example of which is shown in Figure 7.8.

7.3.6 Overall use of photogrammetry

In the past decade, measurement of some areas of the body has undoubtedly been achieved by laser scanning: measurements of the entire body, the trunk and limbs, backs, head, face, jaw, and feet. However, photogrammetry retains some advantages. Because of its basic simplicity for the layperson and its cheapness and its adaptability to a variety of situations, it has found many applications for short-term projects. It entails obtaining suitable equipment and processing software to suit the peculiar needs of the application. To generate results of sufficient accuracy for medical purposes, photogrammetry has the advantage of needing only cameras, software, and a typical computer, as is suggested by the photogrammetric system shown in Figure 7.9. These components are relatively cheap and commonly available. Photogrammetry is adaptable to one-off situations. Because of the inflexibility of scale of commercial laser scanners, photogrammetry can still find uses for objects such as the teeth and eyes. Most importantly, because photogrammetry can be used for moving objects, it has found extensive use in motion studies. These applications include commercial products in both medicine and sport.

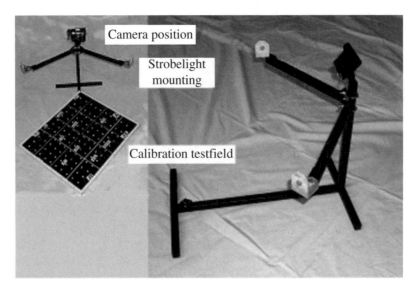

Figure 7.8 *Purpose-built calibration frame used by Majid et al. (2005) in their facial studies which used both photogrammetry and scanning. Reproduced with permission of Blackwell Publishing.*

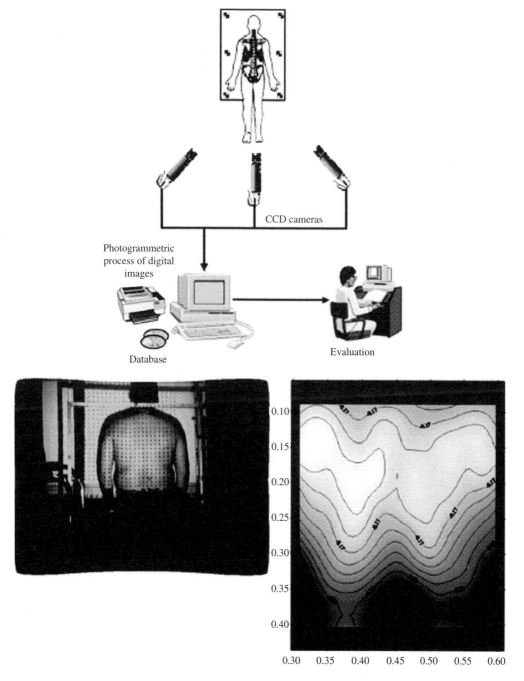

Figure 7.9 *Example of use of optical digital images for 3D back reconstruction (after Sechidis et al., 2000).*

The photogrammetric techniques which have been used in sport and medicine exhibit the following wide variety:

- measurement of either surfaces and individual targeted points;
- photography with and without added optical texture;
- objects as small as teeth to those larger than one person;
- measurement of motionless persons and high-speed measurement of moving people;
- slow turnaround (for many cases) and immediate output (for some rare clinical and surgical applications); and
- expensive commercial systems as well as systems which are fabricated for a specific purpose and which are characterised by their cheapness.

The photogrammetric characteristics of the more common medical applications are shown in Table 7.1.

Table 7.1 Most widely reported trials and ongoing applications of photogrammetry in medicine

Medical application	Special photogrammetric characteristics
Eye (numerous ophthalmic problems, especially relating to retinal region)	Special cameras likely to be used; medical assistance with eye function
Live surgical monitoring and recording	High-speed requirements
Tooth (wear, erosion and abrasion)	Small object may demand microscope; access to mouth by casting teeth
Entire body, trunk, limbs for mass/fat distribution studies	May need all-round coverage
Head, face, jaw reconstruction	The need for detail and the occurrence of complex surfaces may create difficulties
Back for scoliosis detection (screening) and treatment	Suited to photogrammetry in terms of surface shape, size and accuracy Skin on back surface may need added texture
Motion: walking, especially for rehabilitation; sports	High-speed imaging required; problems with occluded targets; targets must not distract patient
Skin sores, ulcers, wounds, melanomas	Small object but not usually suited to microscope

It is worth realising that measurement finds application with medical conditions that are not immediately anticipated. For example, measurements of the face can be useful in psychological studies; gait measurement can be relevant to diabetics because of the lack of blood flow to the extremities of the foot. Measurements may be carried out on healthy people as well as the sick, for research purposes. The location can include large and small hospitals, clinics and other medical institutions.

7.4 Solutions options

The method of data reduction which is adopted depends on a combination of factors such as the data acquisition system, the accuracy required, and the type and amount of data needed, as well as the equipment available. Completely automated computer processing (i.e. including automated image point selection) is usually possible where targets are used, or when the topography of the medically relevant surface is simple and not too convoluted. This can preclude automated measurement of the face, but automated back shape measurement is relatively easy.

In some cases, the medical systems have to be made usable within hospitals and other institutions, by medical staff rather than by photogrammetrists, but commercial photogrammetry software packages require some familiarity with photogrammetric concepts.

There are a few specialised, sophisticated and relatively expensive hardware and software systems which offer fully automated digital processing and are commercially available for motion studies in both medicine and sport (see Section 7.6.3).

7.5 Alternatives to photogrammetry

7.5.1 Allied optical methods

In many of the medical measurement problems which involve surface measurement (i.e. excluding the applications which involve targets, surgical recording etc.), the benefits of photogrammetry's quick, non-contact recording of external surfaces are also offered by a number of other non-photogrammetric techniques, which are often optical, perhaps even being based on the camera. Indeed, some optical techniques can be so closely allied to photogrammetry that they are of interest to users of photogrammetry. Photogrammetry has sometimes been supplanted by such techniques in those cases in which uncomplicated surface measurement is required, and if the scale is suited to existing scanners (such as for the back). As with photogrammetry, the concepts have sometimes been implemented in commercially available scanners. Whilst being accurate and easy to use, they can be relatively expensive, and suited to only those object sizes and accuracies for which they have been designed, and they do not necessarily image instantaneously. Again like photogrammetry, the concepts can also be adapted to short-term projects.

Experimenters who have photogrammetric experience can easily adapt to other optical techniques such as these, because of the common concepts of intersecting rays, and common imaging hardware and image processing software. The problems of occlusions are also common. Like photogrammetry, they compete with laser scanning, but equally they can share photogrammetry's advantages of being cheap and adaptable. The techniques are not widespread, and saw most of their development at least 20 years ago. Despite this, the work deserves to be borne in mind by anyone seeking a simpler solution to a measurement problem involving humans, even though the references may not be recent.

7.5.2 Light sectioning

One technique which has more in common with photogrammetry than most other techniques is that which uses "structured light". In its simplest form, it involves the use of a projector and a camera, rather than the two cameras as in simple photogrammetry. The projector is used to cast a pattern, typically a series of parallel lines as in Figure 7.10, onto the object, which is then imaged from another point. The appearance of the projected pattern can portray immediately to an observer the shape of the object, as indicated in the figure.

Clearly, the projector is equivalent to the second camera used in photogrammetry and similar principles of light ray intersections define the position of the point in space. Any point of light projected in a known direction by the projector and identified on the photograph can then be positioned in three dimensions. The determination of the three-dimensional position involves

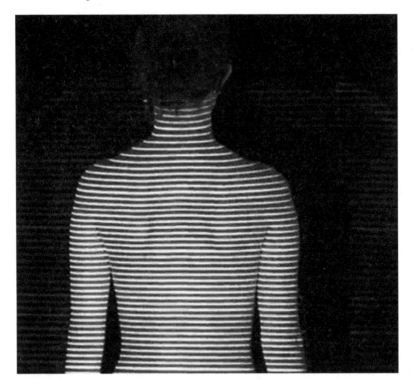

Figure 7.10 *Structured light measurement applied to a human back using the Formetric system from DIERS (adapted from file from DIERS, 2006).*

the intersection of light rays by analogue or analytical reconstruction of the camera/projector relationship, so considerations of lens distortions and principal point positions (for the cameras and projectors), point identification etc., are the same as in photogrammetry. Securing certain simple orientations can simplify the processing and, with a simple camera/projector relationship, three-dimensional coordinates are easily deduced.

In its early days, the technique was valuable because of its portrayal of the object shape. With digital processing to provide coordinates describing the surface, any human perception of the shapes became irrelevant, and automated line identification became a greater problem. This can be overcome by projecting a varying pattern, such as light and heavy lines. Because the structured light technique is best suited to simple surfaces where cast lines are not interrupted, it finds its best medical application in the measurement of backs. Commercial systems which automate the procedure have been made available over the past decade or so (e.g. DIERS, 2006).

Another option to facilitate automation was to move a projected dot or line across the surface, and analyse the individual frames from a video camera. With a single spot or single line, the image position is easily identified immediately, and the position is easily deduced. With a single projected spot, the technique no longer depends on any analysis of the image. The term "light sectioning" is then not appropriate, and the technique reduces to laser triangulation as described in Section 2.7, and for which there are commercial suppliers. Differentiation between laser triangulation, light sectioning and photogrammetry becomes blurred, but the definitions are presumably irrelevant to the user, who will select the most appropriate combination of theory and practice to achieve the required ends.

In medical applications, one real disadvantage for the projection of a moving spot or line can be the finite length of time needed to move the spot or line around the relevant medical surface.

7.5.3 Moiré topography

Interference patterns which are created by the superimposition of two similar patterns in slightly different positions have been used to great effect in some medical surface studies, particularly in relation to the back. With a suitable choice of light and grating geometry, the fringe pattern appearing on the surface will represent "contours" which indicate surface shape, as in Figure 7.11.

The fringes, known as moiré fringes, can be created in two ways, known as the shadow moiré technique and the projection moiré technique. In both cases, a grating of parallel lines is used.

The moiré method is well suited for use by a medical practitioner because, like photogrammetry, it is convenient and safe. Posing of the patient is straightforward since the fringes are visible and the patient can be rotated relative to the grating until a satisfactory orientation and fringe pattern is achieved. Moreover, it is possible to create a system involving no costly equipment, just one camera, a grating and a light source. Windischbauer and Neugebauer (1995) have reported digital adaptation of the moiré technique for human measurement.

Interpretation of the photographs is relatively simple and the method can be used to obtain quantitative information, but there are some complications. Although the fringes represent contours, there is no indication whether the values of adjacent contours are rising or falling. Also

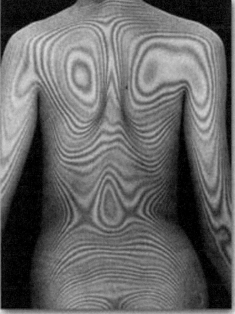

Figure 7.11 *A back shown without and with superimposed moiré interference fringes, courtesy of Gigi Williams, Educational Resource Centre, Royal Children's Hospital, Melbourne, Australia (see Royal Children's Hospital, 2005)*

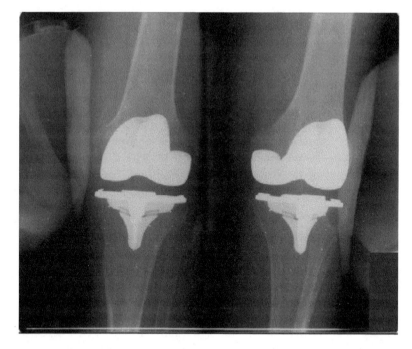

Figure 7.12 *A pair of radiographic images of a total knee prosthesis from two synchronised Roentgen tubes (from Valstar et al., 2002). Reproduced with permission of Elsevier.*

the contour interval is not constant when using a simple point source and this limits the accuracy of the method.

The moiré technique is ideal for back shape measurement, and is usable for mass screening work as well as clinical studies.

7.5.4 Radiography

Unlike photogrammetry, radiographic techniques are used to measure internally. They also require relatively more expensive equipment. Even so, radiographic images can be used in the same manner as surface images to determine three-dimensional coordinates by following photogrammetric theory, a concept that is suggested by the pair of radiographic images shown in Figure 7.12. However, X-rays emanate from a single point and pass through the object and then reach the screen. Although the object lies between the point source (which is equivalent to the centre of the lens) and the film or digital sensor plane, the projection theory for radiographs is similar to that for cameras. Once taken, the radiographs may then be evaluated according to established photogrammetric procedures. The growing use of digital X-ray images and hence of digital image processing assists this evaluation (Ross *et al.*, 1995; Merolli *et al.*, 1995). Further information on X-ray methodologies is given by Stokes (1995).

7.6 Case studies

7.6.1 Case study 1: teeth shape

An interesting and instructive photogrammetric application to measurement involving humans lies in the recording of tooth shape. Being small and generally inaccessible, the teeth can be difficult to measure by any technique at all. They have been the subject of numerous photogrammetric measurements to assist dental studies and treatment. While some tooth measurements have been obtained from photographs taken inside the mouth, the problem of access to the teeth is more usually overcome using casts. Accuracies of 0.01–0.05 mm have been sought and achieved. Chadwick (1992) provides a review and discussion of photogrammetric measurements in dental research applications including assessments of wear on artificial dentures, temporary restorations and restorative materials, and measurements of the palate.

In a dental project reported by Chadwick *et al.* (1991), the measurement was assisted by the use of a microscope, a complication which however serves to indicate the versatility of photogrammetry, and can make the solution informative for the novice photogrammetrist. The goal was to measure the chewing surface of molar teeth with fillings, to accuracies of the order of ±0.01 mm, in order to evaluate the effectiveness of restoration (filling) materials which were under development. Castings were made of the teeth and, in turn, replicas were made of these casts, so that the measurements were being made on objects which were of the same shape as the teeth. Because the measurements were not made on live patients, it can appear that the basic premise that medical photogrammetry involves live humans is not satisfied in this work. However, the use of castings and replicas is an example of the procedures which may be adopted to cope with measurement on live patients, whose discomfort was minimised in this case by carrying out measurements on "copies" of their body parts. Although dental casting and replicating materials are made specially for accurately reproducing teeth, the solution offered by replicating parts can be relevant to other medical measurements. The project retains the fundamental criterion that the measurement tasks relate to live patients.

Technical details relating to the photogrammetric procedures have been reported by Mitchell *et al.* (1989). Pairs of photographs of the replicas were taken through a commercial stereoscopic microscope of the Greenough pattern. This features two converging optical trains which utilise a common objective lens nearest to the object under view. Like many commercial microscopes, this one had an existing facility to mount a camera to photograph objects through the microscope. The body of the 35 mm SLR film camera, i.e. with the lens removed, was attached to the microscope so that a single image was obtained down one of the lens trains. Although a film camera was used in this work, the use of a digital camera is not seen to be fundamentally different.

A pair of photographs could be obtained from different exposure points by rotating the microscope about a vertical axis. The two photographs obtained in this way provided 100% overlap (once one photograph was rotated to align with the other). Being imaged from two different points, photogrammetric intersection could be computed. Replicas were arranged under the microscope so that the chewing surface faced upwards towards the observer and camera. Figure 7.13 shows a pair of images obtained during this project.

The distinctive feature of the technique was the use of the microscope, which could simply be regarded as an extension of the camera. Its relevance lay in the need for the optics of the imaging system to be unravelled. Because both photographs were obtained down the same optical train by rotating the entire microscope body, the optical characteristics such as distortion and principal distance should therefore have remained the same for both images. Even so, the challenge was to uncover the appropriate theory which enabled the measured film coordinates to be converted into (X,Y,Z) coordinates (in a coordinate system which was related to the microscope but which was essentially arbitrary) on the teeth. The Z values thus obtained were more useful than X and Y coordinates, because it was the Z coordinates which could be used to assess wear in the teeth surface. The theory which was developed was specific to convergent photography,

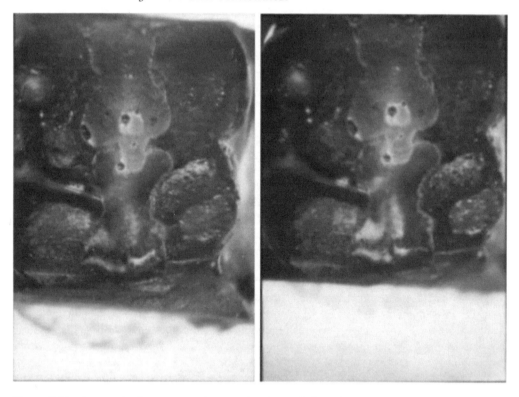

Figure 7.13 *Stereopair of convergent photographs of a tooth through a microscope, from Mitchell et al. (1989).*

but nevertheless its basis was still the depiction of the chief ray path, as explained in Chapter 2. The details of the microscope model are not relevant here but are given by Mitchell *et al.* (1989) and are illustrated in Figure 7.14. It should be noted that a major difficulty with measurement of teeth in dental materials research is to find or establish control. Accordingly, it is crucial that a fixed camera relationship be used, and that this relationship is determined by images of a suitably accurate calibration object.

Photographs covered an object area about 6 mm × 4 mm, and at a scale of about 6:1 (i.e. the film images were actually larger than the original objects). To assist the photographic process, the tooth replicas were sputter-coated with gold palladium to provide them with texture sufficient for imaging (monochrome film was used).

Image coordinates were read off the film by an operator using an optical–mechanical coordinate measuring machine which is usable with film but which is uncommon today in the digital era. The accuracy of this reading was estimated to be 0.01 mm for x coordinates, and 0.001 mm for the differences between x coordinates and differences between y coordinates. The 0.01 mm accuracy of the image measurement represents about 1/3500 of the distance across a 35 mm film. Coordinates were calculated for each of the points whose image coordinates had been measured manually.

To determine the unknown parameters of the optical train, photographs were taken of a metal plate, which was known to be flat and which also carried machined grooves of known thickness. However, certain error sources were ignored. Fiducial marks were not used but frame corners

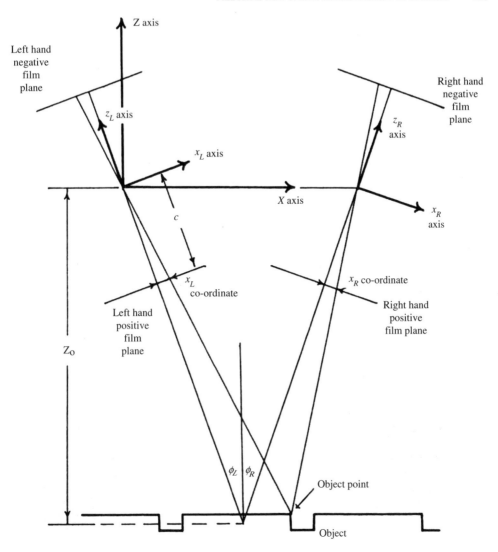

Figure 7.14 *Microscope optics, indicating how the two images have been obtained from different positions, and the principles of ray intersection can be used to recover surface shape data.*

were measured to define an assumed principal point of the photograph. Lens distortion and film flattening errors were ignored because of the low relative accuracy requirements.

To provide accuracy tests, metal feeler gauges of known thickness were measured using the microscope and camera set-up. It was concluded that when points were close together, accuracies for measurements of depth of the chewing surface were of the order of 0.01 mm.

A later dental measurement project reported by Grenness *et al.* (2002, 2005) differed from that described above because a camera with a bellows extension was used, providing a 100 mm focal length, instead of a microscope. Moreover, it used a contemporary commercial photogrammetric system, VirtuoZo (Supresoft, 2005), to achieve automated measurement, rather than manual image measurement which was available for Chadwick *et al.* (1991) and for this reason the work is informative. However, certain fundamental photogrammetric features were common: the use of replicas to overcome patient access problems; the adoption of a technique to provide texture on the replicas (in this case, by adding a suitable substance to the replica material) (see Figure 7.15); the use of calibration techniques to determine optical parameters (in this case, principal distance, principal point, lens distortions and a scaling factor were determined); and the use of a test device to confirm accuracies (see Figure 7.16). Teeth shapes were depicted by contours.

Alternatives to photogrammetry lie in precise but very expensive optical and mechanical dental measuring equipment. Photogrammetry is cheap and simple enough to enable it to be adopted for short projects.

There are features of these similar projects which are novel and interesting. The lessons which arise from them lie not in the techniques or ideas which can be used in other medical cases, but rather the general lessons relating to the benefits of understanding photogrammetry

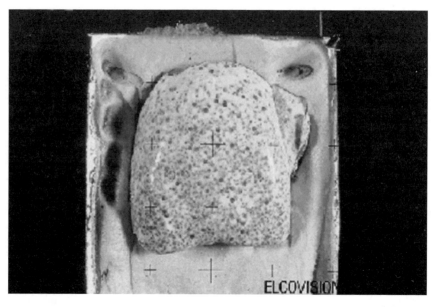

Figure 7.15 *Scanned image of tooth replica (from Grenness et al., 2005). Reproduced with permission of Blackwell Publishing.*

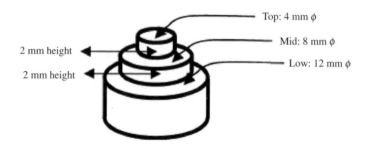

Figure 7.16 *Test object used by Grenness et al. (2005).*

theory and recognising that photogrammetry is surprisingly adaptable and flexible. Overall, the projects had a number of characteristics which may be regarded as typical for medical and sporting photogrammetry:

- The projects draw attention to the unusual and unexpected but beneficial applications of measurement which can be faced in medicine and sport.
- The object sizes are extraordinary; in this case they are very small, and required the use of special photographic techniques, but nevertheless photogrammetry was able to cope and was usable to achieve the required goals.
- The measurements were carried out by health professionals, not photogrammetrists. Those carrying out the measurement were not all photogrammetrically trained so it was necessary during the development stages for a considerable amount of contact between the photogrammetric scientist and the dental investigators.
- The projects faced the problem of interacting with a patient; in this case, the movement and discomfort of the patient was overcome by taking facsimiles of the parts of the body in question.
- The simplicity offered by a photogrammetric solution was emphasised by the ability to use equipment which was widely available (Mitchell *et al.*, 1989, p. 289). In this case, the cameras were of a common type, and the microscope and bellows were widely available commercially. The image coordinate reading machine was not cheap, but the work in the later project showed that the same outcomes could now be achieved by modern digital photogrammetric software, which is comparatively cheap.
- The coordinate accuracies of 0.01 mm are very high in absolute terms; however, in relative terms, the final accuracies were only about 1/500 of the object size, in conformance with the observation made earlier in this section, that relative accuracy requirements of medical measurement are often surprisingly modest.
- The earlier project was developed for the measurement of about 500 objects for a single project. This aspect is again typical for much medical photogrammetry: it is crucial for it to be developed for a given project, but the procedures may not be used again. The cheapness and essential simplicity of photogrammetry make it ideal for relatively short-term applications.
- The use of magnification in the microscope and other aspects of the microscope geometry meant that the basic photogrammetric theory had to be recognised.
- Although some parameters were ignored, calibration was undertaken for selected unknown parameters of the optical arrangement in both cases.
- The projects incorporated the crucial stage of testing.
- The objects had to be treated for effective imaging.

One aspect in which this project was not typical of medical/sports photogrammetry was that there was no need to provide medically informative output. In this case, the users were able to interpret the depth information themselves from basic quantitative output. But overall, the projects showed that photogrammetry was able to achieve the required ends, without great complexity or expense, while the alternatives were generally expensive or otherwise not really available.

7.6.2 Case study 2: face measurement

Medical applications of facial measurement usually relate to assisting the planning of facial reconstruction surgery, which might be necessitated by injury or by abnormality, especially if it affects eating effectiveness or appearance. It is an interesting area of application because there is also a number of applications which involve human subjects but which, unlike medicine and sport, are not related to human health: new uses of face measurement include facial recognition for security purposes, facial feature characterisation for anthropomorphic related studies and others related to animation within the film and entertainment industries.

Facial measurement can sometimes be carried out by laser scanning, but scanning can have the distinct disadvantage of a finite and noticeable scan period, as well as cost and availability disadvantages. Moreover, the provision of fine detail of facial features is often crucial because it is that detail which gives a person their characteristic appearance and realism. Shape therefore is not always sufficient, and an imaging technique which provides spectral texture as well as shape is advantageous. Accordingly, the output can also be a crucial element of the photogrammetric system. D'Apuzzo and Kochi (2003) emphasised feature extraction as an important adjunct to photogrammetric measurement of faces.

There have been many attempts to measure the face by photogrammetric procedures, the studies by Burke and Beard (1967) probably being among the earliest and most famous. However, photogrammetric developments relating to face measurement are still current (e.g. D'Apuzzo, 2002a; Majid *et al.*, 2005). The project reported by D'Apuzzo and Kochi (2003) typifies work on facial measurement: its goals are high accuracy and automation; cameras are synchronised; texture is added to the face; consideration is given to the most useful format for results; and the processing is executed on an ordinary computer. An arc of five monochrome digital cameras is used to obtain extensive coverage of the subject's face; an extra central colour digital camera is used to enable an image of the face to be draped across the surface model. The face is texturised by casting a pattern of random dots as shown in Figure 7.17; two projectors are used in order to provide texture coverage across the entire facial area being imaged producing the effect shown in Figure 7.18.

Figure 7.17 *Random dot pattern as projected by D'Apuzzo (2002a,b) for facial studies. Reproduced with permission of Elsevier.*

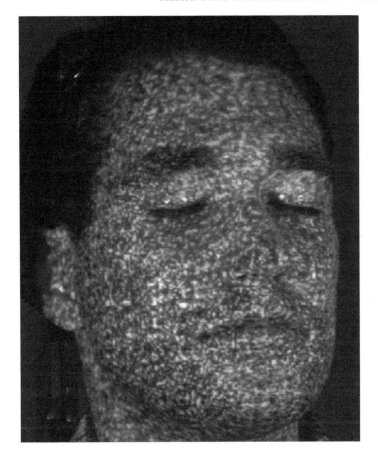

Figure 7.18   *One of five images of the face, showing random pattern projection (from D'Apuzzo, 2002a,b). Reproduced with permission of N. D'Apuzzo.*

The dot pattern is amenable to automated measurement, once a few seed points are selected to within a few pixels, sometimes manually. The software correlates and therefore measures about 2000 points per minute. Data sets, typically of about 45,000 points, are regularised onto an evenly spaced grid; erroneous points are filtered and the data points are thinned to about 10,000 points per face model. Accuracies are claimed to be about 0.2 mm in a direction approximately perpendicular to the surface and 0.1 mm across the surface. Examples of the final form of facial image draped onto the face shape model are shown in Figure 7.19.

The system is calibrated. Locations and orientations of all cameras, principal distance, sensor dimensions (including scaling and shearing) and optical distortion parameters are deduced by using a three-dimensional frame of almost 50 targets.

As with the teeth measurement, this face measurement project exhibits certain characteristics which are typical of medical photogrammetry:

- The project faced the problem of interacting with a patient: in this case, the movement and discomfort of the patient was overcome by synchronising six cameras.
- All equipment was widely available, the software was specialised, this being the nature of the research project being reported.

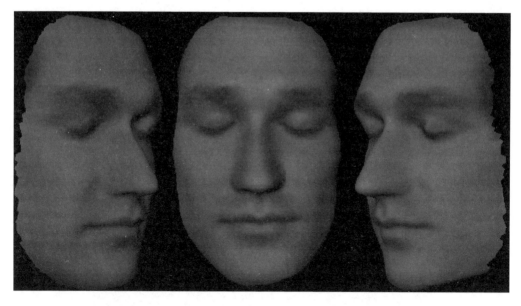

Figure 7.19 *Example of face model with facial image from central camera draped onto the face shape model (from D'Apuzzo, 2002b). Reproduced with permission of N. D'Apuzzo.*

- The accuracies of 0.1 mm are high in absolute terms, although in relative terms, they were only about 1/2000 of the object size.
- Alternatives could not compete in terms of simplicity and cost effectiveness for a relatively small project.
- Calibration was undertaken for selected unknown parameters of the optical arrangement.
- The objects had to be treated for effective imaging.

7.6.3 Case study 3: human motion studies

Human motion analysis by photogrammetry is especially appropriate because the number of applications is so extensive. Primarily, analyses of gait, from whatever source the data derives, have a wide range of medical applications, in studies relating to cerebral palsy (e.g. Gage and Koop, 1995), the design of prostheses, and even diabetes. Motion analysis has uses in sporting applications; Durey and Journeaux (1995) discuss the use of motion analysis for an assortment of sporting purposes: tennis, water-skiing, pole vaulting and table tennis. The uses of motion analysis extend to other health related concerns such as ergonomics (e.g. Lavender and Rajulu, 1995), including the design of acceptable industrial equipment.

There are few alternatives which can match photogrammetry at providing useful data in the case of motion studies. In fact, recording human motion represents the widest and most successful application of medical and sports photogrammetry. There are a number of photogrammetric systems available commercially for dynamic studies of human movement (e.g. Bioengineering Technology and Systems, 2005; Motion Analysis, 2005; Northern Digital Inc., 2005; Qualisys, 2005a; Vicon, 2005). In this case, the simplicity and flexibility of photogrammetry which makes it so suitable for short projects, is not so evident, as the systems are quite expensive. However, the expense simply confirms the extensive use that the systems are expected to undergo.

The requirements of photogrammetric and other optical motion recording systems are described by Pedotti and Ferrigno (1995). Richards (1999) tested seven photogrammetric systems; he also included an electromagnetic system, but it is noteworthy that its accuracy was not commensurate with that of the optical systems. Much of the hardware appears to have developed in the 1980s and 1990s, and many of the technical descriptions of the motion analysis date from that period. An examination of them reveals that motion analysis systems tend to have a number of features in common:

- Primarily, motion studies require imagery at high repetition rates. Accordingly, they use cameras with high rates, up to 1000 Hz (e.g. ProReflex from Qualisys, 2005b).
- They demand full coverage of the moving person, which necessitates a number of cameras, often exceeding ten.
- Measurements are typically based on strategically placed targets. Spherical retro-reflective targets, from 1 mm to 30 mm in size, illuminated by strobe lighting, sometimes infrared, are common.
- The targets are detected by software, which must cope with targets being obscured at various stages of the movement, making their location and identification a difficult task. Positive identification can be achieved from their size and shape and perhaps with assistance from a movement model which anticipates their concealment.
- Software associated with the measurement systems is typically able to analyse the measurements for results relating to the human body, and to provide graphic display of the results, perhaps in the form of body points in motion, and depicting rotations as well as translations of joints.
- The positional results may be used in combination with data from other devices. The opportunity to combine photogrammetric results with output from foot pressure plates and/or from inserted electromyograph (EMG) probes sensing electrical potential is a typical feature of commercial motion measurement systems.
- Accuracies, although hard to typify because they depend on the speed and complexity of the motion being observed, are typically high, are rarely greater than a few millimetres and are characteristically less than a millimetre when conditions make it possible (see Richards, 1999, for some manufacturers' specifications and test results).

Specific applications of human motion analysis are of interest. A commercially available photogrammetric motion analysis system, suited to medical, sporting and other dynamic studies, has been used to examine the action of a bowler in cricket. Lloyd et al. (2000) scrutinised Sri Lankan cricketer, Mutiah Muralitharan, to determine whether his elbow underwent a change of straightness at the time of ball release. This straightening action is not allowed in cricket, the bowler being required to keep the arm holding the ball straight during all stages of delivery until the ball is released.

Recording of the bowler's action required that the measurement be carried out in a hall large enough to allow the bowler to have a run up and slowing distance. A Vicon (Vicon, 2005) motion analysis system, with 50 Hz imaging, was used with six cameras to observe the bowling action of the cricketer. Seven targets were placed on his arm, seven being the "minimum required to carry out a three dimensional kinematic analysis of upper limbs as used in the current model" (Lloyd et al., 2000, p. 976). The arm action was modelled using Vicon's BodyBuilder® software. Although the cricket outcomes are not crucial here, it is useful to note that it was possible to decide that the cricketer's arm was unable to be completely straightened because of a "structural

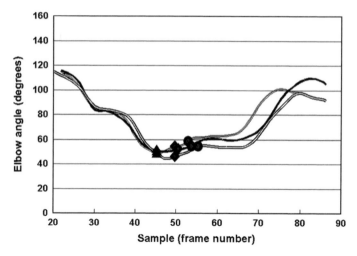

Figure 7.20 *Elbow angles shown for three bowling deliveries: when the arm was vertical, when the arm was horizontal, and when the ball was released. Figure shows an unchanged angle at the three instants of each delivery (marked by symbols). Redrawn from Lloyd et al. (2000). Reproduced with permission of Dr. D.G. Lloyd.*

abnormality at the elbow joint" (ibid, 979), and that his arm retained its lack of straightness during the bowling phase and was no less straight at one crucial time compared to another. What is of interest in the light of discussions in this chapter about the need to provide medically relevant output, is the provision in this case of data defining the elbow angle at different times during the bowler's delivery (Figure 7.20).

7.6.4 Case study 4: live sport studies

Another group of movement studies by photogrammetrists relates to people participating in live sport, notably athletics, golf, tennis and football (Chikatsu *et al.*, 1992; Chikatsu and Murai, 1994). These uses do not seem to have reached commercial levels, but the various applications have notable similarities:

- Although imagery is often collected by professional cameras, it is rarely taken especially for photogrammetry, and users may have to make do with inappropriately positioned cameras, low quality and/or distant coverage of the objects of interest. Moreover, coverage may be provided by just a single image, and the unique analytical procedures which are applicable to monoscopic image geometry may be necessary.
- Because the camera may be inaccessible for various reasons, there may be problems in camera calibration and moreover, changes of internal parameters can occur due to zooming.
- The area of coverage may be equal to the size of a person or larger in the case of moving people and/or team sports.

An interesting study of basketball by Remondino (2003) had the following characteristics. Imagery was only available from old VHS analogue videotapes of sport, the images having been obtained with rotating cameras fixed in position, covering some metres of playing area. Variations in camera principal distance were expected. The images were analysed monoscopically, making use of the height of the basket above floor level and distances between lines marked on the court to provide control (Figure 7.21).

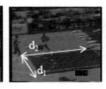

Figure 7.21   *Every second frame of the video sequence analysed by Remondino. The central image shows two distances which were measured on the court. From Remondino (2003). Reproduced with permission of F. Remondino.*

Six distance estimates were derived from the imagery. These were all in the range 1.5–5 m in length, and included:

- the height of the player number 1 at the beginning of the jump;
- the height of the player at the end of the jump;
- the length of the jump;
- the height of the jump given by the ball;
- the height of the jump as indicated at the player's waist; and,
- the height of player number 2.

Accuracies for these distances were estimated from three measurements to be between 2.8 cm and 3.7 cm. In fact, the given and correct heights of player number 1 were 1.71 m and 1.98 m. For player number 2 they were more accurate, being 1.97 m and 2.01 m.

7.7 Conclusions

Photogrammetry finds wide application in various studies related to humans, for medical, sports and other health related purposes, and sometimes for purposes which are not related to health, such as apparel design and fitting. The breadth of the applications can be wider than it may seem to some readers of scientific journals, because the relevant reports of the photogrammetric projects are spread across surveying, photogrammetric, medical, medical physics, medical engineering, sports and other journals. The demands of medical and sport photogrammetry may initially seem to be much the same as they are for other photogrammetric measurement at close range, but in reality, they are distinguished by the involvement of human subjects.

Many of the photogrammetric projects are short term, but this only confirms the flexibility and adaptability of photogrammetry. Those photogrammetric systems used for motion analysis tend to be used on a longer-term basis, but the flexibility of the coverage by cameras is again a strong point in this application. This chapter suggests that, while there are a number of specialised and expensive systems available for various medical measurement purposes, they can be restrictive in their uses. There are other applications for which the specialised systems do not exist. Accordingly, there is a place for photogrammetry's flexibility, especially for short-term projects.

7.8 References

Bioengineering Technology and Systems, 2005. See http://www.bts.it/ (accessed 16 December 2005).

Burke, P.H. and Beard, L.F.H., 1967. Stereophotogrammetry of the face. *American Journal of Orthodontics*, 53(7), 769–782.

Chadwick, R.G., 1992. Close range photogrammetry – a clinical dental research tool. *Journal of Dentistry*, 20, 235–239.

Chadwick, R.G. and Mitchell, H.L., 1999. Presentation of quantitative tooth wear data to clinicians. *Quintessence International*, 30, 393–398.

Chadwick, R.G., McCabe, J.F., Walls, A.W.G., Mitchell, H.L. and Storer, R., 1991. Comparison of a novel photogrammetric technique and modified USPHS criteria to monitor the wear of restorations. *Journal of Dentistry*, 19, 39–45.

Chikatsu, H. and Murai, S., 1994. Application of image analysis to rowing dynamics using video camera. *International Archives of Photogrammetry and Remote Sensing*, 30(5), 35–40.

Chikatsu, H., Turuoka, M. and Murai, S., 1992. Sports dynamics of Carl Lewis through 100m race using video imagery. *International Archives of Photogrammetry and Remote Sensing*, 29(B5), 875–879.

D'Apuzzo, N., 2000. Motion capture by least squares matching tracking algorithm. *AVATARS'2000 Workshop, November 30–December 1, 2000, Lausanne, Switzerland. IFIP Conference Proceedings 196*, Kluwer, Dordrecht, The Netherlands, 2001.

D'Apuzzo, N., 2002a. Surface measurement and tracking of human body parts from multi-image video sequences. *ISPRS Journal of Photogrammetry and Remote Sensing*, 56(5-6), 360–375.

D'Apuzzo, N., 2002b. Measurement and modeling of human faces from multi images. *International Archives of Photogrammetry and Remote Sensing*, 34(5), 241–246.

D'Apuzzo, N. and Kochi, N., (2003). Three-dimensional human face feature extraction from multi images. In Grün, A. and Kahmen, H. (Eds.), *Optical 3D Measurement Techniques VI*, Wichmann Verlag, Heidelberg, Germany, Vol. 1, 140–147.

DIERS, 2006. See http://www.diers.de/ (accessed 6 February 2006).

Durey, A and Journeaux, R., 1995. Application of three-dimensional analysis to sports. In Allard, P., Stokes, I.A.F. and Blanchi, J.-P. (Eds.), *Three-Dimensional Analysis of Human Movement*, Human Kinetics, Champaign, Illinois, USA, 327–347.

Gage, J.R. and Koop, S.E. 1995. Clinical gait analysis: Application to management of cerebral palsy. In Allard, P., Stokes, I.A.F. and Blanchi, J.-P. (Eds.), *Three-Dimensional Analysis of Human Movement*, Human Kinetics, Champaign, Illinois, USA, 349–362.

Grenness, M.J., Osborn, J. and Tyas, M.J., 2005. Stereo-photogrammetric mapping of tooth replicas incorporating texture. *Photogrammetric Record*, 20(110), 147–161.

Grenness, M.J., Osborn, J. and Weller, W.L., 2002. Mapping ear canal movement using area-based surface matching. *Journal of the Acoustical Society of America*, 111(2), 960–971.

Institute of Geodesy and Photogrammetry, ETH Zurich, 2005. See http://www.photogrammetry.ethz.ch/research/ (accessed 6 December 2005).

Lavender, S.A. and Rajulu, S.L., 1995. Application in ergonomics. In Allard, P., Stokes, I.A.F. and Blanchi, J.-P. (Eds.), *Three-Dimensional Analysis of Human Movement*, Human Kinetics, Champaign, Illinois, USA, 311–326.

Lloyd, D.G., Alderson, J. and Elliott, B.C., 2000. An upper limb kinematic model for the examination of cricket bowling: A case study of Mutiah Muralitharan. *Journal of Sports Sciences*, 18, 975–982.

Majid, Z., Chong, A.K., Ahmad, A., Setan, H. and Samsudin, A.R., 2005. Photogrammetry and 3D laser scanning as spatial data capture techniques for a national craniofacial database. *Photogrammetric Record*, 20(109), 48–68.

Malian, A., Azizi, A., van den Heuvel, F.A. and Zolfaghari, M., 2005. Development of a robust photogrammetric metrology system for monitoring the healing of bedsores. *Photogrammetric Record*, 20(111), 241–273.

Merolli, A., Tranquilli Leali, P. and Aulisa, L., 1995. Considerations on the clinical relevance of photogrammetric error in 3-D reconstruction of the spine based on routine X-ray films. In d'Amico, M., Merolli, A. and Santambrogio, G.C. (Eds.), *Three-Dimensional Analysis of Spinal Deformities*, IOS Press, Amsterdam, The Netherlands, 185–190.

Mitchell, H.L., 1994. A comprehensive system for automated body surface measurement. *International Archives of Photogrammetry and Remote Sensing*, 30(5), 265–272.

Mitchell, H.L., Chadwick, R.G. and McCabe, J.F., 1989. Stereomicroscope photogrammetry for the measurement of small objects. *Photogrammetric Record*, 13(74), 289–299.

Motion Analysis, 2005. See http://www.motionanalysis.com (accessed 17 November 2005).

Northern Digital Inc., 2005. See http://www.ndigital.com (accessed 17 November 2005).

Pedotti, A. and Ferrigno, G. 1995. Optoelectronic-based systems. In Allard, P., Stokes, I.A.F. and Blanchi, J.-P. (Eds.), *Three-Dimensional Analysis of Human Movement*, Human Kinetics, Champaign, Illinois, USA, 57–77.

Qualisys, 2005a. See http://www.qualisys.com/images/Sport_AN.pdf (accessed 17 December 2005).

Qualisys, 2005b. See http://www.qualisys.com/proreflex.html (accessed 17 December 2005).

Remondino, F., 2003. Recovering metric information from old monocular video sequences. In Grün, A. and Kahmen, H. (Eds.), *Optical 3D Measurment Techniques VI*, Wichmann, Verlag, Heidelberg, Germany, Vol. 2, 214–222.

Richards, J.G., 1999. The measurement of human motion: A comparison of commercially available systems. *Human Movement Science*, 18, 589–602.

Ross, Y., de Guise, J.A., Polvin, J., Sabourin, R., Dansereau, A. and Labelle, H., 1995. Digital radiography segmentation of scoliotic vertebral pedicles using morphological segmentation. In d'Amico, M., Merolli, A. and Santambrogio, G.C. (Eds.), *Three-Dimensional Analysis of Spinal Deformities*, IOS Press, Amsterdam, The Netherlands, 63–68.

Royal Children's Hospital, 2005. See http://www.rch.org.au/erc/ (accessed 20 January 2006).

Sechidis, L., Tsioukas, V. and Patias, P., 2000. An automatic process for the extraction of the 3D model of a human back for scoliosis treatment. *International Archives of Photogrammetry and Remote Sensing*, 33(B5), 113–118.

Siebert, J.P. and Marshall, S.J., 2000. Human body 3D imaging by speckle texture projection photogrammetry. *Sensor Review*, 20(3), 218–226.

Smith, M.J., Mather, J.S.B., Gibson, K.A.H. and Jowett, S., 1998. Measuring dynamic response of a golf club during swing and impact. *Photogrammetric Record,* 16(92), 249–257.

Stokes, I.A.F., 1995. X-Ray photogrammetry. In Allard, P., Stokes, I.A.F. and Blanchi, J.-P. (Eds.), *Three-Dimensional Analysis of Human Movement*, Human Kinetics, Champaign, Illinois, USA, 125–141.

Supresoft, 2005. See http://www.supresoft.com.cn/english/products/virtuozo/virtuzo.htm (accessed September 2005).

Valstar, E.R., Nelissen, R.G.H.H., Reiber, J.H.C. and Rozing, P.M., 2002. The use of Roentgen stereophotogrammetry to study micromotion of orthopaedic implants. *ISPRS Journal of Photogrammetry and Remote Sensing*, (56)5–6, 376–389.

Vicon, 2005. See http://www.vicon.com/ (accessed 17 December 2005).

Waldhäusl, P., Forkert, G., Rasse, M. and Balogh, B., 1990. Photogrammetric surveys of human faces for medical purposes. *International Archives of Photogrammetry and Remote Sensing*, 28(5/1), 704–710.

Windischbauer, G. and Neugebauer, H. 1995. Digital 3D Moiré-Topography. In d'Amico, M., Merolli, A. and Santambrogio, G.C. (Eds.), *Three-Dimensional Analysis of Spinal Deformities*, IOS Press, Amsterdam, The Netherlands, 409–413.

Applications of 3D Measurement from Images

Key to Colour Section

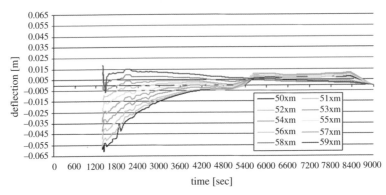

Figure 3.25 *Point displacement versus time in X for the ten targets on beam 5.*

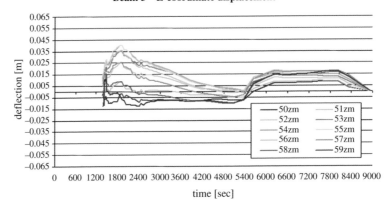

Figure 3.26 *Vertical displacement versus time for the ten targets on beam 5.*

(a)

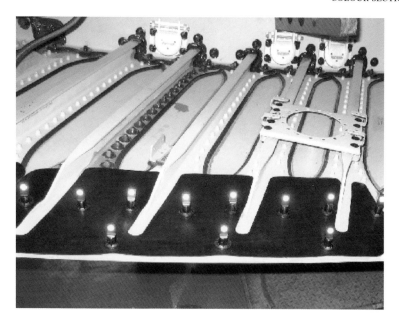

(b)

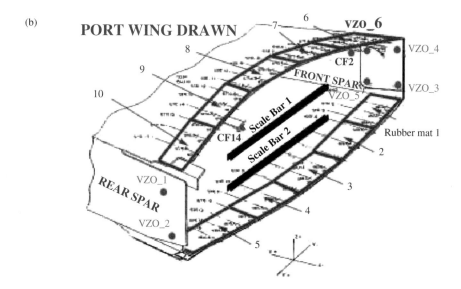

Figure 4.8   *(a) A set of targets located in position using the profiled rubber mat; (b) schematic diagram showing the layout of the various target types and scale bars referenced to the wing root. Reproduced with permission of Airbus UK.*

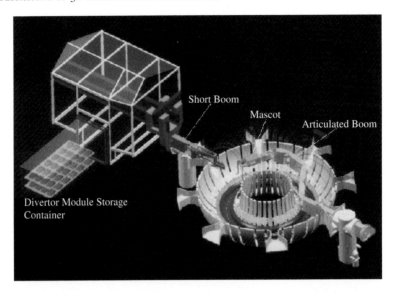

Figure 4.10   *Schematic of the JET Remote Handling system showing the vacuum vessel and manipulator arms. Reproduced with permission of EFD JET.*

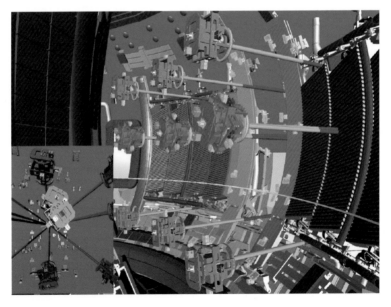

Figure 4.12   *CAD model showing design camera positions and orientations as provided by the virtual robot. Reproduced with permission of EFDA JET.*

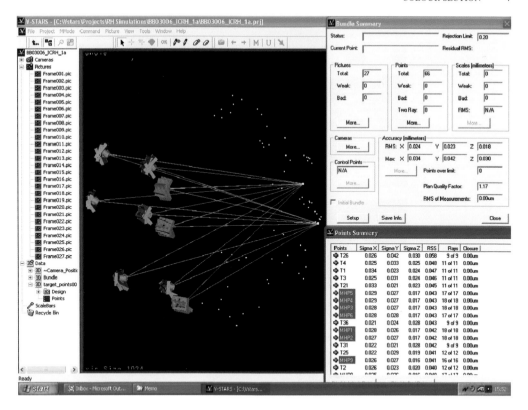

Figure 4.13   *GSI V-Stars network adjustment simulation used to fine tune the camera stations. Reproduced with permission of EFDA JET.*

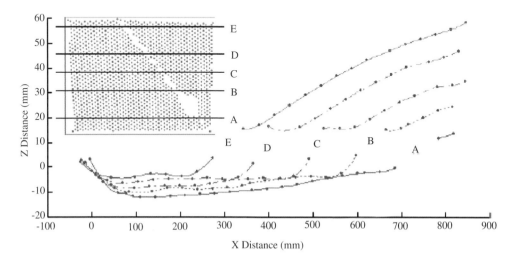

Figure 4.18   *Cross-sectional plots from the corner of a 2 m sail structure. Reproduced with permission of NASA LaRC.*

(a)

(b)

Figure 4.20 *(a) Canon 1Ds digital SLR camera in housing; (b) coded targets located on the supporting table structure. Reproduced with permission of G.W. Johnson (MidCoast Metrology) and S. Laskey (Bath Iron Works).*

(a)

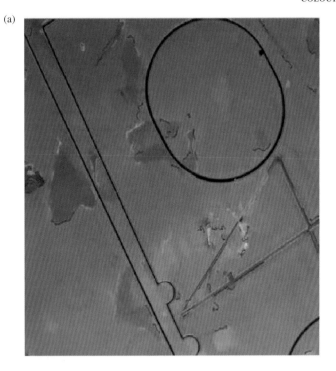

(b)

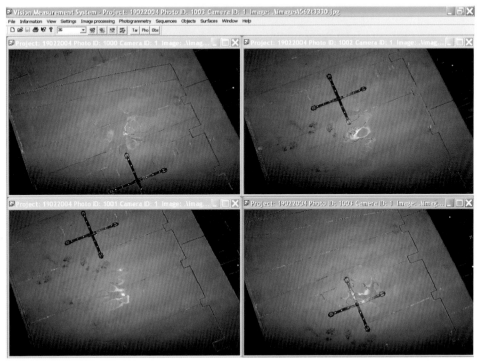

Figure 4.21 *(a) Edge extraction, unfiltered for a typical plate; (b) an image sequence, including a scale artefact used as part of a validation process. Reproduced with permission of G.W. Johnson (MidCoast Metrology) and S. Laskey (Bath Iron Works).*

(a)

(b)

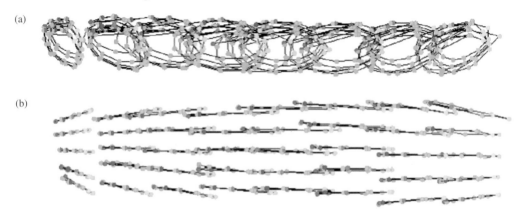

Figure 4.26   *Example of the visualisation of the target movement of the Fresnel lens; (a) out of plane target movement; (b) in plane target movement.*

(a)

(b)

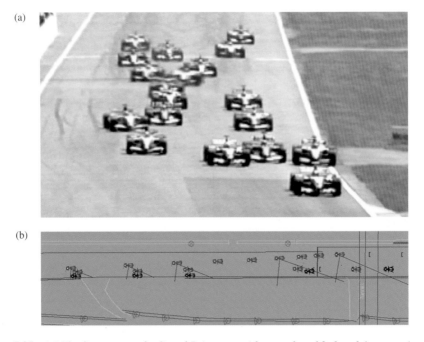

Figure 5.25   *(a) The first corner of a Grand Prix event with cars placed 2, 3 and 4 converging at the first corner, leading to a crash. (b) A CAD reconstruction of the race showing 9 positions of the outside car, the relative positions of the cars involved in the crash and the driver's blind spot. Reproduced with permission of Dr. T. Clarke, Optical Metrology Centre, UK.*

(a)

(b)

Figure 5.26 *(a) Photography of a gunshot wound, noting the metal radiological landmarks; (b) a representation of the wound in 3D via photogrammetry. After Thali et al. (2003b). Reproduced with permission from ASTM International.*

Figure 6.14   *Vertical aerial photograph, Black Ven- 2001. Copyright NERC.*

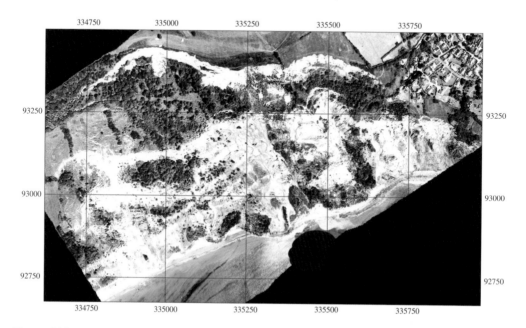

Figure 6.15   *Orthophoto, Black Ven 2001.*

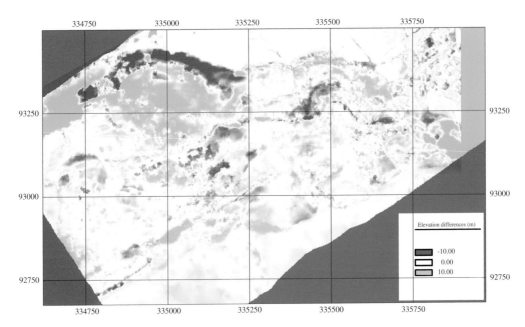

Figure 6.16 *DEM of difference – Black Ven, 2001–1995.*

Figure 6.18 *A perspective view of Mam Tor, created by draping an orthophoto over a DEM, obtained from 1999 images.*

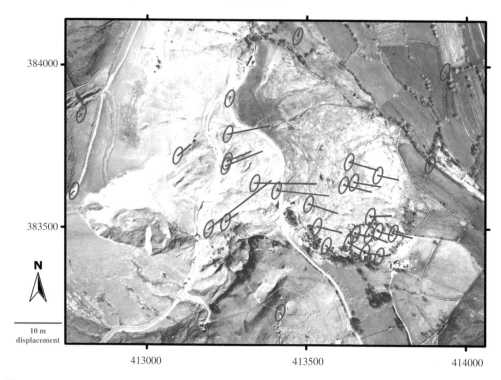

Figure 6.20 *Displacement vectors, representing horizontal displacements between 1973 and 1999. The background orthophoto was created from the 1990 images. The scale of the vectors is 15 times the scale of the background. The displayed error matrices are based on a 95% confidence level.*

(a)

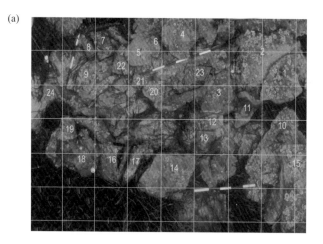

Figure 9.7 *(a) Photomosaic of the entire site of 'Meloria B'; (b) Arpenteur windows for photogrammetric processing; (c) a VRML view of wire-frame model of the site; (d) 3D model textured by photogrammetry mixed with reconstructed wire-frame model.*

(b)

(c)

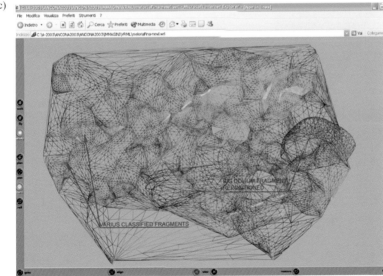

Figure 9.7   *(continued)*

Figure 9.13   *(a) The produced DSM with a contour map of 1 cm contour interval; (b) the produced orthophotomap with a pixel size of 1 cm; (c) the 3D model with draped image; (d) the final 3D GIS output with the archaeological interpretation and the 3D positions of the finds.*

(a)   (b)

(c)   (d)

Figure 9.15   *(a)–(c) Three different types of metric and non-metric images used for the photogrammetric reconstruction; (d) the recovered camera poses and measured points.*

Figure 9.16 *(a) The automated reconstructed 3D surface; (b) the textured 3D model; (c) the anaglyph result; (d) the manually reconstructed folds of the robe; (e) the shaded model; (f) the final textured 3D model of the Great Buddha.*

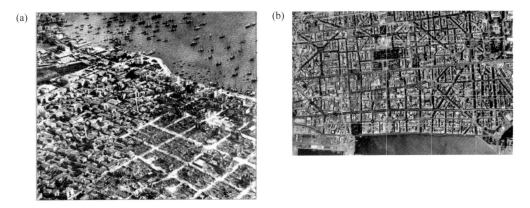

Figure 9.18 *The fire of 5 August 1917 and the resultant damage at the historic centre of Thessaloniki. (a) A close view of the fire damage at the historic centre; (b) a view of the current status. The green line denotes the official borders of the damaged area.*

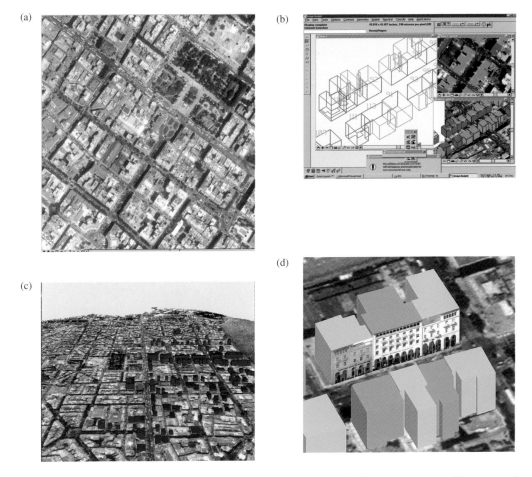

Figure 9.19 *(a) The photogrammetric production of the DTM; (b) the 3D reconstruction of the protected buildings; (c) and (d) the visualization output of the mapping of the city centre.*

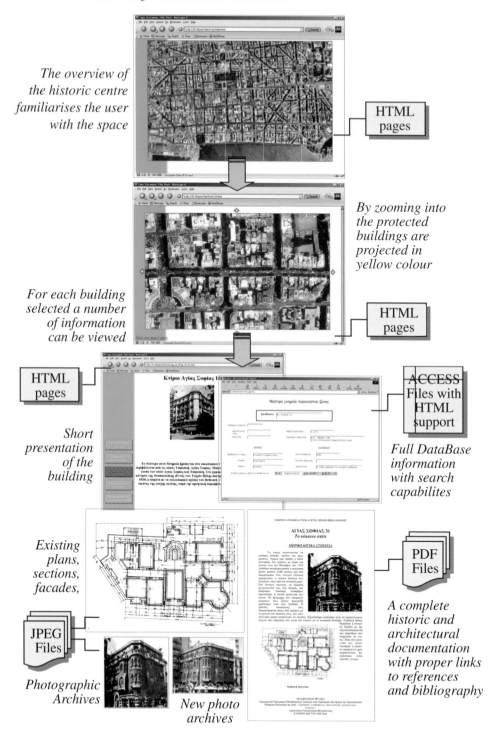

Figure 9.20   *Overview of the web-based information system.*

(a)

(d)

(b)

(e)

(c)

(f)

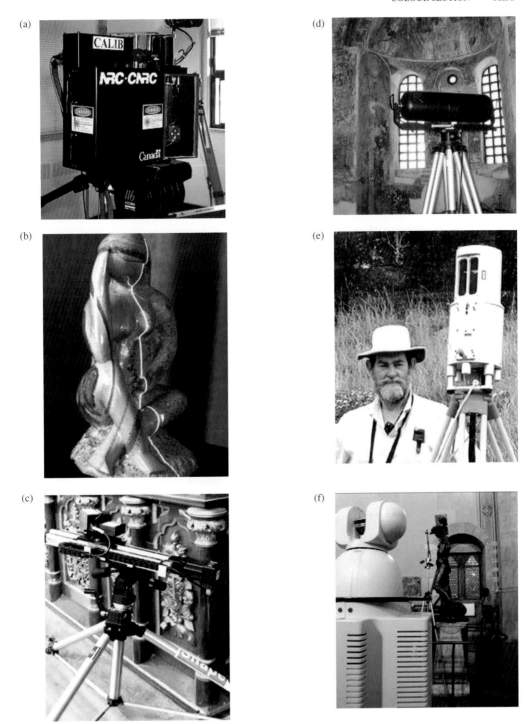

Figure 10.7 *Examples of scanners. (a) Dual axis triangulation-based scanner; (b) projected line; (c) short-base triangulation line scanner on a translation stage; (d) long-base triangulation line scanner rotated by motor; (e) TOF scanner; (f) FMCW scanner.*

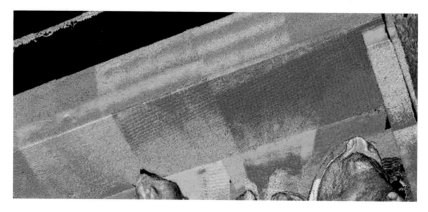

Figure 10.13 *Wave phenomenon created by the motion of the 3D camera.*

(a)   (b)   (c)

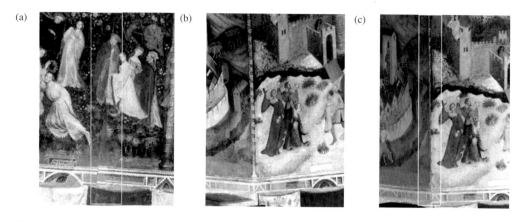

Figure 10.19 *Geometric texture errors: (a) discontinuity error; (b) source image; (c) 3D model using texture from the source image with errors (disappearing textures) at surface intersection.*

(a)

(b)

Figure 10.20   *Two overlapping images: (a) original; (b) after global colour adaptation.*

Figure 10.23   *Detailed model from laser scanner in various representations.*

Figure 10.26   *The final model from all data in wire-frame (left) and with texture.*

Figure 10.29   *Detailed model of the loggia.*

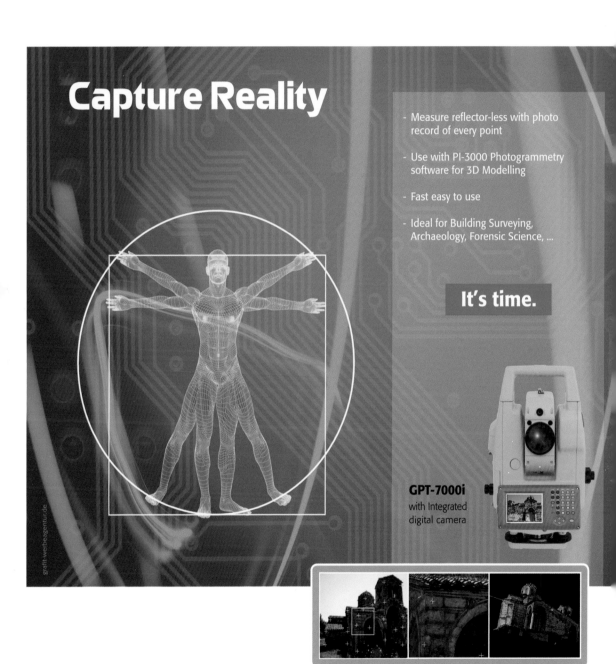

iWitness™
Close Range Photogrammetry

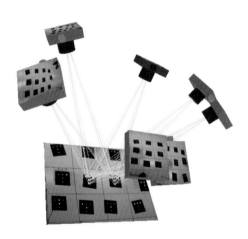

An affordable and exceptionally easy-to-use close-range photogrammetric system that provides fast and accurate 3D measurements from images recorded with both consumer-grade and professional cameras.

- Very intuitive and simple to operate
- Automatic recognition of the camera
- Fully automatic camera calibration, as well as on-the-job self-calibration
- Fully automatic measurement of RGB images via colour coded targets
- Advanced facet-and free-from polyline curve generation
- All photogrammetric orientation processes are initiated automatically
- Robust and reliable error detection through on-line data processing
- Flexible generation of photo-textured 3D models
- Auto-assisted image marking and referencing leading to high-accuracy XYZ coordinates
- Efficient and fast 'stitching' of overlapping networks

iWitness is widely used in engineering, architecture, heritage recording, animation and modeling, and especially in accident reconstruction and forensics.

With its ease of use, **iWitness** is an ideal tool for teaching.

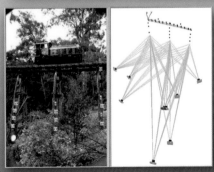

Contact

Photometrix

Photometrix

In North America

DCS

DeChant Consulting Services DCS Inc.

www.Photometrix.com.au

www.iWitnessphoto.com

homometrica
consulting

human body measurements consulting

Homometrica Consulting provides expert assistance
on issues regarding human body measurement,
surface digitization and 3D scanning.

HOMOMETRICA CONSULTING - Dr. Nicola D'Apuzzo

www.homometrica.ch   ·   info@homometrica.ch

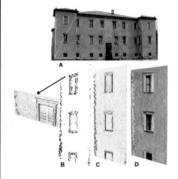

8 Biology and zoology

Albert K. Chong

8.1 General outline of the application

8.1.1 Introduction

Photographs have been used to obtain linear measurement of biological and zoological subjects since the invention of photographic emulsion and the pin-hole camera. However, full-scale research on the use of stereo-photographs and the stereoscopic plotter began around the 1960s (Boyce, 1964). Recently, applications of photogrammetry in biology and zoology have increased notably following the introduction of digital imaging and digital stereo-workstations. Digital stereo-workstations provide flexibility in the use of different imaging sensors. For example, normal and near-infrared (NIR) photography taken with different focal length (principal distance) cameras, and displaying different textures and colours may be viewed together stereoscopically. Improved sensor technology allows flexibility of lighting requirements. Consequently, it is straightforward to capture high-precision NIR images of nocturnal animals in total darkness. Improved photogrammetric software also allows the modelling of refraction of two-media and underwater applications.

8.1.2 General advantages of photogrammetry

Photogrammetry has a number of advantages which make the technique highly suitable for applications in biology and zoology. The crucial advantage is that accurate three-dimensional measurement can be obtained without any physical contact with the subject. Non-contact reduces stress in both animals and humans and eliminates the danger of exposing workers to animals such as tigers, sharks or poisonous insects. This unique feature is utilised in remotely operated vehicles and submersibles which establishes new breadth in the use of photogrammetry in this area. The capability for non-invasive measurement is the second essential feature for biology and zoology applications, so that animals can be studied in their natural states.

Photogrammetry provides a rapid data capture capability with a high sampling rate. This characteristic is a prerequisite for various studies in the fields of biology and zoology. For example, the study of fast moving animals, fishes and birds requires a high recording rate. In addition, the images provide a permanent record which may be accessed later if more measurements are needed. For example, a recent dolphin study showed that the initial measurements obtained from cetacean stereo-photographs did not demonstrate a high correlation between age and length. Subsequently, additional measurements were obtained to achieve a satisfactory result.

8.1.3 General disadvantages of photogrammetry

Measurements obtained from single images are restricted to a two-dimensional plane. As a result, images of two or more perspective views are needed to provide accurate 3D measurements. Moreover, biological and zoological subjects are seldom stationary. In fact, they are often studied during their active time (e.g. insects, fishes, birds and dolphins). Thus, two or more cameras or sensors are needed: they must be synchronised to take images simultaneously, and they must have high shutter speeds.

While it is possible to reduce or to eliminate most systematic errors from images in a digital environment, existing mathematical models must be improved to handle new environmental challenges such as those posed by studies conducted in the depths of the ocean, under adverse temperature and pressure conditions. Accurate modelling of the effect of refraction in the great depths of the ocean and water–air environment often requires *in situ* calibration of imaging systems. Sophisticated laboratory sensor calibration techniques must give way to simple designs to adapt to the field conditions. Consequently, accurate modelling is seldom achievable and the effect is poor quality measurement. In general, photogrammetry requires object–space control to determine the exterior orientation of the imaging system and to provide an accurate scale of the imaged object. In many situations, exterior orientations can be computed without object–space control; however the object-to-sensor distance may still be needed to evaluate the quality of pre-computed exterior parameters.

8.2 Current challenges in close range photogrammetry

Close range photogrammetry is a common tool for the study of land-based animals as there are few constraints for quality imaging. Very large objects (e.g. elephants) or very small (insects) can present some challenges. These challenges include:

- cameras with focal lengths of an appropriate size may be difficult to obtain;
- very short object-to-sensor distances;
- poor geometry of the imagery; and,
- very small object–space control targets.

There are also some reported minor problems involving refraction, and extremes in temperature and pressure. Recent changes in conservation attitudes toward aquatic creatures have resulted in a dramatic shift in the research on aquatic environments. The aquatic environment presents a formidable challenge to researchers who want to apply imaging technology to study living organisms.

Threatened species demand extreme care during study. They must not be disturbed or handled, i.e. they must be free to move in the aquatic environment. For example, there are only about 100–150 New Zealand North Island Hector's dolphins in the wild and there is no captivity breeding program. Study of this cetacean in the wild demands considerable effort and can only be done economically using a remote system involving imaging. Another species of cetacean is the Doubtful Sound Bottlenose dolphin, which lives only in the fiord lands of New Zealand, and it requires similar attention. A unique underwater species of temperate-water red hydrocorals which require similar care while being studied, lives in these fiords. The number of colonies outside the national parks is declining due to seabed trawling and other human activities.

Permits must be obtained before any study can proceed in a national park. There are strict guidelines for photography and this restriction is one challenge that photogrammetrists must overcome. This restriction compromises the accuracy of measurement. Consequently, innovative techniques must be used where extreme care is required and strict guidelines are present.

8.3 Typical project requirements

8.3.1 Measurement accuracies

The accuracy of photogrammetric measurement always depends on the imaging sensor, the geometry of the imagery, the software and the environment. The accuracy of conventional aerial photogrammetry can be estimated accurately because aerial cameras are custom-built for high accuracy. On the other hand, off-the-shelf non-metric film and digital cameras are susceptible to various image distortions which include lens distortion, lens alignment, flatness of film or charge coupled device (CCD), large offsets of the principal point of auto-collimation and variation of the principal distance of the focusable lens. In addition to these common distortions, the portability and remote operation of these cameras could expose them to other adverse factors in the environment such as high pressure in the depths of the ocean, extreme heat and extreme cold.

In stereo-photogrammetry, the stereo-vision of the photogrammetrist becomes an important factor in the measurement, particularly when observing underwater images. Broadly, the highest accuracy practicable is the photogrammetrist's aim, under the constraints of the equipment and the environment. Usually, the measurement accuracy requirement is dependent on the size of animal studied and is typically about 1% of the body size, a comparatively relaxed accuracy specification. In special circumstances such as the study of limb deformity, 0.5% may be needed. So, a large cetacean such as a large cow Gray whale which matures at about 12.4 m length requires a measurement accuracy of approximately 15 cm for growth monitoring (Perryman and Lynn, 2002). Research shows that age determination of the New Zealand Doubtful Sound Bottlenose dolphin requires an accuracy of 5 cm and 2.8 cm for anterior flipper and fluke width measurements, (Chong and Schneider, 2001). In underwater research such as the red hydrocoral study, an accuracy of 0.7 mm was the highest achievable value at 95% confidence level, because of the uncertainty in determining the refractive index of salt water.

8.3.2 Appropriate/available software

Normally, *in situ* self-calibration and on-the-job calibration are needed for calibrating cameras used in biology and zoology. The latter is absolutely essential for underwater and air–water imaging because the temperature and depth of water have an undesirable effect on the principal distance of the imaging cameras. In recent years, camera calibration software, such as Australis (Photometrix, 2005), has become available for both self-calibration and on-the-job calibration. The software provides two essential features: first, high precision digitising of signalised targets; and secondly, bundle adjustment of multiple convergent images. In addition, some software (including Australis) has a relative orientation procedure which can produce a set of coordinates for resection. The technique eliminates the need to input initial estimates for the exposure station or the use of an exterior orientation device for the bundle adjustment.

Stereo-digitising is also a common requirement for applications in biology and zoology. Although it is feasible to project a pattern onto the studied land animal to aid mono-digitising, stereo-digitising enhances the estimation of depth, particularly with a moving animal. Projecting a pattern onto underwater creatures is seldom carried out and, when undertaken, reported results have not been satisfactory. In other words, stereo-digitising is vital for all underwater applications in this field. Most stereo-photogrammetric workstations are suitable for stereo-digitising of close-range images. Commonly used close range focusable cameras require lens modelling (e.g. lens distortion determination and CCD flatness parameters), unlike high quality aerial camera lenses. Consequently, stereo-workstations dedicated to close range applications must have the capability to accept a number of camera lens parameters.

Image processing software has an important role in close range applications where NIR or ultraviolet images are needed. Images must be processed to obtain a composite image for stereo-digitising or bundle adjustment. For example, to obtain a composite image of NIR at 730–780 µm wavelength band it is necessary to subtract the image taken using 780 µm filter from the image taken with 730 µm filter. In addition, the composite image may require edge-enhancement, stretching and brightness adjustment.

8.3.3 Cameras and imaging sensors

In recent years, the application of close range photogrammetry in the biological and zoological fields has become more common because of the availability of low cost off-the-shelf non-metric film and digital cameras and off-the-shelf photogrammetric software. Normally, the application involves remote camera control where the animals must not be disturbed. The remote control may be linked to the camera by a cable or the camera can be activated by an infrared (or laser) sensor. High camera shutter speed is another requirement, however most modern semi-professional SLR digital cameras can be programmed to capture images at speeds up to 1/2000 of a second. For night-time imaging, a night-shot camera and NIR light source are essential. Some applications may require NIR filters which block certain wavelengths from entering the camera lens. Applications in the aquatic environment involve underwater cameras. High shutter speed imaging in water requires a high intensity strobe-light.

8.3.4 Object preparation: targeting and texturising

Photogrammetry, in general, requires a set of control points in the object space to establish the exterior orientations of the image system, scale and datum for the computed measurements and coordinates based on that object–space reference frame. Various types of targets are generally used. A white cross or white circular mark placed against black or grey background on a metal rectangular frame are considered the most economical form of control frame construction. However, retro-reflective targets are considered essential for underwater projects (Figure 8.1). The size of the control target should be comparable with the size of the floating mark of the

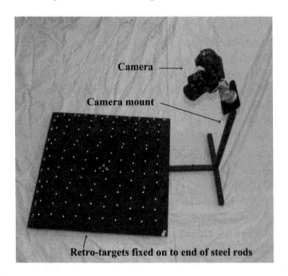

Figure 8.1 *Underwater camera calibration device. The control board may be rotated to obtain four convergent photographs.*

digital stereo-workstation and the type of digitising technique. For example, automated image digitising software such as Australis produces good results when the target covers 16 pixels and manual stereo-digitising software such as DVP requires nine pixels to produce good results. When stereo-orientation is needed, the control targets must be spread around the entire image. Normally, underwater projects require more targets for on-the-job calibration of the sensor, as discussed elsewhere in this chapter.

8.3.5 Products and output

In biology and zoology, the majority of photogrammetric measurements are used for growth monitoring, nutritive condition studies, age prediction and grazing or erosion (e.g. coral). Common photogrammetric outputs are lengths, girth (cross-section) and vectors (angle and distance). Length and girth measurement satisfy most studies which involve growth monitoring, nutritive condition, grazing or age estimation. Vectors are often converted into angular units to determine growth patterns and, occasionally, used to study deformity of growth. Shape is used to identify individual animals. For example, the shape of a fluke or the shape of a body scar may be used to identify individual dolphins from a pod. The shape of flukes varies very slightly between individuals and, consequently, accurate measurements are needed to distinguish between them. Graphic plots are often retained for future identification.

8.4 Case studies

8.4.1 Case study 1: visible light sensor – New Zealand red hydrocoral study

Introduction
New Zealand red hydrozoan corals (hydrocorals) are typically deep-water, temperate species and are not generally found on mainland coastlines. The species can be found in close association with scleractinian corals in tropical environs, but it is more typically found on offshore seamounts and submarine ridges. The red hydrocoral is an endemic species of hydrocoral, *E. novaezelandiae*, which can be found in the fiords of Fiordland, South Western New Zealand.

Much of the current knowledge of hydrocoral biology has been deduced from small, one-of-a-kind samples collected from deep water using dredges. Increased tourism (Grange, 1990; Miller, 1995) and the possibility of souvenir collectors in the Fiordland region heighten the need for research on the biology of this endemic species of hydrocoral. Accurate information on the reproduction and growth of *E. novaezelandiae* will provide valuable insights into the recruitment and population dynamics of the Fiordland population. The knowledge gained will also provide insights into the biology of many offshore species, which are often adversely affected by commercial fishing practices (Probert *et al.*, 1997). Deep-sea trawls frequently bring up deep-water coral species. Regrettably, the long-term effect on these species cannot be studied easily.

Project accuracy requirement
The measurement accuracy from existing single-frame photographs varies between ±2 and ±6 mm. This value is not suitable for the study of New Zealand red coral as the estimated growth rate is less than 6 mm per year and the rate at which it may be grazed can exceed this value. Conventional chemical markers (alizarin/calcein staining) produce accuracies of ±0.1 mm (Ward, 1995). However, the method requires a laboratory procedure which is destructive to the hydrocoral. The study based on the method of chemical staining shows that an alternative non-destructive technique must give a measurement of accuracy ±1 mm to achieve a reliable result.

Stereophotographic system design and calibration

Underwater photography suffers from distortion due to air–water refraction. Consequently, it was essential to calibrate the imaging system underwater. As the hydrocoral species under study live at depths of 16–18 m, a special camera calibration system had to be built for the project. The system needed to be easy to use by divers, easy to assemble underwater; and work in a low-light environment. Subsequently, a simple steel structure which consists of a camera mount and a pin to hold a target board was developed (Figure 8.1). By rotating the board around the pin a set of convergent photographs of the targets could be taken quickly (Chong and Strafford, 2002).

Slide film scanning

Despite the digital age, film-based SLR cameras are still used because of their high image quality. This is particularly true for sub-surface projects because underwater cameras are expensive to replace. To exploit the versatility of digital stereo-workstation in close range photogrammetry, it is necessary, therefore, to scan film to obtain a digital form. Normally the slides are scanned to a resolution of 2500 dpi to obtain the best quality. During interior orientation procedures, the edges of the image are digitised to establish the "four virtual corners", or pseudo-fiducial marks, which are used to determine the image reference coordinate system.

Control frame and exterior orientation of stereo-model

A rigid 250 mm × 300 mm × 25 mm aluminium frame was constructed and bolts of various lengths (10–30 mm) were fastened onto the surface of the frame. Retro-reflectors were placed on the end of each bolt. The required accuracy of the coordinates of the control targets was set at ±0.7 mm in order to achieve an overall project accuracy of ±1.0 mm. Convergent photography and a high precision scale bar were used for the calibration (Figure 8.2).

Figure 8.2 *The underwater stereo-camera system. Note the control frame and the shutter control device.*

Special considerations for underwater application

The Fiordland national park issues strict guidelines which state that the hydrocorals must not be harmed in anyway during photography (Figure 8.3). To avoid damaging the hydrocoral it was necessary to set the control frame at a distance of 50 mm or more in front of the hydrocoral. Initial tests showed that there was a small error in the stereo-measurement when the control frame was placed at a short distance in front of the measured object. Accordingly, it was necessary to study the effect of a change in object distance in addition to the evaluation of the system in an underwater environment. To carry out this test, the underwater test-field was photographed at various distances from the control frame, which was fixed to the camera-mounting frame (Figure 8.2). That is, the object distance to the test field was set at a minimum of 800 mm to a maximum of 1050 mm at increments of 50 mm (Figure 8.4). Five sets of independent stereo-photographs were obtained for each object distance. Each stereo-pair of photographs was

Figure 8.3   *A pair of stereo-photographs of a red hydrocoral colony. Note the identification tag on the colony.*

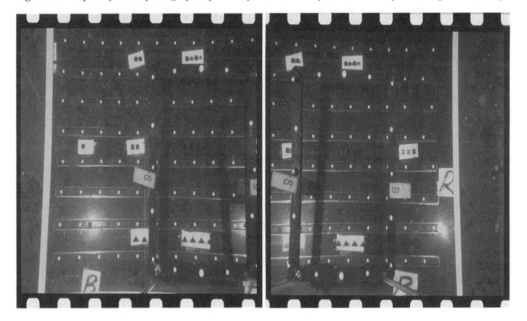

Figure 8.4   *Evaluating the error of underwater object–space control. A tag was used to show the distance of the control frame from the photographed object.*

oriented in a digital stereo-workstation. In each set, a minimum of six three-dimensional vectors between retro-reflectors on the test fields were measured and evaluated.

Results of the study

Table 8.1 shows the results of the system evaluation in sea water. The ratio was obtained by dividing the difference between the measured length and its true length by the true length. The mean ratio and standard deviation (column 2) were computed using 18 sets of measurements. The "zero object distance" row shows the accuracy of the stereo-imaging system. In Table 8.1, this value is 0.003 ± 0.001 mm. A plot of the scale ratio for various distances from the object to the control frame is depicted in Figure 8.5. To work out the accuracy of the system it was necessary to determine the maximum length of a branch, which must be measured in order to compute the growth or linear extension of hydrocoral.

Test samples showed that the maximum measured length of a hydrocoral branch was 128 mm. Multiplying this length by the mean ratio and standard deviation of 0.003 ± 0.001 mm gave 0.4 ± 0.1 mm, which is better than the project specification for mean positional standard error of ±1.0 mm. The majority of the measured lengths for this project were in the order of 40 mm or less. Table 8.1 also shows the results of the linear regression to produce the correction graph shown as Figure 8.5.

Table 8.1   Scale factor for various object distances

Object distance (mm)	Mean ratio and standard deviation	Linear regression
0	0.003 ± 0.001	$s = K + K_1(x)$
50	0.038 ± 0.005	$K = -1.47E\text{-}03$
100	0.088 ± 0.009	$K_1 = 8.65E\text{-}04$
150	0.119 ± 0.005	rms = 5.44E-03
200	0.171 ± 0.021	Where s = mean ratio and
250	0.220 ± 0.023	x = object distance in mm

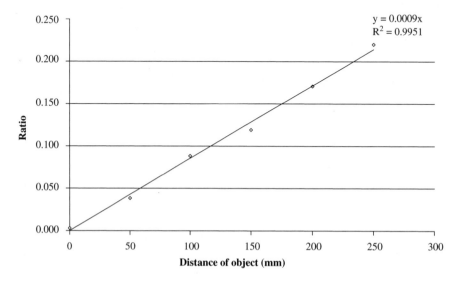

Figure 8.5   *A correction scale factor graph showing the relationship between the correction scale factor and the distance of the test-field from the control frame.*

Significance of the ratio for accurate photogrammetric measurements

As shown in Table 8.1, there was a small change in the ratio for stereo-measurement when the control frame was placed a short distance in front of the measured object. To evaluate the effect of this systematic error source, consider a maximum measured length of 128 mm and an average object distance of 82 mm. Using the ratios in Table 8.1, the computed length of a 128 mm branch photographed at an object distance of 82 mm is actually 120 mm. An 8 mm difference between the measured and computed length is certainly worse than the required measurement accuracy of ±1 mm.

To correct for the error it is necessary to compute the correction using the linear model presented in Table 8.1. The relationship between the correction scale factor and the distance of the test-field from the control frame is depicted in Figure 8.5. Based on the rms of the linear regression shown in the same table, the error for a measured length of 128 mm is 0.7 mm. As a result of the correction the underwater technique can achieve the accuracy required for the project.

Application of the photogrammetric measurement

Linear extension (growth) rates were standardised to annual extension rates. The net extension (growth and erosion) rate was obtained by combining the positive and negative values and the computed values were used to calculate the mean and standard deviation. The age of colonies was computed by dividing the maximum height of a colony by the overall mean extension rate. Some colonies show a substantial amount of erosion or grazing with decreases in branch length of up to 15.6 mm. In the field, small gastropods were observed to live on colonies of *E. novaezelandiae*, but their impact is not known. Corallivorous fish may also graze on the branches of the colonies.

In all colonies, the net growth of the branches was considerably less than the mean growth, with a few colonies showing negative net extension rates, indicating overall colony shrinkage. The study also showed that most colonies at the study site were relatively young (Table 8.2).

Table 8.2 Estimated age of the observed colonies at the Elisabeth Island (Fiordland National Park) study site

Colony No.	1	4	5	7	8	11	14
Total length (mm)	59	107	175	81	139	97	129
Age (years)	27	48	79	36	63	58	58

8.4.2 Case study 2: visible light sensor – medium-size object (dolphin)

Introduction

Bottlenose dolphins are considered to be ecological generalists because they are able to thrive in very diverse marine habitats, from shallow, warm water estuaries to deep, cool open ocean habitats (Schneider, 1999; Scott *et al.*, 1990; Kenney, 1990). Hersh and Duffield (1990) argued that Bottlenose dolphins living in warm, shallow water exhibit relatively small body sizes and proportionately large flippers, perhaps in response to demand for increased manoeuvrability and heat dissipation. They discovered that dolphins living in cool, deep waters appear to have larger body sizes and proportionately smaller flippers than shallow water animals, again perhaps reflecting different thermal and manoeuvrability requirements. It was necessary to obtain body length, fluke width and anterior flipper length of dolphins in the Sounds of Fiordland (Doubtful Sound) to determine whether these cetaceans have the same features as their cousins elsewhere.

There was no record of mass dolphin stranding in the Doubtful Sound, New Zealand because the deep fjords have very few shallow waters and there are very few human inhabitants in the Fiordland. Incomplete morphometric measurements of two dead dolphins were available for this research, but a significant statistical comparison was not possible from the measurement of such a small sample. It was not possible to obtain a licence to capture Bottlenose dolphins for measurement at sea and so a non-contact measurement solution was sought.

Project accuracy requirements

Body measurements and age estimates from 282 dead Bottlenose dolphins stranded on the Texan coast (Chong and Schneider, 2001) were used to determine that the close range photogrammetric technique must achieve a body length measurement accuracy of 10 cm for age estimation. Additionally, the measurement accuracy for fluke width and anterior flipper length comparison was estimated as 4.9 cm and 2.8 cm, respectively. As the project requires accurate age estimation, the accuracy standards must be followed strictly.

Stereo-photographic system design and calibration

Figures 8.6 and 8.7 show the configuration of the stereo-photographic system chosen. Two off-the-shelf Pentax Espio 738 cameras were mounted close to both ends of an aluminium bar. The bar was fastened on to a boom, which was attached to a cabin cruiser mast, 7.0 m above the cabin cruiser's forward deck. Cables were attached to both ends of the bar and mid-point of the boom for positioning, tilting and orientating the cameras correctly at each set-up. The cameras' shutters were triggered electronically from the deck. A video camera was attached at the mid-point of the aluminium bar. The video camera was connected to a TV monitor on the deck.

Figure 8.6 *An 18 ft cabin cruiser for photographing the Doubtful Sound dolphins. Note the electronic camera and video camera on the boom.*

Figure 8.7 *The stereo-camera system. Note the camera cover, the cables for positioning the boom and the video camera.*

Figure 8.8 *Typical Doubtful Sound bow-riding Bottlenose dolphins. Note the object–space control on the deck and railing of the cabin cruiser.*

Control frame and exterior orientation of stereo-model

To test the system, a rigid 100 cm × 100 cm × 25 cm aluminium frame was constructed. Retro-reflective targets were placed on the one side of the frame at 10 cm intervals. The frame was placed on the University of Otago (New Zealand) test-field and four convergent photographs were taken of the range and the control frame. Subsequently, a bundle adjustment of the digitised retro-reflective targets of these photographs established the coordinates of the retro-reflective targets of the aluminium frame. The coordinates of the targets on the bow, front deck and railing of the research vessel were established in a similar manner (Figure 8.8). This was achieved by placing the calibrated aluminium frame on the front deck of the vessel and four convergent photographs were taken of the set-up. A bundle adjustment of the digitised retro-reflective targets on these photographs established the coordinates of the targets on the research vessel. The computed control coordinates would have the same orientation and reference system as the coordinates of the aluminium frame.

Special considerations for two-media applications

Dolphins often bow-ride partially submerged. Consequently, the fluke and flippers are under the surface of the water. Photographing these dolphins for body measurement resulted in image distortions caused by two-media refraction.

In the 1960s research was carried out on two-media photogrammetry (Tewinkel, 1963; Gonenveld, 1969; Rinner, 1969). These experiments assumed a planar sea surface (i.e. a plane water surface). Refraction causes objects in the water to appear closer to the surface, although the effect on the planimetric position of objects is usually insignificant. In the 1980s, Okamoto (1982), Fryer (1984), Fryer and Kniest (1985a,b) and Tan (1989) discussed the effect of waves on stereo-photogrammetric measurements. These authors also concluded that planimetric errors are very small in through-water photogrammetry when ocean waves are present. However, the authors reported that there were depth errors, which peaked near the extremities of the photo-graphs and the magnitude of these errors was symmetrical about the y-axis of the stereopair. Also, Fryer (1984) reported that the effect of waves on close range through-water photography is detrimental to image sharpness, making it impossible to observe accurately.

Nevertheless, it was necessary to evaluate the stereo-camera system to determine its achiev-able accuracy at sea. As mentioned elsewhere, the close range photogrammetric technique must achieve a body length measurement accuracy of 10 cm for age estimation and the measurement accuracy for fluke width and anterior flipper length comparison must be better than 4.9 cm and 2.8 cm respectively. Figure 8.9 shows the design of a wooden dolphin model for this evaluation. Weights were attached to the underside of the four horizontal limbs of the model. Cables were attached to the top of the centre column. Strips of white and black colour paint were applied on the model to provide contrast.

The wooden dolphin model was lowered into the water at the study site. Stereo-photographs were taken of the model at 50 cm depth intervals from the surface to a depth of 350 cm. Trials showed that Doubtful Sound dolphins bow-ride at a depth of 50–300 cm. The photographs were scanned using a commercial high precision scanner and stereo-digitised using a DVP digital photogrammetric workstation. Three-dimensional distances between the limbs were computed using the stereo-digitised coordinates. A scale factor was computed for each measured distance by dividing it by the true distance. At each depth a mean scale factor was obtained from a minimum of four sets of observations. Table 8.3 shows the results of the system evaluation in the fiords of Doubtful Sound. As in Table 8.1, the results of the linear regression to produce corrections are shown.

Significance of the scale factor for accurate measurement

Adult Bottlenose dolphins can reach an asymptotic body length of 320 cm. Some dolphins were photographed at depths of 250 cm. By using the above scale factor the computed length of a 300 cm long dolphin photographed at a depth of 250 cm is 312 cm. A 12 cm difference between the measured and computed body length exceeded the required body length accuracy by 2 cm. Therefore it was necessary to compute a correction for the body length of every photographed dolphin. Based on the rms value of the linear regression in Table 8.3, the error for a body length

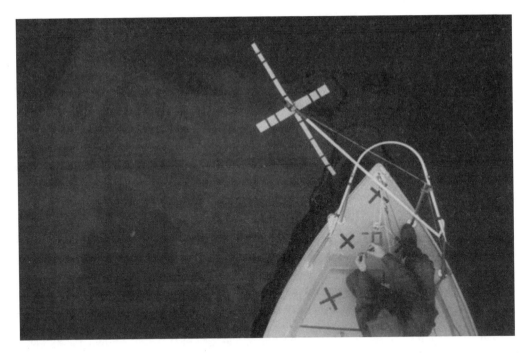

Figure 8.9 *Typical view of the wooden dolphin model under water. Note the cables used to level and manoeuvre the model.*

Table 8.3   Scale factor for various water depths

Water depth (cm)	Mean ratio	Linear regression
50	1.008	$s = K + K_1(x)$
100	1.013	$K = 0.998$
150	1.022	$K_1 = 1.6 \times 10^{-4}$
200	1.031	$\text{rms} = 1.4 \times 10^{-3}$
250	1.040	Where s = mean ratio and
		x = water depth in cm

of 300 cm is 0.4 cm. This value is smaller than the rms of the on-land evaluation provided elsewhere in the paper. As a result of the correction the through-water technique could achieve the accuracy required for the project.

Stereo-orientation and vectorising

Digital stereo-orientation was carried out on a DVP workstation. The software took into account properties of the camera and the radial lens distortion calibration data. The interactive program requires an interior orientation and observation of the fiducial marks. Computed exterior orientation parameters of the stereo-photogrammetric system were also entered into the computer. Vectorising was carried out on each pair of stereo-photographs to determine the body length from tip of the snout to tip of the fluke, the width of the flippers and the width and length of the fluke.

Application of the photogrammetric measurement

A technique for the estimation of the age of photographed dolphins using the photogrammetrically measured body lengths was also developed. The study showed whether the photogrammetrically measured anterior flipper length and fluke width were compatible with the computed Doubtful Sound dolphin age–body-length growth parameters. Additionally, a technique for computing the asymptotic fluke size and anterior flipper size of these dolphins was deduced. The computed values were compared with the values of the stranded Texas dolphins to establish physiological relationships between the two species.

Estimating the ages of Doubtful Sound dolphins

Two dead dolphins have been recovered from the Fiords in recent years. Teeth of the dead animals were used to estimate age using the growth-layer groups technique (Perrin and Myrick, 1980) and the estimated ages were three- and seven-years old. Unfortunately, only the body length of these animals was recorded. A set of age and body measurements of 282 stranded Texas Bottlenose dolphins was also available for the study.

Besides counting the growth layers of teeth, body length is the best body measurement for age estimation of dolphins (Dawson *et al.*, 1995; Bräger and Chong, 1999). Two published growth models were tested on the 282 stranded Texas Bottlenose dolphins (Fernandez and Hohn, 1998). They were the Gompertz equation (Fitzhugh, 1975) and Brody–Bertalanffy equation (Ebert and Russell, 1993) and both models gave a similar growth rate for the Texas dolphins.

A modified Gompertz curve was used to estimate the ages of the photogrammetrically measured Doubtful Sound dolphins. This modification was achieved by assuming that the body length of the seven-year old dolphin was the average asymptotic length of Doubtful Sound dolphins. The body length of the three-year old dolphin was used to provide an anchor point for a modified growth curve. Additionally, the birth size of Doubtful Sound dolphins and the shape of the curve were assumed to be similar to the Texas dolphins because there was no record of Doubtful Sound dolphin birth size. Using the modified growth curve parameters and the photogrammetric body length measurements, the ages of the photographed Doubtful Sound dolphins were estimated.

In a similar fashion, growth parameters for the flippers and flukes of the Doubtful Sound dolphins were estimated from comparison with those involved with the Texas stranding. Fortunately, the shape of the age–body-length growth curve, the age–anterior flipper growth curve and the age–fluke-width growth curve are similar to those of the Texas species. This allowed the use of the age–body-length growth curve of the Doubtful Sound dolphins to determine the asymptotic anterior flipper length and fluke-width of these dolphins. The data shown in Table 8.4 indicate that the photogrammetrically measured anterior flipper length and fluke width did not fit the growth models as well as the measured body-length. Readers desiring more technical details about the curve fitting methods used are referred to the literature cited above.

Table 8.4   Differences in asymptotic size in the morphometric measurements of Texan and Doubtful Sound (DS) Bottlenose dolphins

Measurement (cm)	Texas	DS (estimated)	Difference	Difference (%)
Body length	250	320	70	28
Fluke width	64	75	11	17
Anterior flipper	40	49	9	23

Comments

The adopted photogrammetric technique can provide accurate body length and extremity measurements of Bottlenose dolphins underwater for age estimation, provided that air–water refraction is considered. The system is low cost and simple to operate. Stereo-photographs are excellent for the identification of individual animals because body scars can be viewed in 3D. Additionally, this preliminary work shows that Doubtful Sound dolphins are longer than their Texas cousins, however their fluke width and anterior flipper length are proportionally smaller.

8.4.3 Case study 3: visible light sensor – large objects (Gray whales)

Introduction

Monitoring ocean giants such as Blue, Gray, Humpback or Sperm whales requires a different platform in relationship to the previous case studies. Generally, the length of these animals exceeds 15 m and their girth 3 m or more, making it difficult to photograph them stereoscopically from boats. Consequently, only aerial photography is considered suitable for accurate 3D body measurement of whales. In recent years, whale-watch tourism has prompted a surge in interest in studying the migratory behaviour and physical condition of these giants. One particular research project involves the evaluation of the nutritive condition and reproductive status of both northbound and southbound migrating Gray whales in the Pacific (Perryman and Lynn, 2002). Early migrating southbound Gray whales are more likely to be parturient (i.e. pregnant) than those migrating later. Northbound cows are generally smaller in girth as they fast during their winter south migration. The main objective of the research was to verify findings presented in the 1960s on temporal segregation during migrations, particularly for indications of reproductive status and for changes in shape that may reflect reduction in nutritive condition during winter migration. The nutritive condition may provide clues to the availability of food as a result of climate change. The other objective—and of particular significance for this book—was to determine whether spatial data representing whales could be extracted from high-resolution aerial photographs.

Project accuracy requirements

316 Gray whales were taken under seven scientific research permits issued to the US Bureau of Commercial Fisheries between 1959 and 1969. The research was conducted to provide representative samples from both the southbound and northbound migrations. Other samples were obtained from Russian studies of dead and captured animals. The samples showed that the mature female length was on average 11.1 m long and the average girth was 2.0 m (Figure 8.10). The girth difference between northbound and southbound was 11–25% (Rice and Wolman, 1971). Accordingly, to study the nutritive condition of the cetaceans accurately, the 3D measurement must be better than ±0.2 m or 1% of the body length. Existing data shows that the average length difference between northbound and southbound animals could be less than 0.3 m.

Perryman and Lynn (2002) attempted to determine the measurement accuracy of an aerial photography system for the study of northbound and southbound Gray whales along the Californian coast. The authors used a medium format military reconnaissance camera KA 76 (image size 114 mm × 114 mm) to achieve a favourable accuracy from the aerial photography. In addition, a GPS receiver and a Sperry 300 series twin transducer radar altimeter were installed in the aircraft to provide accurate exterior orientation parameters for the photography. Tests were carried out during aerial photography of the Gray whales. Two lengths of pipe, 12.02 and 6.02 m

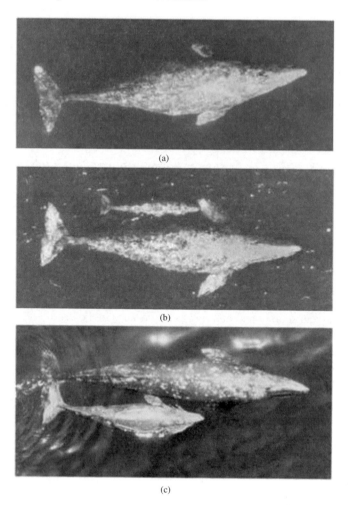

(a)

(b)

(c)

Figure 8.10 *Aerial photos of Gray whale: (a) a near-term female; (b) recent post-partum female with calf, and (c) a cow/calf pair (calf age about 2 months). Note the changes in body shape (Reproduced with permission of W. Perryman).*

long, representing cow and calf were towed to the test site (Figure 8.11). The measured 3D distances were compared to the actual pipe lengths and measurement accuracy was within 1% of the actual length.

Application of the photogrammetric measurement

The main objective of the research is the determination of the shape of the migrating whales. By comparing the shape between the northbound and southbound Gray whales, the nutritive condition of the animals could be studied accurately. Size based on body measurements must be transformed into shape before further analysis could be carried out, and the research utilises a least squares technique developed by Atchley *et al.* (1976). The technique, which utilises the body length (L) and width (W) to determine the shape, removes the effect of size. Consequently, body length, width at the wide point and fluke width (Fw) of the whale were required and these measurements were to be obtained photogrammetrically (Figure 8.12).

Figure 8.11 *Pipes representing adult and juvenile Gray whale for system calibration. Reproduced with permission of W. Perryman.*

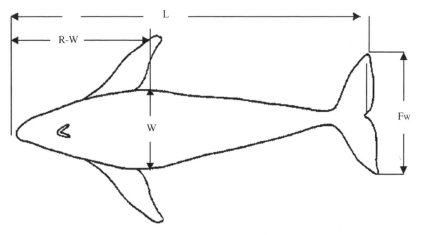

Figure 8.12 *Features measured on vertical aerial photographs of migrating Gray whales. Reproduced from Perryman and Lynn (2002) with permission of the International Whaling Commission, Cambridge, UK.*

Nutritive condition: fatness

Perryman and Lynn (2002) performed a linear regression fit of the measured lengths to width ratio, and the residuals were used to represent shape variables. The means of the residuals (ANOVA) for parturient females, southbound cows, southbound adults, juveniles, northbound cows, northbound adults and juveniles were compared, and the null hypothesis that they were the same was rejected. Figure 8.13 shows a plot of the means of the residuals from the length on the width for all Gray whales for which these two features could be measured accurately. From Figure 8.13, it is apparent that the shape variables decrease steadily during the winter migration and that northbound lactating cows were the narrowest or in the poorest nutritive condition of all categories.

The authors tested the hypothesis that the value R–W (distance from the rostrum, R, (nose) to widest part of body, W) was the same for the five categories of whale. Accordingly, the means of the regression residuals were compared using ANOVA. The location of W was found to vary significantly between whale categories (Figure 8.14). The tests also show that the

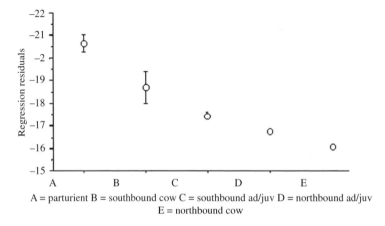

A = parturient B = southbound cow C = southbound ad/juv D = northbound ad/juv
E = northbound cow

Figure 8.13 *Plot of means of residuals from regressions of length (x axis) on the width (y axis) for all Gray whales for which these two features could be measured accurately. Error bars are ±2 standard errors of the mean values. Reproduced from Perryman and Lynn (2002) with permission of the International Whaling Commission, Cambridge, UK.*

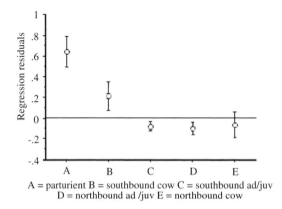

A = parturient B = southbound cow C = southbound ad/juv
D = northbound ad /juv E = northbound cow

Figure 8.14 *Plots of means of residuals from regression of length (x axis) on distance from the tip of the rostrum to the widest point (y axis) on Gray whales. Error bars are ±2 standard errors of the mean values. Reproduced from Perryman and Lynn (2002) with permission of the International Whaling Commission, Cambridge, UK.*

location of greatest width of parturient females and cows was further from their rostrum than all other categories studied. The authors argued that the foetus of near-term pregnant females probably accounts for some of this change in shape. However, for the southbound cows these data indicate that the extra fat carried by the cows was reflected in a change in width that is closer to the fluke.

Comment

The photogrammetric technique was found to be suitable for obtaining accurate measurement of large whales at sea using an aerial platform. It has proved to be an inexpensive technique for monitoring all species. Also, photogrammetric measurements were accurate in the identification of matured and juveniles of migrating Gray whales. The research findings verify the conclusions of the fishery in the 1960s. The conclusions were that:

- There is partial temporal segregation in southbound whales by size and reproductive condition;
- near-term pregnant whales had greater girth to length ratios than any other group migrating south; and,
- south migrating animals were in similar condition to those studied 40 years earlier.

8.4.4 Case study 4: visible light sensor – flying snakes

Introduction

Some species of snakes have the ability to glide from tree to tree. There are many suggestions as to why snakes glide, including:

- less effort is needed to glide in comparison to descending a tree, moving across space and climbing up another tree;
- to avoid predators on the ground;
- to escape from tree-residing predators quickly; and,
- to chase prey which is capable of gliding from tree to tree (or branch to branch).

Socha et al. (2005) and Socha and LaBarbera (2005) argued that gliding flight in snakes is kinematically distinct from gliding in other animals (flying lizard or flying squirrel). The researchers noticed that when a flying snake moves through the air, it passes lateral travelling waves posteriorly along a long and dorsoventrally flattened body and its body moves in a complex motion in three dimensions. Gliding snakes possess a unique body shape which is ideal for flight. In addition, the researchers hypothesise that gliding snakes exhibit aerial behaviour that is the most dynamic among vertebrate gliders. Consequently, an investigation was carried out to determine the actual flight performance of the species. The particulars of flight performance are:

- the snakes produce sufficient lift to glide further;
- the snakes are true gliders; and,
- postural change during flight is critical for supremacy in flight performance.

The research involved a sample of 21 C. Paradise Boie (genus *Chrysopelea*) with equal number of males and females. All were caught in the wild. However, only the flight data of 14 individuals met the criteria of high quality trajectories. To ensure minimal error in photogrammetric digitising, the snakes were marked on the dorsal surface at the head-body junction, body midpoint and vent with a 1 cm band of non-toxic white paint during the study.

Project accuracy

Two sets of gliding flight data were acquired photogrammetrically (Socha *et al.*, 2005). The first set involved trajectory variables such as:

- glide angle;
- glide ratio;
- ballistic dive angle;
- trajectory shallowing rate;
- speed and acceleration; and,
- heading angle.

The second set consisted of postural variables including: excursion (vertical, fore-aft, lateral); horizontal body angle; body angle of attack; wing loading; undulation frequency, undulation wavelength and wave speed; and, undulation wave height.

Detailed discussion of the variables is given by Socha *et al.* (2005) and Socha and LaBarbera (2005). The variables can be easily determined by using the 3D coordinates obtained by photogrammetry. As existing data on this type of research was not available, it was difficult to establish an accuracy requirement for the project. However, tests were conducted to evaluate the accuracy of the photogrammetric system. It was apparent that the low base-to-height ratio of the video cameras and the small amount of height variation in the vertical axis of the control points would produce poor 3D coordinates. Therefore, it was decided that the position of the camera stations should be determined accurately and held fixed in subsequent photogrammetric computation.

Trials show that the mean rms errors were in the ranges 1–4 cm, 2–14 cm and 3–13 cm in the X, Y and Z (vertical) axes, depending on the distance between the snake and the video cameras. The error sizes were considered satisfactory for the computation of trajectory and postural variables.

Methodology

The snakes were launched from a scaffolding tower constructed in an open, grassy field (Figure 8.15). The reptiles propel themselves off a typical tree branch which protrudes 1 m from the edge of the platform at a height of 9.62 m. Each snake was used 11 times and a total of 237 trials were recorded during the study.

Two Sony DCR-TRV 900 digital video cameras were positioned on the top of the tower, 12 m above the ground. They were placed about 2 m apart to ensure a good base-to-height ratio. The cameras were synchronised by short-duration, high amplitude audio signals. Wide-angle mode was used in the video capture. In addition, two Nikon SLR cameras were used to capture specific aspects of the posture throughout the trajectory (see Figure 8.15).

Object–space control system and camera calibration

A grid of reference points which covers a flat grassy area approximately 8 m × 8 m were used as object–space control (Figure 8.15). The coordinates of the control points were determined by conventional surveying techniques and their accuracies were ±0.1 mm in all axes. The direct linear transformation (DLT) technique in ERDAS Imagine Orthobase software, was used in the photogrammetric camera calibration, image measurement and object-space coordinate computation of points of interest on the snake in flight. Detailed discussion on the DLT technique for camera calibration and object–space coordinate computation can be found in Abdel-Aziz and Karara (1971) or Grün (1997).

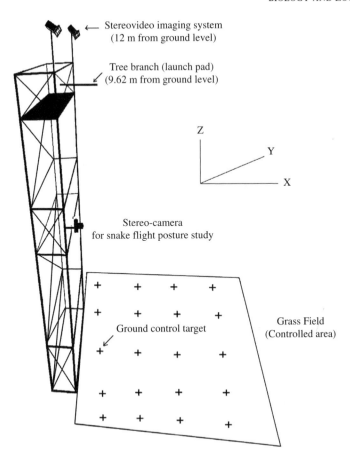

Figure 8.15 *Stereo-imaging system for flying snake flight study. Note that the Nikon cameras were mounted midway up the platform to photograph the snake posture during flight (data obtained from Socha et al., 2005).*

Image processing and photogrammetric digitising

Video clips of the gliding trials were processed to obtain 60 images per second using Adobe Premiere software and were exported in jpg format for use in ERDAS Imagine software. Stereo-pairs of video images were sorted for further photogrammetric work in ERDAS Imagine. Finally, the positions of the marks on the snakes were digitised to obtain X, Y and Z coordinates from the stereo-pairs. About 60 positions could be computed per second epoch of the video clip.

Application of the photogrammetric measurement

As stated earlier, it was hoped that the research could answer the following three questions:

- Do the snakes produce sufficient lift to introduce a significant horizontal component to the trajectory?
- If the snakes are true gliders, then at what position in the trajectory do the snakes reach equilibrium?
- Does the postural change provide any significant aerodynamic forces which may allow the snakes to gain horizontal distance?

To seek evidence for the first question, it was necessary to compute the airspeed, sinking speed and horizontal speed of the flight. Once the marker's position on each stereo-pair (30 pairs per second) was determined using the DLT technique, the speed components could be determined readily. Figure 8.16 shows the computed value of the three speeds which were required to determine whether there was sufficient lift during the flight. The speed at the termination of flight was not recorded by the video cameras as the snake was out of view. The sinking speed began to level off at a speed of 6.1 ± 0.7 ms^{-1} or the instant it began to decrease (position of arrow shown in plot (c). In other words, the lift was higher than the force of gravity.

The computed speed information was also used to answer the second and third questions. It is assumed that to reach equilibrium in flight the three speeds must achieve a constant value simultaneously during the flight. However, Figure 8.16 shows that equilibrium was never reached. Subsequently, images captured by the Nikon cameras were used to determine the postural change during flight. The posture during flight revealed that the snake's cylindrical body becomes flattened, coils into an 'S' shape and the body undulates aerially. Figure 8.17a shows the early stage of an 'S' posture after the initial dive. The next stage (b) involves widening the 'S' shape and simultaneously flattening the cylindrical body. At a later stage of the flight (c), the anterior body begins to move parallel to the ground while the posterior body is angled downward. The aerial behaviour is unique among true gliders. Consequently, it is confirmed that the flying snake's postural change during flight provides additional lift to gain horizontal distance.

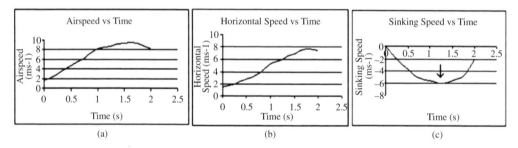

Figure 8.16 *Plots of average flight speed vs time: (a) airspeed; (b) horizontal speed; and (c) sinking speed (data obtained from Socha et al., 2005).*

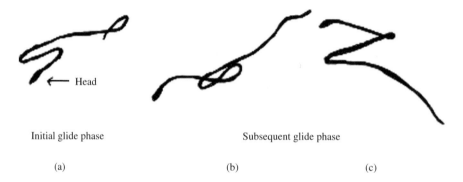

Figure 8.17 *A stick figure of the snake during flight. Note the change in the posture in the initial and subsequent glide phase (data obtained from Socha et al., 2005).*

Conclusions

In previous research, the glide performance of animals was estimated using a few basic measurements, such as take-off height, landing height and horizontal distances travelled and aerial time. Only basic equipment such as a rangefinder, maps and a hand-held stopwatch were used. Consequently, the results of these studies were not able to provide accurate kinematic details which are critical for the evaluation of flight performance. Accurate photogrammetric measurements have certainly provided the necessary information to give a better understanding the flight performance of flying snakes. Certainly, studies of other gliding animals such as flying lizards and flying squirrels could benefit from similar photogrammetric measurement.

8.5 References

Abdel-Aziz, Y.I. and Karara, H.M., 1971. Direct linear transformation from computer coordinates into object space coordinates in close range photogrammetry. In *ASP Symposium on Close Range Photogrammetry*, American Society of Photogrammetry, Bethesda, Maryland, USA, 1–18.

Atchley, W.R., Gaskins, C.T. and Anderson, D., 1976. Statistical properties of ratios. *Systematic Zoology*, 25, 563–583.

Boyce, R.E., 1964. Simple scale determination on underwater stereo pairs. *Deep Sea Research*, 11, 89–91.

Bräger, S. and Chong, A.K., 1999. An application of close-range photogrammetry in dolphin studies. *Photogrammetric Record*, 16(93), 503–517.

Bräger, S., Chong, A., Dawson, S., Slooten, E. and Würsig, B., 2000. A combined stereo-photogrammetry and underwater-video system to study group composition of dolphins. *Helgoland Marine Research*, 53 (2), 122–128.

Chong, A.K. and Schneider, K., 2001. Two-medium stereo-digital photogrammetry for Dolphin study. *Photogrammetric Engineering and Remote Sensing*, 67(5), 621–628.

Chong, A.K. and Strafford, P., 2002. Underwater digital stereo-observation technique for Red Coral study. *Photogrammetric Engineering and Remote Sensing*, 68(7), 745–751.

Dawson, S.M., Chessum, C.J., Hunt P.J. and Slooten, E., 1995. *An inexpensive, stereo-photographic technique to measure sperm whales from small boats*. Report of the International Whaling Commission, 45, 431–436.

Ebert, T.A. and Russell, M.P., 1993. Growth and mortality of sub-tidal red sea urchins (*Strongylocentrotus franciscanus*) at San Nicolas Island, California, USA: problems with models. *Marine Biology*, 117, 79–89.

Fernandez, S. and Hohn, A.A., 1998. Age, growth, and calving season of bottlenose dolphins, Tursiops truncatus, off coastal Texas. *Fishery Bulletin*, 96, 357–365.

Fitzhugh, H.A., 1975. Analysis of growth curves and strategies of altering their shape. *Journal of Animal Science*, 42, 1036–1051.

Fryer, J.G., 1984. Errors in depths determined by through-water photogrammetry. *Australian Journal of Geodesy, Photogrammetry and Surveying*, 40, 29–39.

Fryer, J.G. and Kniest, H.T., 1985a. Errors in depth determination caused by waves in through-water photogrammetry. *Photogrammetric Record*, 11(66), 745–753.

Fryer, J.G. and Kniest, H.T., 1985b. Some strategies for improving the accuracy of depths determined by through water photogrammetry. *Australian Journal of Geodesy, Photogrammetry and Surveying*, 43, 45–60.

Grange, K.R., 1990. *Unique marine habitats in the New Zealand Fiords: A case for presentation*, Report prepared for the Department of Conservation, Wellington, New Zealand.

Groenveld, M., 1969. Formula for conversion of stereoscopically observed apparent depth of water to true depth. *Photogrammetric Engineering*, 30(6), 1037–1045.

Grün, A., 1997. Fundamentals of videogrammetry – A review. *Human Movement Science*, 16, 155–187.

Hersh, S. and Duffield, D., 1990. Distinction between Northwest Atlantic Offshore and Coastal Bottlenose Dolphins based on hemoglobin profile and morphometry. In (Leatherwood, S. and Reeves, R.R. (Eds.), *The Bottlenose Dolphin,* Academic Press, San Diego, California, USA, 129–139.

Kenney, R.D., 1990. Bottlenose dolphins off the northeastern United States. In Leatherwood, S. and Reeves, R.R. (Eds.), *The Bottlenose Dolphin,* Academic Press, San Diego, California, USA, 369–386.

Miller, K.J., 1995. *Size frequency distribution of red corals at Te Awaatu Marine Reserve, Doubtful Sound, Fiordland.* Report to the Department of Conservation, Invercargill, New Zealand.

Okamoto, A., 1982. Wave influences in two-media photogrammetry. *Photogrammetric Engineering and Remote Sensing,* 48(9), 1487–1499.

Perrin, W.F. and Myrick, A.C. (Eds.), 1980. *Age determination of toothed whales and sirenians.* Report of the International Whaling Commission, Cambridge, UK, Special Issue 3.

Perryman, W.L. and Lynn, M.S., 2002. Evaluation of nutritive condition and reproductive status of migrating gray whales based on analaysis of photogrammetric data. *Journal of Cetacean Resource Management,* 4(2), 155–164.

Photometrix, 2005. See www.photometrix.com.au (accessed 23 August 2005).

Probert, P.K., McKnight, D.G. and Grove, S.L., 1997. Benthic invertebrate bycatch from a deep-water trawl fishery, Chatham Rise, New Zealand. *Aquatic Conservation: Marine and Freshwater Ecosystems,* 7, 27–40.

Rice, D.W. and Wolman, A.A., 1971. *The life history and ecology of the gray whale (Eschrichtus robustus).* American Society of Mammalogists, Special Publication No 3, Stillwater, Oklahoma, USA.

Rinner, K., 1969. Problems of two media photogrammetry. *Photogrammetric Engineering,* 35(3), 275–282.

Schneider, K., 1999. *Behaviour and Ecology of Bottlenose Dolphins in Doubtful Sound, Fiordland, New Zealand,* Ph.D. thesis, Otago University, New Zealand.

Socha, J.J. and LaBarbera, M., 2005. Effects of size and behavior on aerial performance of two species of flying snakes (Chrysopelea). *The Journal of Experimental Biology,* 208, 1835–1847.

Socha, J.J., Dempsey, T.O. and LaBarbera, M., 2005. A 3D kinematic analysis of gliding in a flying snake, Chrysopelea paradise. *The Journal of Experimental Biology,* 208, 1817–1833.

Scott, M.D., Wells, R.S. and Irvine, A.B., 1990. A long-term study of bottlenose dolphins on the West Coast of Florida, In Leatherwood, S. and Reeves, R.R. (Eds.), *The Bottlenose Dolphin,* Academic Press, San Diego, California, USA, 235–244.

Tan, W., 1989. Positional errors in analytical through-water photogrammetry. *Australian Journal of Geodesy, Photogrammetry and Surveying,* 50, 73–87.

Tewinkel, G.C., 1963. Water depths from aerial photographs. *Photogrammetric Engineering,* 29(6), 1037–1042.

Ward, S., 1995. The effect of damage on the growth, reproduction and storage of lipids in the scleractinian coral *Pocillopora damicornis* (Linnaeus). *Journal of Experimental Marine Biology and Ecology,* 187, 193–206.

9 Cultural heritage documentation

Petros Patias

9.1 General outline of the application

9.1.1 Introduction and background

Cultural heritage is a testimony of past human activity, and, as such, cultural heritage objects exhibit great variety in their nature, size and complexity, from small artefacts and museum items to cultural landscapes, from historic buildings and ancient monuments to city centres and archaeological sites.

Cultural heritage around the globe suffers from wars, natural disasters and human negligence. The importance of cultural heritage documentation is well recognised and there is an increasing pressure to document our heritage both nationally and internationally. This has alerted international organisations to the need for issuing guidelines describing the standards for documentation. Charters, resolutions and declarations by international organisations underline the importance of documentation of cultural heritage for the purposes of conservation works, management, appraisal, assessment of the structural condition, archiving, publication and research. Important ones include the International Council on Monuments and Sites, ICOMOS (ICOMOS, 2005) and UNESCO, including the famous Venice Charter, The International Charter for the Conservation and Restoration of Monuments and Sites, 1964 (UNESCO, 2005).

This suite of documentation requirements, as stated by the international agreements, imposes important technical restrictions and dictates specifications, which should be always borne in mind when recording strategies are designed and followed. The common features of the above documents are, briefly:

- recording of a vast amount of four-dimensional (i.e. 3D plus time) multi-source, multi-format and multi-content information, with stated levels of accuracy and detail;
- digital inventories in 3D and, as far as available, dated historical images;
- management of the 4D information in a secure and rational way, making it available for sharing and distribution to other users; and
- visualisation and presentation of the information in a user-friendly way, so that different kinds of users can actually retrieve the data and acquire useful information, using internet and visualisation techniques.

Recognising these documentation needs, ICOMOS and the International Society of Photogrammetry and Remote Sensing (ISPRS) joined efforts in 1969 and created the International Committee for Heritage Documentation, CIPA (CIPA, 2005). CIPA's main objective is to provide an international forum and focal point for efforts in the improvement of all methods for surveying of cultural monuments and sites. The combination of all aspects of photogrammetry with other surveying methods is regarded as an important contribution to recording and monitoring cultural heritage, to the preservation and restoration of any valuable architectural or other cultural monument, object or site, and to provide support to architectural, archaeological and other art-historical research.

An important issue to bear in mind is that documentation is a complicated process, and includes a wide suite of activities that include surveying, testing and monitoring, and gathering textual and other information. "The geometry of the object is not the only parameter to be recorded. All specificities making the object unique are meaningful; all potential values—architectural, artistic, historical, scientific and social—are parameters to consider" (D'Ayala and Smars, 2003).

The guiding concepts of cultural heritage documentation have been detailed by D'Ayala and Smars (2003) as:

- Objectivity: "...an objective basis is a guarantee to having a firm ground on which to debate the conservation choices..." and "...the use of any specific set of data necessarily influences any decision-making process. The manner in which a survey is executed significantly influences further actions".
- Values: "The recorder's choices are critical...... What is seen today as uninteresting may appear tomorrow as extremely valuable. The importance of thorough recording is emphasised by the common loss of minor details which may disappear at the moment of new conservation work, leading to loss of integrity or of historical evidence."
- Learning process.
- Continuity: "... Documentation should not be seen as an activity confined within a set time.... Therefore, a basic requirement is that the results of documentation should be available for future use."
- Fabric: "Documentation should not stop at the surface.... Integration with other documentation techniques is necessary."
- Documentation sets: "Information gathered during documentation may be large and manifold.... thus it is critical to organise the available information, for which the metric survey is a natural support. Sets of thematic drawings (geometry, materials, pathologies etc.) can be prepared. A specific set prepared by one specialist can bring insight to other specialists who are working on other sets."
- Redundancy: "....Every piece of information is associated with uncertainty. Documentation data should be supplemented by information about the quality of the data. Control procedures offer a way to assess quality."

In this framework, photogrammetry is called upon to offer its services at a variety of levels and in all possible combinations of object complexities, scientific procedures, quality requirements, usage of final products, time restrictions and budget limitations.

9.1.2 General advantages of photogrammetry

Photogrammetry has always been a technique that provides accurate, detailed, 3D data in a cost-effective way. Compared to other techniques, photogrammetry shows distinct advantages. Recognising that photogrammetry needs to couple its data with data from other sensors and other techniques, it is valid to state that the comparative advantage of photogrammetry consists of the provision of:

- **Large amounts of data:** this can be at various scales and resolutions, referring to whole areas or to single objects, and which are based on photogrammetric measurements or on combinations with other types of measurements.

- **Very accurate data:** under the current technology, this is routinely on the order of $\frac{1}{3}$ to $\frac{1}{10}$ of the camera pixel size. More importantly, the whole procedure is regularly monitored and checked to ensure that the quality of the results (accuracy and precision) is based on sound scientific statistical indicators. The mapping is objective, and the results are repeatable, verifiable, with a consistent overall accuracy.
- **3D data:** this is required whether this refers to a whole area, city, or simply a small archaeological artefact. Photogrammetry, by its nature, reconstructs the 3D surface of objects in a detailed and accurate way. The geometric reconstruction is based either on 3D point or continuous surface determination.
- **Texture data:** this is a natural consequence since the technique is based upon images of the objects to be reconstructed. Spectral or texture data are very important since they give a natural look to the reconstructed 3D objects, thus enhancing the user's cognition. More importantly, this texture also carries the 3D object's geometry, and therefore allows metric characteristics to be matched with those of the vector data.
- **High resolution and detailed data:** this is needed both in vector and texture. Based on the current high rates of advances in technology, photogrammetric sensors are capturing more and more detailed data, which in turn are processed by increasingly effective automatic procedures. Centimetre-level pixel sizes are routinely realised on the object in medium-scale mappings, and may decrease to millimetre-level or less for close range mappings.
- **Geo-referenced data:** data defined above are referred to common reference frames, whether they are global or local coordinate systems. By reference to common ground coordinate systems, the metric characteristics of the data gain one more important advantage: they all refer to real-world geometry and facilitate the ability to extract indirect measurements any time after leaving the monument.
- **Metadata:** this refers to information about the data. Metadata is valuable, since it can be used for tracing down original sources, acquisition times, qualities, metrics, and even ownership.
- **Stereo-viewing capabilities of the 3D data:** the technique frequently uses stereoscopic images, which is a unique characteristic of photogrammetric data.
- In addition to all of the above, two more characteristics of photogrammetry should be highlighted. First, the required equipment is becoming less costly; secondly, it can provide final products under strict time requirements.

9.1.3 General disadvantages of photogrammetry

In general, there are two main disadvantages of photogrammetric methodology. First, photogrammetry requires medium to high-end hardware/software. This has been a traditional impediment but is changing since: (a) digital systems are based on computer technology (in contrast to analogue and analytical instrumentation, which were based on optical–mechanical parts) where costs are rapidly declining; (b) off-the-self "consumer grade" digital cameras are becoming cheaper and more common, and their resolution has increased over the past few years. Even though these cameras are non-metric, extensive literature shows that metric content is recoverable to a high degree from non-metric imagery, the quality of the recovery depending on the knowledge of the internal geometry of the camera (through camera calibration); and (c) a lot of low-end and low-price software has become available during the last decade. Secondly, photogrammetry requires conventional field survey measurements to establish a reference frame, upon which the photogrammetric measurements are based. This seems to be a drawback, since the "remote" character of photogrammetric measurements can lose their major advantage. However, again this disadvantage is becoming less of an issue, since reflectorless instruments have entered the surveying market, alternative systems can be used (e.g. cameras on top of total stations, laser scanners etc.), and in addition, geometric features (straight lines, parallel lines and planes etc.) can be used in order to relax control point requirements.

9.1.4 History and brief literature review

Architectural photogrammetry is nearly as old as photography itself. In 1858 the German architect A. Meydenbauer developed photogrammetrical techniques for the documentation of buildings and established the first photogrammetric institute in 1885 (Royal Prussian Photogrammetric Institute). In the same year, the first application of photogrammetry proved its potential. The ancient ruins of Persepolis were the first archaeological objects recorded photogrammetrically.

Since then, photogrammetric techniques have been constantly developing (Patias and Peipe, 2000; Patias, 2001; Atkinson, 1996; Karara, 1989; Carbonnell, 1989; Ogleby and Rivett, 1985) on all fronts: hardware, software and procedures. Over the recent past, there has been a movement from optical–mechanical to purely digital hardware and from high-end to low-budget software (Hanke, 2001; Mills *et al.*, 2000; Glykos *et al.*, 1999; Dequal *et al.*, 1999). New procedures are constantly developing: the use of wide angles or even panoramic cameras (Pöntinen, 2004; Haggrén *et al.*, 2004; Reulke *et al.*, 2005); fusion of photogrammetric data with data from other sensors, especially laser scanners (Böhler, 2005; Balletti *et al.*, 2004; Tucci *et al.*, 2003; Lingua *et al.*, 2003; Guarnieri *et al.*, 2004a,b; Barber *et al.*, 2002); use of non-metric sensors like off-the-self digital cameras (e.g. Patias *et al.*, 1998; Peipe and Stephani, 2003; Stephani *et al.*, 1999) or video recorders; the production of specialised products like digital rectification (e.g. Hemmleb, 1999; Wiedemann *et al.*, 2000); monoplotting (Karras *et al.*, 1996; Karras and Petsa, 1999); unwrapping developable surfaces (Karras *et al.*, 1996, 1997, 2001); true orthophoto production (e.g. Boccardo *et al.*, 2001; Grammatikopoulos *et al.*, 2004; Balletti *et al.*, 2003); augmented reality products (e.g. Koussoulakou *et al.*, 1999, 2001a) and virtual museums (e.g. Boulanger and Taylor, 2000) etc.

9.2 Distinctive aspects of heritage recording

In contrast to most conventional photogrammetric applications, cultural heritage documentation is much more challenging, due specifically to the varying size of the objects to be documented as well as the different quality and resolution requirements.

It must be stressed that the documentation of cultural heritage is not an aim in itself: it aims at making information accessible to other users. The reasons for this have been identified by Böhler (2005) and include the cases:

- when the object is not accessible to interested parties;
- when the object is too large or complicated;
- when the object (or just a part of it) is visible only for a short period of time in context or at its original location (as in many archaeological excavations);
- when persons living far from the object cannot afford to visit it; and,
- when the object is in danger of slow deterioration (from environmental factors) or sudden destruction (from earthquakes or other natural disasters, war or vandalism).

Choosing the appropriate technology (sensor, hardware, software), the appropriate procedures, designing the production workflow and assuring that the final output is in accordance with the set technical specifications is a challenging matter in cultural heritage documentation. For this, the leading parameters are the size and the complexity of the object and the level of accuracy required. These two are the major factors, which crucially influence the procedure to be followed and, sometimes, even the viability of photogrammetry itself.

Other secondary factors, which may have an impact (and are sometimes crucial) are speed/time limitations for the documentation and budget limitations. The reason that these factors are secondary in nature is that they do not influence the viability of photogrammetry as the technique to be followed (since photogrammetry already has a distinct advantage over other techniques concerning the above factors) but they may influence the technicalities of the procedure which should be followed.

9.3 Typical requirements of cultural heritage documentation

9.3.1 Accuracies

When discussing accuracy requirements, one cannot avoid classifying the different applications and final products, since requirements vary widely.

Architectural and archaeological applications of photogrammetry and vision techniques can be classified in many ways and according to different parameters (McGlone, 2004). Probably the most reasonable way to classify applications is by the purpose of the documentation. Inherent in this type of classification are the parameters of object size and complexity, time and cost requirements, and type of final product.

Architectural analysis of monuments

The purpose is to show both the principal structural elements and the principal stylistic and technical peculiarities (Carbonnell, 1989). Typical scales are in the range 1:100 for the first and 1:10–1:20 for the second, with degrees of accuracy in the range 0.5–5 cm. The survey rarely concerns the whole monument, whereas the required output is normally ground plans and vertical sections in 2D-vector form. Details are frequently shown by planar texture maps (e.g. rectified photography). Time constraints are rarely imposed but the required budget is limited (e.g. Gemenetzis et al., 2001; Monti et al., 1999; Guerra and Balletti, 1999a; Pomaska, 2000; Nour El Din et al., 1999).

Conservation and restoration of monuments

The purpose is to reveal architectural structures, type of materials, cracks and indications of damage. Typical scales are from 1:50 to 1:10, and the required accuracies up to 1 cm. Although the surveys are normally partial, wider photographic coverage for future use is advisable (Carbonnell, 1989). Time and budget restrictions normally are not imposed and the required output generally consists of 2D vector and texture maps.

Studies of artifacts

The purpose is an exhaustive 3D representation of an object with high accuracy (0.1–2 mm) and large scales (1:1–1:5). 3D reconstructions in vector or solid form as well as 3D texture representations are the normal outputs, while visualisation and GIS archiving issues are important (e.g. Chikatsu and Anai, 1998; Gemenetzis et al., 2001).

Special studies

These include studies of damage and structural stability, surface deteriorations due to external factors such as chemicals, humidity, environment, etc. (e.g. Monti et al., 2001). Scales and accuracy requirements are varying and often time limitations are imposed. In many cases multi-sensor as well as multi-resolution data are involved.

Archaeological documentation

The main purpose is the documentation of archaeological excavations (Doneus, 1996; Georgiadis *et al.*, 2000; Koistinen, 2000; Koussoulakou *et al.*, 2001a,b). When the excavation is on-going, time limitations are strict. The scales involved are normally 1:20–1:100 and the required accuracies 1–5 cm. In these cases 2D representations do not suffice and 3D texture maps are generally required. In this group one should also add special kinds of surveys, like underwater photogrammetric ones.

Studies of city centres and settlements

These applications involve both aerial and close range photography at multi-scales and resolutions. The objects of interest vary from whole settlements to historic centres, to streetscapes and façades (e.g. Ogleby, 1996; Nakos and Georgopoulos, 2000; Patias *et al.*, 2002). Both scales and accuracies have a wide range (1:500–1:20) and normally 3D texture maps are the most preferable outputs. Visualisation issues and GIS development are also important.

Importations to GIS, visualisation purposes and virtual museums

The purpose of these surveys is the production of 2D and 3D texture maps, which will be used either in GIS environments or for visualisation purposes (Bartolota *et al.*, 2000; Boulanger and Taylor, 2000; Guerra and Balletti, 1999b; Ito *et al.*, 2000; Ogleby, 2001; Pomaska, 1999). Scales vary, while accuracies are relaxed. Data volume handling, administration and delivery are the most important issues.

A good reference about technical specifications, especially for documentation of monuments, which the interested reader can consult, is the English Heritage Guidelines (English Heritage, 2005).

9.3.2 Hardware and software requirements

Typical hardware and software requirements for such applications include:

- digital photogrammetric workstations or low-end hardware, either with or without stereo-viewing capabilities;
- digital rectification software;
- digital othorphoto production software;
- digital 3D reconstruction software (plus digital surface model collection);
- convergent multi-photo photogrammetric processing software; and,
- the combination of photogrammetric data with other sensors (e.g. laser scanners).

To this list one can also add visualisation and GIS software, if such output is also sought.

9.3.3 Sensors

While generally medium format (6 cm × 6 cm) metric film cameras are used, the use of non-metric medium and small format cameras is increasing. In this case a proper calibration of the camera focal length to 0.01 mm, plus a compensation for lens distortions is required. Of course, analogue images are scanned (at a pixel size of 25 μm or less), in which case photogrammetric or even desktop scanners can be used, provided these are properly calibrated.

9.3.4 Object preparation

Except in situations where high accuracy is required, targeting on the object is not essential although this does provide many advantages. When targets are used, they should be visible but not overly obtrusive and not be destructive. Natural features such as corners of windows, distinctive marks on rocks etc., sometimes suffice as targets.

When smaller objects are documented, portable calibration fields usually provide all the necessary targeting.

9.4 Solutions options

9.4.1 Existing commercial systems

Table 9.1 gives only an indicative list of the currently available commercial systems.

Table 9.1   Commercially available software systems

Indicative list of digital photogrammetric stations		
Company	Software	Web reference
Digi	Digi21	http://www.digi21.net
DVP-GS	DVP	http://www.dvp-gs.com
BAE Systems	Socet Set	http://www.vitec.com
DAT/EM	Summit Evolution	http://www.datem.com
Z/I Imaging	ImageStation	http://www.ziimaging.com
NETCAD	FOTOCAD	http://www.netcad.com.tr
TopoLSoftwareLtd. and Atlas Ltd.	PhoTopoL Atlas	http://www.topol.cz/photopol
Indicative list of convergent multi-photo software		
Company	Software	Web reference
PhotoMetrix	iWitness	http://www.photometrix.com.au
ShapeQuest Inc.	ShapeCapture	http://www.shapcapture.com
Eos Systems	PhotoModeler	http://www.photomodeler.com
RealVis	ImageModeler	http://www.realviz.com
Vexcel	Fotog	http://www.vexcel.com
RolleiMetric	RolleiMetric CDW	http://www.rollei.de
Technet Gmbh	Pictran	http://www.technet-gmbh.com
Indicative list of digital rectification and monoplotting software		
Company	Software	Web reference
Z/I Imaging	IRAS/C	http://www.ziimaging.com
Bentley	Descartes	http://www.bentley.com
Rollei	RolleiMetric MSRPlan	http://www.rollei.de
CIGRAPH	ArchiFacade	http://www.cigraph-store.com

9.5 Alternatives to photogrammetry for this application

In Figure 9.1, possible techniques for metric surveying are arranged according to the scale of the final document which in turn is a function of object size and achievable representation of details. The complexity of the survey can be expressed by the number of points to be recorded. This ranges from one single point describing the geographic location of an artefact, to some thousands of points, typical for a CAD drawing of a building or a topographic situation, or to about a million points or more for the description of the whole surface of a sculpture or a digital elevation model (Böhler, 2005).

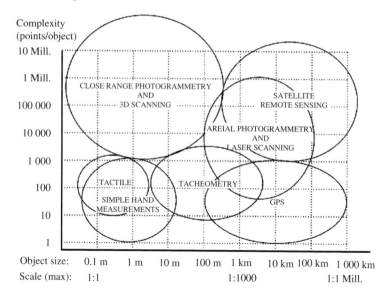

Figure 9.1 *Selection of appropriate surveying methods depending on object size and complexity (after Böhler, 2005).*

Besides size and complexity, other factors may influence the optimal method to be chosen. These include: the necessary accuracy and resolution; accessibility of the object and availability of ideally located vibration-free observation stations; availability of instruments and power supply; and the possibility of touching the object and permission to use the selected method.

For large objects (monuments, sites etc.), architectural mapping, regular topographic surveys and laser scanning are used along with photogrammetric techniques. Table 9.2 summarises the advantages and disadvantages of these techniques with respect to photogrammetry. In many cases a combination of methods is used, especially the coupling of photogrammetry with laser scanning. For the latter, the user can refer to Böhler (2005), and to Balletti *et al.* (2004), for a valuable comparison between the two techniques.

In addition, alternative systems have been proposed: for example, hybrid geodetic–photogrammetric stations coupled with laser ranging and digital cameras (Kakiuchi and Chikatsu, 2000; Scherer, 2004; Tauch and Wiedemann, 2004).

For small objects such as coins, tools and instruments, everyday items, jewellery, cloths, furniture, ornaments, weapons, musical instruments, paintings and frescos, and statues, alternative vision-based technologies are:

- shape from silhouettes (Tosovic *et al.*, 2002);
- shape from structured light (Tosovic *et al.*, 2002; Salvi *et al.*, 2004; Rocchini *et al.*, 2001; Gühring, 2001);
- shape from motion (video cues) (Enciso *et al.*, 1996; Chiuso *et al.*, 2000; Pollefeys *et al.*, 2000);
- shape from shading (Zhang *et al.*, 1999);
- shape from texture (Forsyth, 2002); and,
- shape from focus/defocus (Schechner, 2000; Favaro, 2002).

Table 9.2   Advantages and disadvantages of various measurement techniques

	Pros	Cons
Architectural mapping	Low level instrumentation and processing Low experience Adequate accuracy Quick	Requires an accurate reference frame (topography – photogrammetry) Mainly good only for objects of limited extent (e.g. interior spaces, excavation holes) and of low complexity Requires more time for field work than photogrammetry Point-wise mapping, no texture mapping
Topographic surveys	High accuracy Homogeneous overall accuracy Sound scientific indicators for quality assurance of the final product Good for plans and cross-sections	Mainly good only for objects of low complexity, otherwise not cost/time effective Requires more time for field work than photogrammetry Point-wise mapping, no texture mapping
Laser scanning	Good for complex continuous surfaces Good for surface analysis and visualisation	Edges cannot be extracted Line drawings cannot be derived Huge amount of data to handle

9.6 Case studies

9.6.1 3D mapping of cultural heritage artefacts

Kuzu (2003) reports on the reconstruction of a vase which is a museum item. A full 3D reconstruction serves as a permanent record of heritage artefacts. Such a reconstruction can be used to detect the changes required for conservation purposes. The models may also serve as manufacturing blueprints for machine production of replicas for exhibitions. Highly precise reconstructions of the artefacts or replica can be made accessible to scholars and visitors. Finally, another application area is virtual museums where artefacts and information resources of the museum can be viewed locally or on the internet.

An off-the-shelf CCD video-camera for acquisition of still images of the artefact was used. Additionally, a portable calibration frame was used in order to calibrate the interior orientation parameters of the camera. The calibration frame consisted of three perpendicular square planes and 25 control points on each side. The object was placed inside the calibration frame in order to aid definition of some natural control points.

Image segmentation was required to recover the vase's external shape. In order to segment object pixels from background pixels the object was placed in front of a homogeneous blue background. Multiple views of the object were captured, requiring a circular camera set-up. The focus of the camera was held fixed. A bundle block adjustment of all the images produced the coordinates of points on the vase, which then served as control points in the space carving processes. This bundle adjustment also determined the camera calibration parameters.

Obtaining the shape from the silhouette is a well-known approach for recovering the shape of objects from their contours. This approach is popular in computer vision and in computer graphics due to its fast computation and robustness. As a pre-condition of volume intersection algorithms, the contour of the real object must be extracted from input images. For this, the monochromatic blue background was used.

The output
The produced output is shown in Figure 9.2.

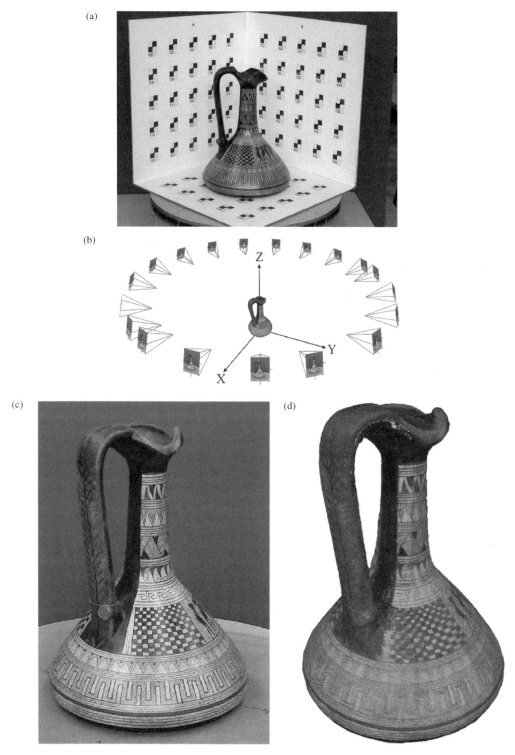

Figure 9.2   (a) *Camera calibration and control point determination; (b) camera set-ups; (c) visualisation of a replica of a Greek vase, Geometrical Period 900–800 BC, original image; (d) 3D reconstruction.*

Conclusions

For 3D mapping of artefacts, many techniques have been proposed (see Figures 9.3–9.6). Almost all of them are based on imaging techniques. Some of them are based purely on photogrammetric procedures and others use other sensors as well (e.g. laser scanners). The above example is based on the shape-from-silhouette vision technique, whereas reference is given below to other approaches as well. All in all, it can be concluded that photogrammetric vision techniques are very valuable for the accurate 3D mapping of small artefacts.

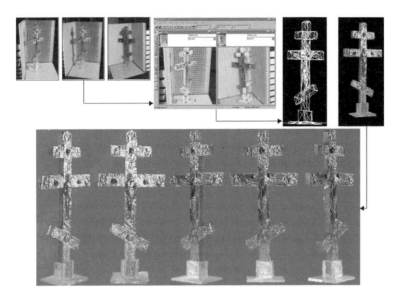

Figure 9.3   *3D reconstruction using photogrammetry (after Gementzis et al., 2001).*

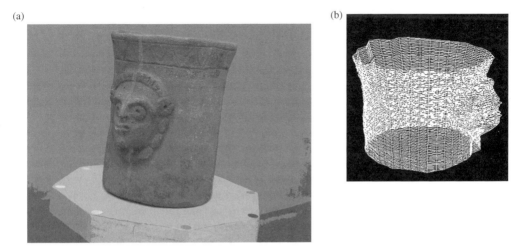

Figure 9.4   *The optical scanning system consists of a laser unit, a rotation unit and an imaging unit. (a) The mapping principle, (b) the output wire-frame. Expected accuracy 1/10,000 or better than 20 μm (after Tsioukas et al., 2004).*

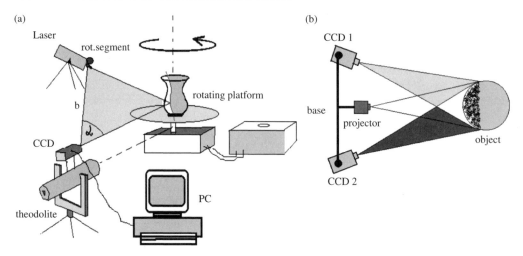

Figure 9.5   *The mapping system is similar to that of Figure 9.2. (a) The mapping principle; (b) the principle of two cameras (after Pavelka and Dolanský, 2003).*

For further discussion on documentation of objects of similar size and/or complexity, see also Böhler *et al.* (2003, 2004); Gemenetzis *et al.* (2001); Hanke and Böhler (2004); Pavelka and Dolanský (2003) and Tsioukas *et al.* (2004). The reader is urged to also consult other related very interesting applications like epigraphic surveys (Meyer *et al.*, 2004) or cave paintings (Fryer *et al.*, 2005).

9.6.2 3D mapping of underwater archaeological sites and artefacts

Location and historical background

Some 2000 years ago, *dolia* were large, almost spherical, ship-borne containers which had diameters up to 170 cm and a capacity of up to 2000 l. Unlike *amphorae* which were designed as removable containers, the *dolia* were placed on the keel during the construction of small ships. Situated underwater at the centre of the Secche della Meloria, near Livorno, Italy, there is a concentration of *dolia* wall and lip fragments spread out over an area of approximately 9 m × 6 m, part of which is on a sandy bottom but part is hidden by *posidonia* (*Posidonia oceanica*) bushes at depths of 5–6 m (Canciani *et al.*, 2000). The larger concentration of fragments is located near the *posidonia*, whose roots extend over 2 m and may be hiding part of the cargo load; other large *dolia* fragments have been found under a thin sand layer, close to the *posidonia*.

Purpose of the documentation

Only stratigraphic excavation of the site could bring definitive results: from preliminary data the wreck seemed to belong to the kind of trading ships specialised in carrying wine with very big *dolia*, during the first imperial age. The first layer of fragments now visible cannot be restituted with a traditional graphic survey because of the enormous quantity of fragments and the position of the site in the open sea. To gather the data, numerous and successive dives would be needed to take underwater measurements with respect to an underwater origin point. This process would be subject to human error.

A photogrammetry and 3D modelling programme on the site was started in 2001 by an international team of researchers from Nucleo Operativo Subacqueo, of Soprintendenza per i Beni Archeologici della Toscana (SBAToscana, the regional division of the Italian Ministry of

(a)

(b)

Figure 9.6 *Mapping of Maximilian I tomb using photogrammetry and laser scanning. (a) Virtual model from scanned data; (b) results from stereo plotting (after Böhler et al., 2003; Hanke and Böhler, 2004).*

Cultural Heritage which overseas the archaeological heritage of Tuscany), University of Rome 3 and Centre National de la Recherce Scientifique (CNRS) at Marseille, France. The purpose was to obtain more precise models than traditional graphic surveys, which are costly and rarely used for underwater sites now.

Instrumentation

Low cost instruments were used:

- Nikon Coolpix 990 digital camera in a Ikelite housing;
- PhotoModeler software for the orientation and plotting phase; and,
- Arpenteur software for photogrammetric restitution (Drap and Grussenmeyer, 2000);

A set of rulers to provide scale and several buoys for vertical reference were used. (The buoys were made with perforated table tennis balls.)

Documentation

In the summer of 2001, in only three days of diving, about 100 good quality digital photographs were obtained while the water was of crystal transparency. The camera was calibrated with the PhotoModeler calibration module in order to compensate for lens distortion and the effects of refraction in the water. The underwater housing for the camera was considered rigid and non-deformable. The radial distortion (pin-cushion shape) was modelled by an increase in the focal length and the usual parameters for radial distortion.

To simplify and organise the photogrammetric restitution, a photomosaic was made as a first approximation for a photographic map covering the complete site. About 400 model points were traced from this restitution, with an accuracy of 1%, establishing an average of six points on at least four adjoining photos.

For the photogrammetric restitution, the Arpenteur software used a wire-frame model imported from PhotoModeler and utilised the computed exterior orientations for the underwater camera positions. The model was further enhanced to obtain a dynamic visualisation, adapting the textures derived by the underwater shootings to provide a more complete 3D representation.

In 2003, a survey of two large fragments discovered in 2000 in the "*dolia*" wreck site, was carried out photogrammetrically. The restitution was very accurate with respect to the depth survey (*ca.* 1400 points, with a precision of 0.075%), an exact definition of the fragment's geometry (rotation axis, maximum and minimum diameter, internal and external diameter of the lip), and a characterisation of the surface through the textures acquired by the photographs.

Output

A comparison has been made between the textured fragment and the model defined by a reconstructive hypothesis using a direct survey of a generic section. In this way it has been possible to verify the congruence of the model reconstructed with the photogrammetric survey directly with the final mixed model. Software was used to allow virtual reality modelling in accordance with standard predefined views (frontal, lateral, planimetric etc.). The photogrammetric output from the Meloria is illustrated in Figure 9.7 (see colour section).

Conclusions

Mapping underwater sites and artefacts is a very difficult job. Even if divers are able to approach the site, staying in deep water is difficult and potentially dangerous. Therefore diving time should be kept as short as possible (normally measured in minutes). It is thus essential that the data collection time is kept very short. This makes other surveying techniques impossible to use. Also, reduced accuracy could usually be expected at such depths. Photogrammetry has long been proved to be a viable technique in such situations.

Related references which discuss the documentation of objects of similar size and/or complexity include Drap and Grussenmeyer (2000), Drap and Long (2001) and Drap *et al.* (2002).

9.6.3 Mapping the excavation site of "Toumba", Thessaloniki, Greece

Historical background

"Toumba" is today one of the most well-known neighbourhoods in Thessaloniki. The name of the neighbourhood comes from an impressive wide and high hill which commands the whole area. Today, this area is 3 km from the sea and its highest point is at an altitude of 80 m above mean sea level. The archaeological site is on the hill occupying 18,000 m^2 and rising to a height of 22 m (Figure 9.8). It has been a settlement with various uses through time. All the findings indicate that the area has been inhabited since at least the Late Bronze Age (starting from the fifteenth century BC) and possibly even in the Middle Bronze Age (in the fifteenth to twentieth century BC).

The excavation process on the site has continued for more than a decade, with about 7% of the whole area having been investigated. The two main areas that have been excavated cover part of the top and part of the western hillside. The excavation proceeds on a cellular (excavation ditches) basis. Each cell is a 4 m × 4 m square and between the cells a corridor of 1 m width is left. As the excavation proceeds, the corridors are eventually destroyed and the neighbouring cells are merged to form larger cells.

For further information see: Patias *et al.* (1999), Georgiadis *et al.* (2000), Pateraki *et al.* (2002) and Koussoulakou *et al.* (2001b).

Purpose of the documentation

There is a wide range of excavation types, each of which is for a differing purpose and requires differing approaches (e.g. University of Vienna, 2005b):

- rescue excavations: very fast solutions, accuracy is a secondary consideration;
- stratigraphic excavations: 3D models of strata;
- excavations of graveyards and burial sites: sometimes there are several thousands of skeletons; standardised recording procedure is needed;
- excavations of settlements without stone walls: recording procedures can change depending on features; any system used must be adaptable;

Figure 9.8 *Aerial image of the "Toumba" archaeological site.*

- excavations of settlements with stone walls: to go on with excavation, walls and pavements must be removed after they have been documented (excavations in towns can be very complicated);
- documentation of wall-paintings;
- excavations in moist areas: organic materials (tools, wooden pavements, wooden walls etc.) have to be recorded instantly;
- excavations in remote areas without infrastructure (desert, jungle, mountains etc.);
- excavations underwater: there are special problems, which require sophisticated solutions; and,
- excavation of hunter-gatherer sites that require different approaches.

In this case, full documentation of the ditches during the on-going excavation (i.e. almost in real-time) is required. This includes the 3D mapping of the ditches, the documentation of the 3D position of any finds and their archaeological interpretation. A special requirement is that the documentation should be quick, accurate to 1–2 cm and should not interfere with the archaeological work on site. Finally, the delivered digital products will be used as a digital 3D record of the excavation period and thus should be part of a GIS system, serving as a digital archive.

Instrumentation

For control point measurements, a total station has been used and regular surveying procedures adopted. Imagery was acquired, using the Kodak DCS420 digital camera which uses a Nikon N90 camera body as shown in Figure 9.9 and which features a 1.5 million pixel sensor with 9 μm pixel size and 36-bit colour. Its advantages are: rapid image acquisition, ease of use by non-specialists, low price and the ability to check the quality of the images on site. The disadvantages are that the camera is "amateur" and thus calibration is required, and due to the small sensor format, many images are required in order to cover the area of a single ditch.

The documentation process

Mapping of the excavation continues throughout the day, and contains all the steps shown in the workflow in Figure 9.10. Early in the morning, when an excavation day starts, control points are placed and images acquired.

Figure 9.9 *The Kodak DCS 420 digital camera, featuring a Nikon N90 camera body, 1.5 million pixels sensor, 9 μm pixel size, with 36-bit colour. (Photo by Jarle Aasland, NikonWeb.com)*

It should be mentioned that the control points are measured only once, at the beginning of the excavation period which usually lasts for one month. They are targeted and are placed mainly in the corridors between the ditches, as well as at the paries of the ditches, and in other places, where they are less likely to be disturbed. Control points are also placed in the ditches, and replaced if destroyed (Figure 9.11). These points are surveyed and their 3D coordinates determined relative to the coordinate system, generally to an accuracy of 5 mm.

The basic problem in the photogrammetric process is that due to excavation a number of the control points are frequently lost, mainly those which lie on the ditch floor. This sometimes leads to the use of tie points from a previous day's solution as secondary control points in the next day's images.

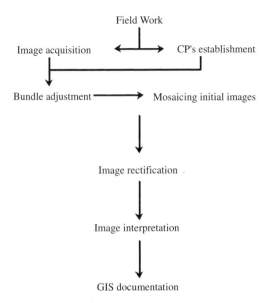

Figure 9.10 *Workflow of the documentation procedure.*

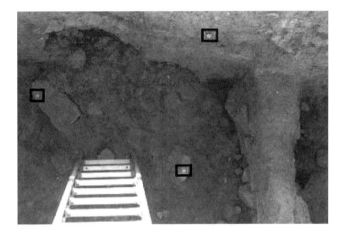

Figure 9.11 *Example of control points in the excavation ditch.*

The images are normally acquired in the early morning, to avoid the generation of saturated spots that can appear under strong noonday sunlight. Colour images were originally acquired, but greyscale was finally used. As mentioned previously, due to the small sensor size and the narrow field-of-view, the number of images was large and additionally the geometry of the block of images had some deficiencies. The base-to-height ratio and the camera rotations were not optimally configured as the bases between the camera stations were rather small and differences in camera axes angles were up to 30°. An average number of ten images were acquired per ditch, from a height of approximately 5 m, using a small ladder.

A bundle adjustment procedure with simultaneous camera self-calibration using the parameters proposed by Brown (1971) was performed in order to improve the accuracy of the orientation through modelling the lens distortion parameters. The overall theoretical precision computed by the bundle adjustment was given by standard deviations of 17 mm, 14 mm and 21 mm in the X, Y and Z directions, respectively.

The procedure followed thereafter aimed at reducing data storage and making the handling easier by combining the acquired images into large virtual images (Figure 9.12). Consequently, a virtual image could be substituted for a strip of images and be used for further processing, such as digital surface model (DSM) generation and 3D mapping. The generation of virtual images through projective transformation has been discussed by Stephani (1999). Pöntinen (2004) and Koistinen (2000) use concentric image sequences, which can be projected onto a cylindrical surface and adjacent frames combined to a panoramic image. Detailed information on the procedure can be found in Pateraki *et al.* (2002).

The radiometry of the overlapping virtual images was optimised through filtering with a Wallis filter in order to reduce noise, while preserving fine detail such as one-pixel wide lines, corners and line end-points. Additionally, Canny filtering was used to extract edge features and to help the subsequent automatic matching. On average 18,000 points per model were

(a)

(c)

(b)

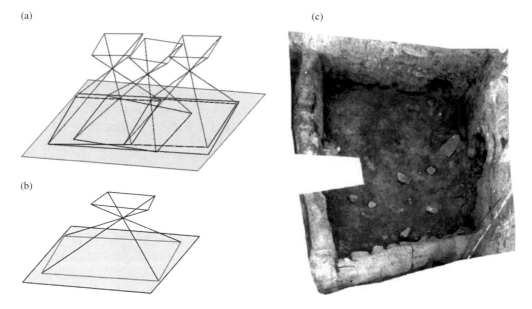

Figure 9.12 *(a) Corners of sub-images are projected on the predefined plane and the enclosed rectangle in object space is found; (b) back projections of the corners of the rectangle define the size of the virtual image; (c) the final virtual image.*

successfully matched. The matched 3D edges were finally converted to 3D grid points through interpolation, with a 2 cm grid spacing and an accuracy of 3–5 cm in height.

The computed DSM was used for the generation of the orthorectified images with an accuracy of 2–3 cm. These orthoimages were transferred back to the site, at about midday, so that archaeologists were able to proceed with the interpretation and the information collection for the GIS development.

The photo-interpretation vector information was digitised, firstly by hand and later in the office, overlaid on the images and saved in digital format. Additional archaeological information was marked on the orthophotomap. As at many archaeological sites, the hill of Toumba has been created by continuous layered deposits through time, being settlements built one on the top of the other. During the excavation process it was found that in most cases, parallel layers were not discovered in reverse time order. This means that finds or elements of earlier layers were found in later layers. Archaeologists, in order to define the relationship between findings and layers, established the term "stratigraphy".

Stratigraphy is the recording of the successive deposition of layers and findings in 3D space. So, there was a clear need to build a GIS that manipulates 3D drawings. A time-based classification has been created by archaeologists, called "phases", to describe findings and artefacts as belonging in a specific epoch. Phase can be assigned to a prehistoric object more clearly than its exact chronology and is generally assigned to it by means of the object's characteristics (from appearance, comparison etc.).

The developed 3D GIS consists of different layers to accommodate the six different "phases" for the two different categories of finds (i.e. architectural constructions, e.g. houses, storing places etc., and mobile elements, stone, metal, bone, pottery and shell remains).

Output
Figure 9.13 (see colour section) depicts the full range of delivered output.

Conclusions
It is concluded that photogrammetry, even with amateur cameras, can provide full and accurate documentation of excavation sites. With the use of sophisticated techniques the procedure can further be simplified and can even be adopted by non-experts. The 3D digital products produced can serve a variety of aims including digital 3D archiving. Readers in need of further details on the advantages and disadvantages of photogrammetry in archaeological excavations are referred to University of Vienna (2005a).

For related references for documentation of objects of similar size and/or complexity, see Arias *et al.* (2004), Brown (1971), Guidi *et al.* (2002), Koistinen (2000), Skarlatos *et al.* (2004) and Stephani (1999).

9.6.4 3D reconstruction of the Great Buddha of Bamiyan

Historical background
In the centre of Afghanistan, the town of Bamiyan is situated about 200 km north-west of Kabul at an altitude of approximately 2500 m. Bamiyan is considered an oasis in the centre of a long valley that separates the big chains of the Hindu Kush Mountains, and the city functioned as one of the greatest Buddhist centres for nearly five centuries. In the great valley of Bamiyan, two large standing Buddha statues and a small seated Buddha were carved out of the sedimentary rock of the region. They were begun in the second century AD under the patronage of Emperor Kanishka and probably finished around the fourth and fifth centuries AD.

The two massive Buddhas, which stood about one kilometre apart, were carved out of a high stretch of cliff facing the widest part of the valley. These colossal statues were the largest Buddhist sculptures in the world. The larger statue was 55 m high and was carved at the western end of the cliff face. It was painted red and it is thought to represent Vairocana, the "Light Shining throughout the Universe Buddha". The smaller statue was about 38 m high and it was situated at the eastern end of the cliff. It was painted in blue and probably represented Buddha Sakyamuni, although the local Hazara people believe it depicted a woman.

The two colossal Buddha statues carved into the sandstone cliffs of Bamiyan, as well as many other Buddhist statues of the area, were demolished by the Taliban militia on 8 March 2001 (see Figure 9.14). After the destruction, a campaign was launched to reconstruct the Buddhas in their original shape, size and place (see http://www.photogrammetry.ethz.ch/research/bamiyan/buddha/index.html). A virtual reconstruction is based on archived metric and non-metric images, using photogrammetric techniques. For a series of detailed publications on this project, see Grün *et al.* (2002a,b; 2003a–c; 2004), Grün and Remondino (2005) and Remondino (2004).

(a)

(b)

Figure 9.14 *(a) The Green Valley of Bamiyan with the two large standing Buddha statues and a smaller one in the middle; (b) the explosion of March 2001; (c) the empty niche where the Great Buddha once stood, as seen in August 2003.*

(c)

Figure 9.14   *(continued)*

Data used

Different 3D computer models of the Great Buddha have been created, derived from three different data set samples of which are shown in Figure 9.15 (see colour section).

The first data set consists of four low resolution images of the big statue found on the internet. These images have different resolutions and size, unknown camera constants and were acquired at different times. The second data set consists of high resolution metric photographs taken in 1970 (see Grün *et al.*, 2002a). The reconstructed model of the statue, in particular the one obtained with the manual measurements, can be used for the physical reconstruction of the Great Buddha in Bamiyan. The third data set consists of some tourist images taken between 1965 and 1969.

The documentation process

The reconstruction process for all three data sets consists of:

- image calibration and orientation;
- image matching to extract the 3D point cloud;
- point cloud editing and surface triangulation; and,
- texture mapping and visualisation.

For the three metric images, the camera parameters were recovered using a self-calibrating bundle adjustment, measuring the tie points semi-automatically by means of least squares matching. The final average standard deviations of the object coordinates in the X, Y and Z directions were 0.014 m, 0.017 m and 0.021 m, respectively.

After the establishment of the adjusted image block, the 3D reconstruction of the statue was performed with automated procedures and manual measurements. The automated tie point generation methods provide dense point clouds (in the presence of good texture) but they can fail to extract the very fine details of the statue, like the folds of the robe. Only detailed manual measurements allow the generation of 3D models which are precise, reliable and complete enough to serve as a basis for a physical reconstruction. Three stereo-models were set up and points measured along horizontal profiles at 20 cm intervals, while the main edges were measured as break-lines. A triangulated irregular network for the surface model was generated using the method of Delaunay triangles by dividing the measured point cloud into separate parts. A mesh was created for each single point cloud and then all surfaces were merged together. The various products are illustrated in Figure 9.16 (see colour section).

The modelling of the empty Buddha niches was also performed using five digital images acquired with a Sony Cybershot F707 during a field campaign in August 2003. The image sizes were 1920 × 2560 pixels and the pixel size is about 3.4 µm. After image orientation three stereo-models were set up and points were manually measured along horizontal profiles, while the main edges were measured as break-lines. Thus a point cloud consisting of approximately 12,000 points was generated.

Output

The 3D model, generated with manual measurement of the metric images, can be used for a physical reconstruction of the Great Buddha of Bamiyan, due to its high accuracy (1–2 cm relative accuracy). The final 3D model of the Great Buddha was used for the generation of different physical models of the Great Buddha as illustrated in Figure 9.17. In particular, a 1:25 scale model was generated for the Swiss pavilion of the 2005 EXPO in Aichi, Japan.

(a)

Figure 9.17 *(a) The produced physical model (1:200); (b) the 1:25 physical model realised for the EXPO 2005 in Japan.*

(b)

Figure 9.17 *(continued).*

Conclusions

It has been demonstrated that photogrammetry can accurately reconstruct medium-size objects based on either metric images or non-metric images. The final output can be used for many purposes, including this case of recreating destroyed objects either in replicas or even in 1:1 scale physical models. Readers interested in related references to this case study should consult Beraldin *et al.* (2002), Bitelli *et al.* (2002), Dorffner *et al.* (2000), Drap *et al.* (2000), El-Hakim *et al.* (2002), Monti *et al.* (1999) and Ruther *et al.* (2001).

9.6.5 3D mapping and documentation of historical city centres

Historical background

Thessaloniki, the second most populated city in Greece, has a long history. Traces of the first inhabitants in the area have been found, which date back to 3000 BC. The city's development went through different historic periods, such as Roman, Byzantine, Ottoman and finally its liberation in 1912. During its history it has repeatedly suffered from invasions by Germans, Slavs,

Normans, Arabs, Francs, Catalans and Venetians. The beginning of the twentieth century found a wealthy Thessaloniki, with an important harbour, a much developed public transportation system, a gas delivery system, and an electric power factory.

On 5 August, 1917, a fire destroyed a large part of the inner city; an area of about 1,000,000 m². Due to the many wooden buildings and a strong wind, the fire burnt for three days destroying whole blocks. The extent of the damage is obvious from aerial photography of that time (see Figure 9.18 (see colour section)). After the fire, a new urban plan was designed by the French architect E. Hebrard, to which the city owes its current form.

The historic centre has since been reconstructed and the few buildings that have been salvaged are protected by the Greek law as sites of architectural and historic value. Due to its high cultural value, it has been proposed that the city archives, consisting of engineering plans, façade drawings, old photos and construction details be put under the protection of UNESCO, as an item belonging to world cultural heritage. This application concerns the historic part of the city, and offers a management solution to architectural heritage information.

Purpose of the documentation

The purpose of the documentation is: first, to develop a 3D digital archive for the city's architectural heritage; secondly, to develop a web-based system for the dissemination of information to all involved parties and to the general public and finally, to develop an information management system in order to support a range of uses (see Patias *et al.,* 2002).

Instrumentation

The instrumentation used to create this documentation consists of:

- Ashtec-L1 GPS receivers for control point determination;
- digital photogrammetric station for aerial restitution;
- image rectification software;
- visualisation software;
- GIS system; and,
- HTML programming.

The documentation process

In order to support a proper and accurate geo-referencing of the information system, the first phase was the production of an accurate and detailed orthophoto representing the urban structure. Emphasis was put on the core part of the historic centre, which had been destroyed by the fire of 1917.

For this reason, recent aerial photography at 1:20,000 scale was used. A total of 39 images were scanned at a resolution corresponding to a pixel size of 15 μm on the image plane. The necessary ground control points were measured by an Ashtec-L1 GPS receiver to a 3D accuracy of 3 cm. A total of 26 ground control points, aligned to the national reference system, were measured to support the aerial mapping required.

The aerial images were photogrammetrically oriented and triangulated, and the terrain surface was modelled by measuring DTM points. These were photogrammetrically collected using manual methods, due to the harsh conditions (high buildings, shadows etc.) in the historic urban centre of this densely populated city. A total of approximately 400 points was measured to an estimated height accuracy of 40 cm.

Subsequently, an ortho-image was produced at a scale of 1:5,000 with a planimetric accuracy of 35 cm and a ground pixel size of 50 cm. Finally the ortho-image was draped on the underlying DTM in order to produce a 3D visual effect. In addition, and at a much larger scale, a geometric database of 250 officially protected buildings, has been developed. These buildings have been reconstructed and rendered in 3D, by standard stereoscopic photogrammetric techniques from aerial images. Each building has been "cropped" at the existing DTM level, in order to reduce any visual "floating" effect, an undesirable feature caused by inaccuracies in DTM modelling.

For each building, the ground plan and at least one roof point (to produce the building height) have been stereoscopically measured on the aerial images. From these measurements, a control point frame was derived for the reconstruction of the main façades. This frame will be used later for the rectification of the digital close range image of each building façade.

Each building's main façade has been visually documented by close range terrestrial digital images with the use of amateur digital off-the-shelf 3 megapixel cameras at an average scale of 1:200. All the images were taken in winter time, in order to reduce the extent of the areas that might be obscured due to tree canopies and shadows. A total of 250 images were acquired.

Using the above mentioned control point frame (i.e. two points at the ground level and two points at the roof level, all at the façade corners), the façade images have been rectified, mosaiced, and draped on the 3D building model for a full view of the façade at a scale of 1:50, and at a ground pixel size of 2 cm.

Finally, for each building, a thorough search of official and individually owned archives, as well as a literature survey revealed a wealth of historic information, concerning engineering drawings (floor plans, façade construction details etc.), old photographs, ownership records, protection status and plans etc. All of this information, has been collected, transformed to digital form and "intelligently" archived in a database, with proper links to old and new images, document files and a bibliography. Furthermore, the information system has been designed and developed for internet access, through the use of HTML pages.

The developed mapping material aims to familiarise and orientate the user through space at different scales and resolution. It also aims at giving a visual impression of the whole urban area and of the historic centre in particular. The user can choose to fly over the city through virtual video sequences, in which case an overview of the protected sites can be provided. These can also be viewed as 3D rendered blocks. The user can further choose to go "on-the-ground" and walk through the city centre via the virtual walk-through video option. Through choosing one of the available routes the user can examine the protected buildings in more detail, complete with their surrounding area, and view their façades in 3D. The full photogrammetric process is illustrated in Figure 9.19 (see colour section) while the web-based information is shown in Figure 9.20 (see colour section).

Output and conclusions

The study process has produced:

- a DTM consisting of about 400 points with a height accuracy of 40 cm;
- an orthophotomap at a scale of 1:5,000 with a planimetric accuracy of 35 cm and a ground pixel size of 50 cm;
- 3D reconstruction of 250 buildings with full-view facades at a scale of 1:50, and with a ground pixel size of 2 cm; and,
- GIS and visualisation products.

This case study shows how image analyses can be integral, not only for historical documentation over a large and complex area, but also to provide the basis for future preservation and visualisation. For details of documentation projects of objects of similar size and/or complexity, see Agosto *et al.* (2004), Bouroumand and Studnicka (2004), Gatti *et al.* (1999), Grün and Beutner (2001), Grussenmeyer and Yasmine (2003), Hanke (2004), Lianos *et al.* (2004), Lohr *et al.* (2004), Nour El Din *et al.* (1999), Ogleby (1996), Pomaska (2000) and Streilein and van den Heuvel (1999).

9.7 References

Agosto, E., Porporato, C. and Rinaudo, F., 2004. Stereo photomap and 3D realistic model of city centers. *Proceedings of the International Workshop on Vision Techniques applied to Rehabilitation of City Centres"*, Lisbon, Portugal, 25–27 October 2004: Workshop CD.

Arias, V., Kuntz, T.R., Richards, H., Watson, R.P. and Van Der Elst, J., 2004. Documentation and virtual modeling of archaeological sites using a non-metric extreme wide-angle lens camera. *International Archives of Photogrammetry and Remote Sensing*, 35(B5), 824–827.

Atkinson, K.B. (Ed.), 1996. *Close Range Photogrammetry and Machine Vision*, Whittles Publishing, Scotland.

Balletti, C., Guerra, F., Lingua, A. and Rinaudo, F., 2003. True digital orthophoto of the San Marco basilica in Venice. International Workshop on Vision Techniques for Digital Architectural and Archaeological Archives, Ancona, Italy, 1–3 July, *International Archives of Photogrammetry, Remote Sensing and Spatial Information Sciences*, 34(5/W12), 43–48.

Balletti, C., Guerra, F., Vernier, P., Studnicka, N., Riegl, J. and Orlandini, S., 2004. Practical comparative evaluation of an integrated hybrid sensor based on photogrammetry and laser scanning for architectural representation. *International Archives of Photogrammetry and Remote Sensing*, 35(5), 536–541.

Barber, D., Mills, J. and Bryan, P., 2002. Experiences of laser scanning for close range structural recording. *Proceedings of CIPA WG 6: International Workshop on Scanning for Cultural Heritage Recording*, Corfu, Greece, 1–2 September, 121–126.

Bartolota, M., Di Naro, S., Lo Brutto, M., Misuraca, P. and Villa, B., 2000. Information systems for preservation of cultural heritage. *International Archives of Photogrammetry and Remote Sensing*, 33(B5/1), 864–870.

Beraldin, J.-A., Picard, M., El-Hakim, S., Godin, G., Latouche, C., Valzano, V. and Bandiera, A., 2002. Exploring a Byzantine crypt through a high-resolution texture mapped 3D model: Combining range data and photogrammetry. *Proceedings of CIPA WG 6: International Workshop on Scanning for Cultural Heritage Recording*, Corfu, Greece, 1–2 September, 65–72.

Bitelli, G., Capra, A. and Zanutta, A., 2002. Digital photogrammetry and laser scanning in surveying the "Nymphaea" in Pompeii. *Proceedings of CIPA WG 6: International Workshop on Scanning for Cultural Heritage Recording*, Corfu, Greece, 1–2 September, 115–120.

Boccardo, P., Dequal, S., Lingua, A. and Rinaudo, F., 2001. True digital orthophoto for architectural and archaeological applications. International Workshop on Recreating the Past – Visualization and Animation of Cultural Heritage, Ayutthaya, Thailand, 26 February–1 March, *International Archives of Photogrammetry and Remote Sensing,* 34(5C15), 50–55.

Böhler, W., 2005. Comparison of 3D scanning and other 3D measurement techniques. *Proc. International Workshop on Recording, Modeling and Visualization of Cultural Heritage*, Centro Stefano Franscini, Monte Verità, Ascona, Switzerland, 22–27 May 2005, 89–99.

Böhler, W., Bordas Vicent, M., Hanke, K. and Marbs, A., 2003. Documentation of German Emperor Maximilian I's Tomb. *International Archives of Photogrammetry and Remote Sensing*, 34(5/C15) and *CIPA International Archives for Documentation of Cultural Heritage*, 19, 474–479. See http://cipa. icomos.org/fileadmin/papers/antalya/127.pdf (accessed 30 October 2006).

Böhler, W., Bordas Vicent, M., Heinz, G., Marbs, A. and Müller, H., 2004. High quality scanning and modeling of monuments and artifacts. *Proc. FIG Working Week 2004*, Athens, Greece, 22–27 May. See http://www.fig.net/pub/athens/programme.htm (accessed 30 October 2006).

Boulanger, P. and Taylor, J., 2000. The virtual museum – data generation and visualization. *International Archives of Photogrammetry and Remote Sensing*. 33(*Special Sessions, Supplement*), 4–9.

Bouroumand, M. and Studnicka, N., 2004. The fusion of laser scanning and close range photogrammetry in Bam laser-photogrammetric mapping of Bam citadel (Arg-E-Bam)/Iran. *International Archives of Photogrammetry and Remote Sensing*, 35(5), 979–985.

Brown, D.C., 1971. Lens distortion for close-range photogrammetry. *Photogrammetric Engineering*, 37(8), 855–866.

Canciani, M., Gambogi, P., Romano, F., Cannata, G. and Drap, P., 2000. Low cost digital photogrammetry for underwater archaeological site survey and artifact insertion. The case of the Dolla wreck in Secche della Meloria, Livorno, Italy. International Workshop on Vision Techniques for Digital Architectural and Archaeological Archives, Ancona, Italy, 1–3 July. *International Archives of Photogrammetry, Remote Sensing and Spatial Information Sciences*, 34(5/W12), 95–100.

Carbonnell, M., 1989. *Photogrammetrie appliquee aux releves des monuments et des centres historique.* ICCROM (International Centre for the Study of the Preservation and the Restoration of Cultural Property), Rome, Italy.

Chikatsu, H. and Anai, T., 1998. Relics modeling and visualization in virtual environment. *Proceedings of the Conference on Real-Time Imaging and Dynamic Analysis*, Hakodate, Japan, 2–5 June. *International Archives of Photogrammetry and Remote Sensing*, 32(5), 528–532.

Chiuso, A., Jin, H., Favaro, P. and Soatto, S., 2000. MFm: 3D motion and structure from 2D motion causally integrated over time: implementation. In Vernon, D. (Ed.), *Computer Vision – ECCV 2000*, Lecture Notes in Computer Science 1843, Springer-Verlag, Berlin, Heidelberg, Germany, 734–750.

CIPA, 2005. See http://cipa.icomos.org/ (accessed 20 September 2005).

CIPA-Documentation underwater photogrammetric task group, 2005. http://cipauwp.gamsau.archi.fr (accessed 20 September 2005).

D'Ayala, D. and Smars, P., 2003. *Minimum requirement for metric use of non-metric photographic documentation*, University of Bath Report. See http://www.english-heritage.org.uk/server/show/ conWebDoc.4274 (accessed 30 October 2006).

Dequal, S., Lingua, A. and Rinaudo, F., 1999. A new tool for architectural photogrammetry: The 3D Navigator. *Proc. XVII CIPA Symposium, Mapping and Preservation for the New Millennium*, (Olinda, Brazil, 3–6 October).

Doneus, M., 1996. Photogrammetrical applications to aerial archaeology at the Institute for Prehistory of the University of Vienna, Austria. *International Archives of Photogrammetry and Remote Sensing*, 31(B5), 124–129.

Dorffner, L., Kraus, K., Tschanner, J., Altan, O., Külür, S. and Toz, G., 2000. Hagia Sophia – photogrammetric record of a world cultural heritage. *International Archives of Photogrammetry and Remote Sensing,* 33(B5/1), 172–178.

Drap, P. and Grussenmeyer, P., 2000. A digital photogrammetric workstation on the web. *ISPRS Journal of Photogrammetry and Remote Sensing*, 55(1), 48–58.

Drap, P. and Long, L., 2001. Towards a digital excavation data management system: the "Grand Ribaud F" Estruscan deep-water wreck. *Proceedings of Conference on Virtual Reality, Archeology, and Cultural Heritage, 2001* (Glyfada, Greece), 17–26.

Drap, P., Bruno, E., Long, L., Durand, A. and Grussenmeyer, P., 2002, Underwater photogrammetry and xml based documentation system: The case of the 'Grand Ribaud F' Estruscan wreck. ISPRS Commission V Symposium: Close-range Imaging, Long-range Vision, Corfu, Greece, 2–6 September, *International Archives of Photogrammetry and Remote Sensing*, 34(5), 342–347.

Drap, P., Gaillard, G., Grussenmeyer P. and Hartmann-Virnich, A., 2000. A stone-by-stone photogrammetric survey using architectural knowledge formalised on the Arpenteur photogrammetric workstation. *International Archives of Photogrammetry and Remote Sensing*, 33(B5/1), 187–194.

El-Hakim, S., Beraldin, J.A. and Picard, M., 2002, Detailed 3D reconstruction of monuments using multiple techniques. *Proceedings of CIPA WG 6: International Workshop On Scanning For Cultural Heritage Recording*, Corfu, Greece, 1–2 September, 58–64.

Enciso, R., Zhang, Z. and Viéville, T., 1996. Dense reconstruction using fixation and stereo cues. *Project RobotVis, World Automation Congress* 1996, Montpellier, France.

English Heritage, 2005. Metric survey specifications for English Heritage. See http://www.english-heritage.org.uk/ (accessed 20 September 2005).

Favaro, P., 2002. *Shape from Focus/Defocus*. Washington University Department of Electrical Engineering Electronic Signals and Systems Research Lab, June 25th, 2002. See http://homepages.inf.ed.ac.uk/rbf/CVonline/LOCAL_COPIES/FAVARO1/dfdtutorial.html (accessed 30 October 2006).

Forsyth, D.A., 2002. Shape from texture without boundaries. *Proceedings of the 7th European Conference on Computer Vision*-Part III, 225–239.

Fryer, J., Chandler, J. and El-Hakim, S., 2005. Recording and modelling an aboriginal cave painting: with or without laser scanning? ISPRS Comm. V. Workshop on 3D Virtual Reconstruction and Visualisation of Complex Architectures, 22–24 August 2005, Mestre-Venice, Italy. *International Archives of Photogrammetry and Remote Sensing*, XXXVI (SW 17) CD-ROM.

Gatti, M., Govoni, C., Pellegrinelli, A. and Russo, P., 1999. A 3-D urban model from aerial and terrestrial photographs. ISPRS WG V/2 and WG5 Workshop, Photogrammetric Measurement, Object Modeling and Documentation in Architecture and Industry, Thessaloniki, Greece, 7–9 July. *International Archives of Photogrammetry and Remote Sensing,* 32(5W11), 141–145.

Gemenetzis, D., Georgiadis, H. and Patias, P., 2001. Virtuality and Documentation. International Workshop on Recreating the Past – Visualization and Animation of Cultural Heritage, Ayutthaya, Thailand, 26 February–1 March. *International Archives of Photogrammetry and Remote Sensing,* 34(5C15), 159–164.

Georgiadis, Ch., Tsioukas, V., Sechidis, L., Stylianidis, E. and Patias, P., 2000. Fast and accurate documentation of archaeological sites using in-the-field photogrammetric techniques. *International Archives of Photogrammetry and Remote Sensing*: 33(Supplement B5), 28–32.

Glykos, T., Karras, G. and Voulgaridis, G., 1999. Close-range photogrammetry within a commercial CAD package. ISPRS WG V/2 and WG5 Workshop: Photogrammetric Measurement, Object Modeling and Documentation in Architecture and Industry, Thessaloniki, Greece, 7–9 July. *International Archives of Photogrammetry and Remote Sensing,* 32(5W11), 103–106.

Grammatikopoulos, L., Kalisperakis, I., Karras, G., Kokkinos, T. and Petsa, E., 2004. On automatic ortho-projection and texture-mapping of 3D surface models. *International Archives of Photogrammetry and Remote Sensing*, 35(5), 360–365.

Grün, A. and Beutner, S., 2001. The geoglyphs of San-Ignacio – New results from the Nasca project. *Proceedings of the International Workshop on Recreating the Past – Visualization and Animation of Cultural Heritage*, Ayutthaya, Thailand, 26 February–1 March.

Grün, A. and Remondino, F., 2005. Die Rueckker der Buddhas – Photogrammetrie und kulturelles Erbe in Bamiyan, Afghanistan. *PFG - Photogrammetrie, Fernerkundung, Geoinformation*, 1(9), 57–68 and 2(10), 157–164.

Grün, A., Remondino, F. and Zhang, L., 2002a. Reconstruction of the Great Buddha of Bamiyan, Afghanistan. *ICOMOS International Symposium*, Madrid, 15 December 2002.

Grün, A., Remondino, F. and Zhang, L., 2002b. Reconstruction of the Great Buddha of Bamiyan, Afghanistan. *International Archives of Photogrammetry and Remote Sensing,* 34(5), 363–368.

Grün, A., Remondino, F. and Zhang, L., 2003a. Automated Modeling of the Great Buddha Statue in Bamiyan, Afghanistan. IC II/IV, WG III/4, III/5, III/6 Workshop, Photogrammetric Image Analysis, Munich, Germany, 17–19 September. *International Archives of Photogrammetry, Remote Sensing and Spatial Information Sciences*, 34(3/W8), 11–16.

Grün, A., Remondino, F. and Zhang, L., 2003b. Computer Reconstruction and Modeling of the Great Buddha of Bamiyan, Afghanistan, 18th CIPA International Symposium 2003, Antalya, Turkey, 30 September – 4 October. *International Archives of Photogrammetry Remote Sensing*, 34(5/C15), 440–445. See http://cipa.icomos.org/fileadmin/papers/antalya/120.pdf (accessed 30 October 2006).

Grün, A., Remondino, F. and Zhang, L., 2003c. Image-based reconstruction and modeling of the Great Buddha of Bamiyan, Afghanistan. International Workshop on Vision Techniques for Digital Architectural and Archaeological Archives, Ancona, Italy, July 1–3. *International Archives of Photogrammetry, Remote Sensing and Spatial Information Sciences*, 34(5/W12), 173–175.

Grün, A., Remondino, F. and Zhang, L., 2004. Photogrammetric reconstruction of the Great Buddha of Bamiyan, Afghanistan. *Photogrammetric Record*, 19(107), 177–199.

Grussenmeyer, P. and Yasmine, J., 2003. The restoration of Beaufort Castle (South-Lebanon): A 3D restitution according to historical documentation. 18th CIPA International Symposium 2003, Antalya, Turkey, 30 September–4 October. *International Archives Photogrammetry Remote Sensing*, 34(5/C15), 322–327.

Guarnieri, A., Vettorea, A., El-Hakim, S. and Gonzo, L., 2004a. Digital photogrammetry and laser scanning in cultural heritage survey. *International Archives of Photogrammetry and Remote Sensing*, 35(5), 154–158.

Guarnieri, A., Vettore, A. and Remondino, F., 2004b. Photogrammetry and ground-based laser scanning: Assessment of metric accuracy of the 3D model of Pozzoveggiani Church. *FIG Working Week 2004*, Athens, Greece, 22–27 May, 15 pages. See http://www.fig.net/pub/athens/programme.htm (accessed 30 October 2006).

Guerra, F. and Balletti, C., 1999a. 3D Reconstruction for the representation of the church of S. Martino: Solid modelling, mapping, CAD, photogrammetry. ISPRS WG V/2 and WG5 Workshop: Photogrammetric Measurement, Object Modeling and Documentation in Architecture and Industry, Thessaloniki, Greece, 7–9 July. *International Archives of Photogrammetry and Remote Sensing,* 32(5W11), 210–214.

Guerra, F. and Balletti, C., 1999b. Survey and virtual reality: the Guggenheim museum in Venice. ISPRS WG V/2 and WG5 Workshop: Photogrammetric Measurement, Object Modeling and Documentation in Architecture and Industry, Thessaloniki, Greece, 7–9 July. *International Archives of Photogrammetry and Remote Sensing,* 32(5W11), 160–165.

Gühring, J., 2001. Dense 3D surface acquisition by structured light using off-the-shelf components. Videometrics and optical methods for 3D shape measurement. *SPIE Proceedings 4309*, 220–231.

Guidi, G., Tucci, G., Beraldin, J.A., Ciofi, S., Damato, V., Ostuni, D., Costantino, F. and El Hakim, S.F., 2002. Multiscale archaeological survey based on the integration of 3D scanning and photogrammetry. *Proceedings CIPA WG 6: International Workshop on Scanning for Cultural Heritage Recording*, Corfu, Greece, 1–2 September, 13–18.

Haggrén, H., Junnilainen, H., Järvinen, J., Nuutinen, T., Laventob, M. and Huotarib, M., 2004. The use of panoramic images for 3D archaeological survey. *International Archives of Photogrammetry and Remote Sensing,* 35(5), 958–963.

Hanke, K., 2001. *Accuracy Study Project of Eos Systems' PhotoModeler*. Final Report. See http://www.photomodeler.com/pdf/hanke.pdf (accessed 30 October 2006).

Hanke, K., 2004. Restitution and visualization of the medieval fortress Kufstein. *Proceedings International Workshop on "Vision Techniques applied to Rehabilitation of City Centres"*, Lisbon, Portugal, 25–27 October 2004.

Hanke, K. and Böhler, W., 2004. Recording and visualization of the cenotaph of German emperor Maximilian I. *International Archives of Photogrammetry and Remote Sensing*, 35(5), 413–418.

Hemmleb, M., 1999. Digital Rectification of historical images. *Proceedings of XVII CIPA Symposium, Mapping and Preservation for the New Millenium*, Olinda, Brazil, 3–6 October. See http://cipa.icomos.org/fileadmin/papers/olinda/99c806.pdf (accessed 30 October 2006).

Institute of Geodesy and Photogrammetry, ETH Zurich, 2005. See http://www.photogrammetry.ethz.ch/research/bamiyan/buddha/index.html (accessed: 20 September 2005).

ICOMOS, 2005. ICOMOS Charters, Resolutions and Declarations. See http://www.international.icomos.org/charters.htm (accessed 20 September 2005).

Ito, J., Tokmakidis, K. and Inada, K., 2000. Reconstruction of the stadion area of ancient Messene using 3D computer graphics and analysis of its townscape. *International Archives of Photogrammetry and Remote Sensing*, 33(B5/1), 395–400.

Kakiuchi, T. and Chikatsu, H., 2000. Construction of stereo vision system for 3D objects modelling. *International Archives of Photogrammetry and Remote Sensing,* 33(B5/1), 414–421.

Karara, H.M., (Ed.), 1989. *Non-Topographic Photogrammetry,* 2nd edn, American Society for Photogrammetry and Remote Sensing, Bethesda, Maryland, USA.

Karras, G. and Petsa, E., 1999. Metric information from single uncalibrated images. *Proc. XVII CIPA Symposium, Mapping and Preservation for the New Millenium,* Olinda, Brazil, 3– 6 October. See http:// cipa.icomos.org/fileadmin/papers/olinda/99c802.pdf (accessed 30 October 2006).

Karras, G., Patias, P. and Petsa, E., 1996. Monoplotting and photo-unwrapping of developable surfaces in architectural photogrammetry. *International Archives of Photogrammetry and Remote Sensing,* 31(5), 290–294.

Karras, G., Patias, P., Petsa, E. and Ketipis, K., 1997. Raster projection and development of curved surfaces. *International Archives of Photogrammetry and Remote Sensing,* 32(5C1B), 179–185.

Karras, G., Petsa, E. Dimarogona, A. and Kouroupis, S., 2001. Photo-textured rendering of developable surfaces in architectural photogrammetry. *Proceedings of International Symposium on Virtual and Augmented Architecture (VAA01),* Dublin, 21–22 June 2001. See http://www.survey.ntua.gr/main/labs/ photo/staff/gkarras/Karras_Vaa01.pdf (accessed 30 October 2006).

Koistinen, K., 2000. 3D documentation for archaeology during Finnish Jabal Haroun Project. *International Archives of Photogrammetry and Remote Sensing,* 33(B5/1), 440–445.

Koussoulakou, A., Sechidis, L. and Patias, P., 1999. Virtual inverse photogrammetry. ISPRS WG V/2 and WG5 Workshop: Photogrammetric Measurement, Object Modeling and Documentation in Architecture and Industry, Thessaloniki, Greece, 7–9 July. *International Archives of Photogrammetry and Remote Sensing,* 32(5W11), 111–117.

Koussoulakou, A., Patias, P. and Stylianidis, E., 2001a. Documentation and visual exploration of archaeological data in space and time. *Proceedings of Third International Conference on Ancient Helike and Aigialeia,* Nikolaika of Diakopto, Greece, 6–9 October.

Koussoulakou, A., Patias, P., Sechidis, L. and Stylianidis, E., 2001b. Desktop cartographic augmented reality: 3D mapping and inverse photogrammetry in convergence. *Proceedings 20th International Cartographic Association Conference,* Beijing, China, 6–10 August, 2506–2513.

Kuzu, S., 2003. Volumetric reconstruction of cultural heritage artefacts. 18th CIPA International Symposium, 2003, Antalya, Turkey, 30 September–4 October. *International Archives of Photogrammetry and Remote Sensing,* 34(5/C15), 93–98.

Lianos, N., Spatalas, S., Tsioukas, V. and Gounari, E., 2004. 3D modeling of the old town of Xanthi in Greece. *International Archives of Photogrammetry and Remote Sensing,* 35(B), 542–545.

Lingua, A., Rinaudo, F., Auditore, G., Balletti, C., Guerra, F. and Pilot, L., 2003. Comparison of surveys of St. Mark's square in Venice. 18th CIPA International Symposium 2003, Antalya, Turkey, 30 September–4 October. *International Archives of Photogrammetry and Remote Sensing,* 34(5/C15), See http://cipa.icomos.org/fileadmin/papers/antalya/71.pdf (accessed 30 October 2006).

Lohr, U., Hellmeier, A. and Barruncho, L., 2004. Precise lidar data – An efficient way to build up virtual 3D models. *Proceedings of International Workshop on "Vision Techniques applied to Rehabilitation of City Centres",* Lisbon, Portugal, 25–27 October 2004.

McGlone, J. (Ed.), 2004. *Manual of Photogrammetry,* 5th edn., American Society of Photogrammetry and Remote Sensing, Bethesda, Maryland, USA.

Meyer, E., Grussenmeyer, P., Tidafi, T., Parisel, C. and Revez, J., 2004. Photogrammetry for the epigraphic survey in the great hall of Karnak temple: A new approach. *International Archives of Photogrammetry and Remote Sensing,* 35(B5), 377–382.

Mills, J., Peirson, G., Newton, I. and Bryan, P., 2000. Photogrammetric investigation into the suitability of desktop image measurement software for architectural recording. *International Archives of Photogrammetry and Remote Sensing,* 33(B5/2), 525–532.

Monti, C., Brumana, R., Fregonese, L., Savi, C. and Achille, C., 2001. 3D photogrammetric restitution of carvings and GIS applies. The case of the fontana of Nettuno at Conegliano Veneto. *Proceedings of International Workshop on Recreating the Past – Visualization and Animation of Cultural Heritage*, Ayutthaya, Thailand, 26 February–1 March.

Monti, C., Guerra, F., Balletti, C. and Galeazzo, G., 1999. The survey of Palazzo della ragione in Padova. *Proceedings of XVII CIPA Symposium, Mapping and Preservation for the New Millenium*, Olinda, Brazil, 3–6 October. See http://cipa.icomos.org/fileadmin/papers/olinda/99c314.pdf (accessed 30 October 2006).

Nakos, B. and Georgopoulos, A., 2000. The potential of applying 3D visualization methods for representing destroyed settlements. *International Archives of Photogrammetry and Remote Sensing,* 33 (B5/1), 285–290.

Nour El Din, M., Koehl, M. and Grussenmeyer, P., 1999. 3-D geometric modelling of the Aquest El Ghuri. ISPRS WG V/2 and WG5 Workshop, Photogrammetric Measurement, Object Modeling and Documentation in Architecture and Industry, Thessaloniki, Greece, 7–9 July. *International Archives of Photogrammetry and Remote Sensing,* 32(5W11), 146–152.

Ogleby, C., 1996. A reconstruction of the ancient city of Ayutthaya using modern photogrammetric techniques. *International Archives of Photogrammetry and Remote Sensing*, 32(B5), 416–425.

Ogleby, C., 2001. Olympia – Home of the ancient and modern Olympic Games. A VR 3D experience. *International Archives of Photogrammetry and Remote Sensing,* 34(5/W1), 97–102.

Ogleby, C. and Rivett, L., 1985. *Handbook of Heritage Photogrammetry.* Australian Heritage Commission, Special Australian Heritage Publication Series No. 4, Australian Government Publishing Service, Canberra, Australia.

Pateraki, M., Baltsavias, E. and Patias, P., 2002. Image combination into large virtual images for fast 3D modelling of archaeological sites. *International Archives of Photogrammetry and Remote Sensing*, 34(5), 440–445.

Patias, P., 2001. Caring for the past, aiming at the future: plans and policy of ISPRS Commission V. Keynote address, International Workshop on Recreating the Past – Visualization and Animation of Cultural Heritage, Ayutthaya, Thailand, 26 February–1 March. *International Archives of Photogrammetry and Remote Sensing,* 34(5C15), 1–4.

Patias, P. and Peipe, J., 2000. Photogrammetry and CAD/CAM in Culture and Industry – An ever changing paradigm. *International Archives of Photogrammetry and Remote Sensing*, 33(B5), 599–603.

Patias, P., Karapostolou, G. and Simeonidis, P., 2002. Documentation and visualization of historical city centers: A multisensor approach for a new technological paradigm. *International Archives of Photogrammetry and Remote Sensing*, 34(5), 393–399.

Patias, P., Stylianidis, E. and Terzitanos, K., 1998. Comparison of simple off-the-shelf and of wide-use 3D modelling software to strict photogrammetric procedures for close-range applications. *International Archives of Photogrammetry and Remote Sensing*, 32(5), 628–632.

Patias, P., Stylianidis, E., Tsioukas, V. and Gemenetzis, D., 1999. Rapid photogrammetric survey and GIS documentation of pre-historic excavation sites. *Proceedings of XVII CIPA Symposium, Mapping and Preservation for the New Millenium*, Olinda, Brazil, 3–6 October, printed in symposium CD. See http://cipa.icomos.org/fileadmin/papers/olinda/99c503.pdf (accessed 30 October 2006).

Pavelka, K. and Dolanský, T., 2003. Using of non-expensive 3D scanning instruments for cultural heritage documentation. 18th CIPA International Symposium 2003, Antalya, Turkey, 30 September–4 October. *International Archives of Photogrammetry and Remote Sensing,* 34(5/C15), 534–536. See http://cipa. icomos.org/fileadmin/papers/antalya/140.pdf (accessed 30 October 2006).

Peipe, J. and Stephani, M., 2003. Performance evaluation of a 5 megapixel digital metric camera for use in architectural photogrammetry. *International Archives of the Photogrammetry, Remote Sensing and Spatial Information Sciences*, 34(5/W12), 259–261.

Pollefeys, M., Vergauwen, M. and Van Gool, L., 2000. Automatic 3D modeling from image sequences. *International Archives of Photogrammetry and Remote Sensing*, 33(B5), 619–626.

Pomaska, G., 1999. Documentation and internet presentation of cultural heritage using panoramic image technology. *Proceedings of XVII CIPA Symposium, Mapping and Preservation for the New Millenium*,

Olinda, Brazil, 3–6 October. See http://cipa.icomos.org/fileadmin/papers/olinda/99c405.pdf (accessed 30 October 2006).

Pomaska, G., 2000. Reconstruction of the appearance of Schloss Herborn back to 1540 and its multimedia presentation. *International Archives of Photogrammetry and Remote Sensing*, 33(B5/1), 627–634.

Pöntinen, P., 2004. On the geometrical quality of panoramic images. *International Archives of Photogrammetry and Remote Sensing*, 35(5), 82–87.

Remondino, F., 2004. The Bamiyan project: 3D object reconstruction and landscape modeling for cultural heritage documentation. In *Commemorative Volume for the 60th birthday of Prof. Dr. Armin Grün*, Institute of Geodesy and Photogrammetry, ETH, Zurich, Switzerland, 213–219.

Reulke, R., Scheibe, K. and Wehr, A., 2005. Integration of digital panoramic camera and laser scanner data. *Proceedings of International Workshop on Recording, Modeling and Visualization of Cultural Heritage*, 22– 27 May 2005. Centro Stefano Franscini, Monte Verità, Ascona, Switzerland.

Rocchini, C., Cignoni, P., Montani, C., Pingi, P. and Scopigno, R., 2001. A low cost 3D scanner based on structured light. *Computer Graphics Forum*, 20(3), 299–308.

Ruther, H., Fraser, C., Grün, A. and Bührer, T., 2001. The recording of Bet Giorgis, a 21st century rock-hewn church in Ethiopia. *Proceedings of International Workshop on Recreating the Past – Visualization and Animation of Cultural Heritage*, Ayutthaya, Thailand, 26 February–1 March.

Salvi, J., Pages, J. and Batlle, J., 2004. Pattern codification strategies in structured light systems. *Pattern Recognition*, 37(4), 827–849.

Schechner, Y., 2000. Depth from defocus vs. stereo: How different really are they? *International Journal of Computer Vision*, 89, 141–162.

Scherer, M., 2004. How to optimise the recording of geometrical data and image data for the purpose of architectural surveying. *International Archives of Photogrammetry and Remote Sensing*, 35(B5), 228–231.

Skarlatos, D., Theodoridou, S. and Glabenas, D., 2004. Archaeological surveys in Greece using radio-controlled helicopter. *FIG Working Week 2004*, Athens, Greece, 22–27 May. See http://www.fig.net/pub/athens/programme.htm (accessed 30 October 2006).

Stephani, M., 1999. Digitale Abbildung durch projektive Transformation eines fiktiven Bildes. *Festschrift für Prof. Dr.-Ing. Heinrich Ebner zum 60. Geburtstag*, (Heipke, C. and Mayer, H., Eds.), Technische Universität München, Lehrstuhl für Photogrammetrie und Fernerkundung, 303–311.

Streilein, A. and van den Heuvel, F., 1999. Potential and limitation for the 3D documentation of cultural heritage from a single image. *Proceedings of XVII CIPA Symposium, Mapping and Preservation for the New Millenium*, Olinda, Brazil, 3–6 October. See http://cipa.icomos.org/fileadmin/papers/olinda/99c804.pdf (accessed 30 October 2006).

Tauch, R. and Wiedemann, A., 2004. Generation of digital surface models for architectural applications with Archimedes 3D. *International Archives of Photogrammetry and Remote Sensing*, 35(B5), 350–353.

Tosovic, S., Sablatnig, R. and Kampel, M., 2002. On combining shape from silhouette and shape from structured light. In Wildenauer, H. and Kropatsch, W. (Eds.), *Proceedings of the 7th Computer Vision Winter Workshop*, Bad Aussee, Austria, 108–118.

Tsioukas, V., Patias, P. and Jacobs, P., 2004. A novel system for the 3D reconstruction of small archaeological objects. *International Archives of Photogrammetry and Remote Sensing*, 35(B5), 815–818.

Tucci, G., Algostino, F., Bonora, V. and Chiabrando, F., 2003. 3-D modeling and restoration: from metric to thematic survey. The case study of San Frascesco al Prato in Perugia. 18th CIPA International Symposium 2003, Antalya, Turkey, 30 September–4 October. *International Archives of Photogrammetry and Remote Sensing*, 34(5/C15). See http://cipa.icomos.org/fileadmin/papers/antalya/142.pdf (accessed 30 October 2006).

UNESCO 2005. Venice Charter, 1964. See http://www.international.icomos.org/charters/venice_e.htm (accessed 20 September 2005).

University of Vienna, 2005a. See http://www.univie.ac.at/Luftbildarchiv/wgv/procon.htm (accessed 30 October 2005).

University of Vienna, 2005b. Notes on excavation. See http://www.univie.ac.at/Luftbildarchiv/wgv/appl. htm (accessed 20 September 2005).

Wiedemann, A., Hemmleb, M. and Albertz, J., 2000. Reconstruction of historical buildings based on images from the Meydenbauer archives. *International Archives of Photogrammetry and Remote Sensing*, 33(B5/2), 887–893.

Zhang, R., Tsai, P.S., Cryer, J.E. and Shah, M., 1999. Shape from shading: A survey. *IEEE Transactions on Pattern Analysis and Machine Intelligence*, 21(8), 690–706.

10 Sensor integration and visualisation

Sabry F. El-Hakim and Jean-Angelo Beraldin

10.1 Introduction

The process of digitally reconstructing real objects or sites for documentation and virtual reality simulation is an application-oriented undertaking that entails thoroughly planned selection of the appropriate sensing technologies. This 3D reconstruction is usually referred to as *modelling from reality* (Ikeuchi and Sato, 2001) to distinguish it from computer-graphics creations of artificial-world models such as in games and movies. The resulting model is usually an information-rich surface representation, a mesh of triangles with texture and lighting. Figure 10.1 displays a typical modelling and visualisation, or rendering, pipeline. Although in the literature the steps are often addressed separately, in practical applications it is essential to tackle them all together as a single system. Unfortunately there is no turnkey system available for the whole pipeline from the data capture to rendering. While this may change in the future, training and experience remain necessary to design and operate each of the system components. The expertise draws on numerous foundations including: photogrammetry (3D from images), computer vision (procedure automation), physics and electronics (laser scanning), computational geometry (3D modelling), and computer graphics (rendering).

This chapter is intended to review and analyse 3D capture, modelling, and visualisation techniques with an emphasis on using multiple sensors and methods. There is a large amount of published source material, so this chapter will focus on citing the most recent papers and include reviews in each step when possible. Although the subject is fast evolving and parts of the chapter will inevitably soon be out of date, most of the discussed issues are persistent and will remain challenging until breakthrough solutions are discovered.

10.1.1 3D modelling from reality: motivations and requirements

The most obvious motive for 3D modelling from reality is the truthful documentation of objects and sites for reconstruction, restoration, or modification purposes. Other motives include creating education and research resources, visualisation from viewpoints that are impossible in the real world due to size or accessibility, interaction with objects without risk of damage, design simulation, reverse engineering, training, remote tourism, and virtual museums. The models should have high geometric accuracy, all the important details, realistic appearance, and efficient size. The models should, if at all possible, be created with easy to use or automated methods, and at low cost. The order of importance of those requirements depends on the application but, as a rule, all should be pursued.

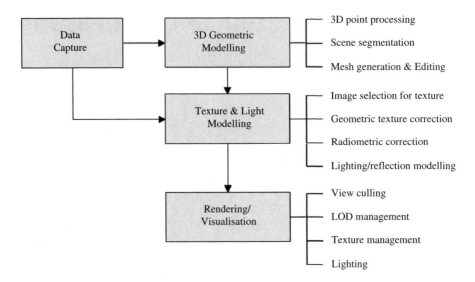

Figure 10.1 *Overview of the modelling and rendering pipeline.*

10.1.2 3D capture and modelling techniques

The techniques can be subdivided into three main categories:

- geometric modelling by a CAD modeller based on surveying data, and existing engineering diagrams or architectural drawings such as floor plans;
- image-based modelling or structure from motion, with photogrammetry and computer vision approaches; and,
- active 3D or range capture, primarily with laser scanners.

The first category, which is the classical approach, is undoubtedly time consuming and impractical for relatively large or complex sites and the resulting models do not look realistic or include all the fine details. The second category, although it can produce accurate and realistic-looking models, remains highly interactive since current computer-vision automated methods are still unproven in real applications. It also cannot capture details on unmarked or featureless surfaces. Sensors in the third category can capture relatively accurate and complete details with a high degree of automation, but at least for now they are costly, usually bulky, hard to use, and influenced by surface properties. Also, as a laser scanner is intended for a specific range and volume, one designed for close range is not suitable for medium range or long range. It is therefore apparent that, except for simple objects or sites, a single sensor or approach is not sufficient and an integration of data from multiple sources will be needed. In Sections 10.2–10.4 the 3D capture and modelling techniques will be reviewed and the potentials and limitations of each will be discussed. Section 10.5 will cover the integration of data from different sources.

10.1.3 Photo realism and interactive visualisation

There are several ways of representing a digitised object or site, but a 3D model that allows interactive visualisation and manipulation is the most desirable. This offers the user the freedom to choose any viewpoint with a variety of lighting conditions, unlike pre-rendered animations or movies where the viewpoints and lighting are pre-determined and invariable. On the other hand, a movie has the advantage of affording off-line rendering, which can make use of the highest level of details offered by the data. Visualisation issues can be divided into those affecting photo-realism, defined as having no difference between the rendered model and a photograph taken from the same viewpoint (Fleming, 1998), and those affecting the interactivity or real-time to ensure smooth navigation through the model. Due to the large size of most detailed models, it is usually not feasible to achieve both photo-realism and smooth navigation with current computer and graphics hardware without simplification and/or creative rendering techniques. The hardware is, of course, continuously improving, but model sizes are also increasing at a high rate to satisfy the ever-rising need for realistic details. Realism and interactivity may also be reached with image-based rendering (IBR), which eliminates the need for the creation of a geometric model (Shum *et al.*, 2003; Evers-Senne and Koch, 2005). This may suffice for some applications, but the lack of geometric information makes IBR unsuitable for documentation or reconstruction purposes. Section 10.6 covers methods for achieving photo-realism and interactive visualisation of 3D models. Other forms of representation such as movies, IBR, Quick-Time VR, panoramas, orthophotos, and cross-sections will not be discussed.

10.1.4 Selecting the proper approach

The task of designing and executing a 3D modelling project requires clear understanding of the performance of available sensors and methods. However, for digital 3D modelling techniques, there is historically no tradition of rigorous and systematic validation for the whole 3D modelling pipeline, apart from the photogrammetric 3D point measurement step (Fraser, 2005). In no particular order, the key performance criteria to be considered with respect to the application requirements and constraints are:

- geometric fidelity of the final model (the accuracy or uncertainty);
- generated level of detail (spatial resolution);
- model completeness (percentage of object or site captured);
- visual quality (photo-realism);
- impact of the environment (surface type, accessibility etc.); and
- user interaction and level of automation (ease of creation).

This list is by no means complete and the reader may add other criteria, for example cost and time. There is a need for comprehensive comparison of all sensors and techniques based on controlled experiments with the specific objective of evaluating the above criteria. Some efforts, albeit in their early stage or covering only some criteria for a specific application, are becoming available (Roy-Chowdhury and Chellappa, 2002; Eid and Farag, 2004; Böhler, 2005). This chapter will try to discuss each criterion for every 3D technique. Based on this discussion, as well as the examples given in Section 10.7, Section 10.8 will provide suggestions on how to select the right approach.

10.2 3D modelling from surveying and existing drawings

Although sensor-based 3D measurement and modelling techniques have been showing promising results for over a decade, building models with a CAD modeller using only surveying or engineering drawings remains the most widespread (Haval, 2000). These methods cannot capture the fine details, and necessitate hypotheses about surface shapes and surface relationships, which can be inaccurate. Synthetic textures, which yield a computer-generated look, are

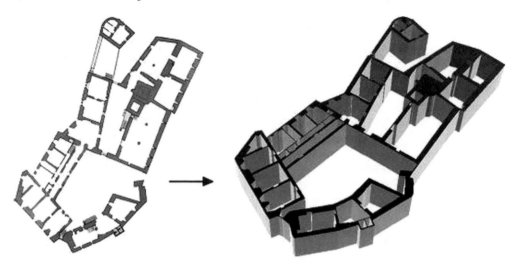

Figure 10.2 *3D model extruded from a floor plan.*

usually applied, but textures from images have sometimes been used for realistic appearance (DeLeon, 1999; Foni *et al.*, 2002). The approaches are typically manual although automated model generation techniques, based on libraries of predefined 3D building blocks, have been developed (Dikaiakou *et al.*, 2003). The focus here is on issues concerning 3D modelling from 2D architectural drawing (Figure 10.2).

10.2.1 Utilising existing architectural drawings

Several issues need to be addressed when modelling from existing drawings or floor plans. In the case of drawings created directly in digital form from surveying data, file formats such as AutoCAD or DXF allow the information to be arranged in separate layers, each containing a different element, which facilitates the modelling process. When the drawings are only available on paper, it becomes more difficult due to the variety and complexity of the drawing notations and the presence of noise and unwanted information. Therefore, significant research has been devoted to converting paper drawings, after they have been scanned or digitised, to CAD representation. Kernighan and Van Wyk (1996) developed a technique that can extract lines and poly-lines from digitised floor plans and remove the details not needed for 3D modelling. Several add-ons to standard commercial software are now available to ease this process. Once the data contains only the appropriate information, some adjustments to obtain consistent local geometry and layout topology are needed (Nguyen *et al.*, 2005). Creating 3D models may be done interactively on simple floor plans, but can be time consuming and error prone on large complex structures. This motivated the development of a few automatic and semi-automatic techniques (e.g. Lewis and Sequin, 1998).

10.2.2 Combining architectural drawings and sensor data

Integrating architectural drawings with sensor data can be divided into three categories:

- The drawings are the prime modelling source while sensor-based techniques are used to add missing details, that are either too small for the drawings scale factor or have shapes too complex to be modelled from 2D representation.

- Sections are entirely modelled from drawings and others are entirely modelled from sensor data. Sections are then connected together using joint elements or portals.
- Sensor-based data is the prime modelling source while drawings cover unsensed or destroyed parts. The drawings may also be used to help assemble different models.

The first category is sensible if most of the site lacks interesting details and accurate drawings already exist. The third category is the most desirable because it gives the truly "as built" representation, or the most faithful documentation, and also provides the photo-realistic aspect. The second category is practical if some sections are difficult to access with 3D sensors, or there are entire sections that are simpler to model from floor plans, such as plane surfaces without details of interest.

10.3 3D capture and modelling from images

A few effective interactive photogrammetric modelling techniques have been developed (Debevec *et al.*, 1996; El-Hakim, 2002; Grün *et al.*, 2004) and have been applied to many projects in the past few years. Also some commercial software packages are available and can be used for at least uncomplicated objects. Heritage sites like castles were the most frequently modelled from images (Hanke and Oberschneider, 2002; Bacigalupo and Cessari, 2003; Gonzo *et al.*, 2004; Kersten *et al.*, 2004), sometimes with surveying and CAD software to fill in missing parts. Automation of the techniques is a very active area of research in computer vision but to our knowledge no large complex site model has been based purely on automated methods. As a matter of fact, only selected sections of structures have been modelled automatically (Pollefeys *et al.*, 2004).

10.3.1 Automated techniques

Efforts to completely automate camera orientation, self-calibration, 3D modelling are, while scientifically promising, thus far not often practical. The fully automated procedure, reported in computer vision (Rodriguez *et al.*, 2005; Pollefeys *et al.*, 2004; Fusiello, 2000; Liebowitz *et al.*, 1999), starts with an image sequence taken by an uncalibrated camera. The system automatically extracts interest points like corners and matches them across views, then computes camera parameters and 3D coordinates of matched points using robust techniques. Since the key to the success of this procedure is that successive images must not vary significantly, the images have to be taken at short intervals, both in space and time. Two images are selected to initialise the sequence. The formulation is done using projective geometry (Faugeras *et al.*, 2001) and is followed by a projective bundle adjustment. Self-calibration to compute the intrinsic camera parameters, usually only the focal length, follows in order to upgrade the projective construction into a metric one, to scale. The next step, the creation of the 3D model, is more difficult to automate (Gibson *et al.*, 2003) and is usually done interactively to define the topology between components (adjacency, connectivity, intersection or containment) and to add points suitable for modelling. Automated feature extraction often creates areas with too many points that are not all needed for modelling and areas with too few points to produce a full model. Also since image-based methods rely on features that can be seen and matched in multiple images, occlusions and untextured surfaces are obviously problematic. Figures 10.3 and 10.4 show typical classical architectures, which are difficult to model automatically since they have occlusions and limited or no texture on the surfaces.

Some attempts have been made to automate the modelling step once the 3D points are acquired (Dick *et al.*, 2001; Ruggeri *et al.*, 2001; Werner and Zisserman, 2002; Cantzler, 2003; Schindler and Bauer, 2003; Wilczkowiak *et al.*, 2003). Such techniques rely on constraints of

Figure 10.3   *Untextured ceiling.*

Figure 10.4   *Courtyard with occlusions.*

surface shapes and assumed relationships between surfaces. Without any knowledge about the object, the modelling procedure may still be automated using constraints derived recursively from the data such as visibility constraints (Hilton, 2005).

The effectiveness of all fully automated techniques remains to be proven in real application environments. Since closely spaced images are needed for robust matching, large numbers of images are required for large structures and scenes, which may be difficult or impractical. Also accuracy becomes an issue on long sequences due to error propagation since points are usually tracked in only two or three images. Another important issue is that since automating the modelling step requires constraints, the final model accuracy is dependent on the validity of the assumptions made for those constraints. Also the generated model will usually require manual editing to correct errors resulting from the automated procedure. Therefore, automated techniques may be considered, at least for now, only on selected parts of a site where a short image sequence is possible and there are suitably located well-defined features and clearly identifiable surface constraints.

10.3.2 Semi-automated techniques

The most impressive results remain those achieved with interactive approaches. Rather than full automation, a hybrid easy to use system known as *Façade* (Debevec *et al.*, 1996) has been developed to create 3D models of architectures from small numbers of photographs. The basic geometric shape of a structure is first recovered using models of polyhedral elements. In this interactive step, the actual size of the elements and camera pose are computed assuming that the camera's intrinsic parameters are known. The second step adds geometric details by automated stereo-matching, constrained by the model constructed in the first step. The approach proved to be effective in creating relatively accurate and realistic models. However, it requires a high level of interaction and, since assumed shapes determine camera poses and all 3D points, the results are only as accurate as the extent to which the structure elements match those shapes. A semi-automatic approach has been designed to model regular shapes such as blocks, arches, columns, doors and windows (El-Hakim, 2002). On such shapes, a small number of seed points are interactively measured in all the images in which they appear. Camera positions and the 3D coordinates of seed points are determined using a bundle adjustment. Then, the remaining points completing those shapes are automatically added.

Lee and Nevatia (2003) developed another semi-automatic technique to model architectures. The camera was calibrated using the known shapes of the buildings being modelled. The models are created in hierarchical manner by dividing the structure into basic shape, façade textures, and detailed geometry such as columns and windows. The procedure requires the user to interactively provide shape information such as width, height, radius and spacing then the shape is completed automatically. Grossmann and Santos-Victor (2005) used planarity, orthogonality, parallelism, and symmetry constraints to semi-automatically construct man-made objects from one or more images and provided precision measures to verify construction quality.

10.3.3 A semi-automatic modelling pipeline

Figure 10.5 shows an example of an image-based modelling pipeline. It indicates which steps are interactive and which are automatic (interactive steps are light grey) with an option of taking a closely spaced sequence of images to increase the level of automation on selected parts. The calibration and bundle adjustment steps are well known and will not be discussed here. The texturing and lighting step will be described in Section 10.6. Overlapping images are first acquired at proper geometric configuration to cover the object. The procedure then starts by extracting several features in multiple images, followed by image registration and 3D coordinate computation with bundle adjustment. The modelling operation starts by dividing the scene into connected segments to define the surfaces' topology and boundaries. To add more points to any segmented region, automatic feature extraction and stereo-matching may follow. The matching will require constraints such as template colours, the epipolar condition, and disparity range set-up from the 3D coordinates of the initially measured points. Points can also be added automatically to surfaces of known shapes, like cylinders or quadrics, whose parameters can be computed using a proper number of interactively measured seed points. When some surface boundaries are not

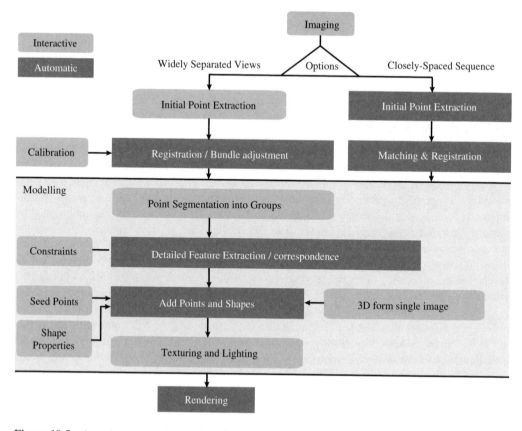

Figure 10.5 *A semi-automatic image-based modelling pipeline.*

visible in any image they can be computed by surface intersections. The general rule for adding points on known elements, as well as for generating points in occluded parts, is to do the work in the 3D space to add points to complete the shape then project them on the images using the known camera parameters. It is important to include an approach to obtain 3D coordinates from a single image since some parts of the scene will appear in only one image, mainly due to occlusions (van den Heuvel, 1998; Liebowitz *et al.*, 1999; El-Hakim, 2002). Constraints for surface shape such as being plane or cylindrical, and surface relations such as being parallel or orthogonal to one another are required to compute 3D coordinates from a single image.

10.4 3D capture and modelling from laser scanners

Laser scanners promise to provide highly detailed and accurate representation of most shapes. Combined with colour information, either from the scanner itself or from a digital camera, a realistic-looking model can be created. Laser scanners have been used for many years to acquire the entire geometry for small objects (Godin *et al.*, 2002), industrial parts for inspection and reverse engineering (Boulanger *et al.*, 1992; Varady *et al.*, 1997), and for city modelling (Fruh and Zakhor, 2003). The technology has also been used in some large-scale heritage sites (Allen *et al.*, 2003; Beraldin *et al.*, 2002). Levoy *et al.* (2000) modelled Michelangelo's 5 m tall statue "David" with a large number of high-resolution scans. Ikeuchi *et al.* (2003) developed geometric and photometric modelling techniques and applied them to constructing a multimedia model of the 15 m tall statue of the Great Buddha of Kamakura. Laser scanners typically capture range images (range organised in 2D grids) and come with widely different accuracies, speeds and capabilities. For all scanners, the accuracy depends not only on the technology used but also on the range, angle of incidence and surface characteristics. Without an understanding of at least the basic physics behind each type of scanner and knowing their measurement limits, it will be difficult to select the appropriate instrument for a given task within an application. In this section, range measurement uncertainty and spatial sampling considerations are addressed for the current scanning techniques.

The goal of this overview is not to survey commercial systems but instead to present some basic theory about 3D sensing, along with data processing examples, to help the reader decide which scanner to use. The focus is mainly on single-spot scanning methods rather than full-field systems such as fringe projection or moiré-based sensors.

10.4.1 A 3D modelling pipeline with laser scanners

Figure 10.6 outlines the main steps required to create a 3D model, using laser scanning data. Other technologies like photogrammetry or GPS may augment this data by providing alignment or registration information in a common framework.

Due to object size, shape and occlusions, it is usually necessary to use multiple scans from different locations to cover every surface at the desired resolution or level of detail. The resolution is defined as the smallest distance between points captured by the scanner. Aligning and integrating all the scans (the registration step) requires significant effort and affects the final accuracy of the 3D model. Since the measured points vary significantly in accuracy, for example due to variation in range and angle of incidence, understanding the factors affecting the quality of the measured points is required to properly clean up and integrate the different views. Device-based or data-based (manual or automatic) initial global alignment followed by the

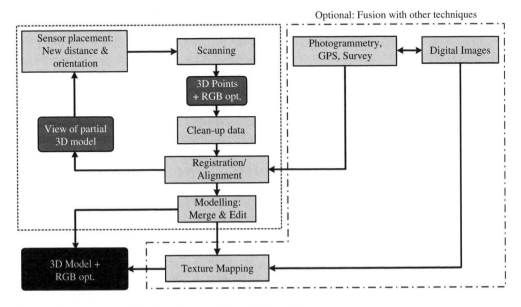

Figure 10.6 *A general pipeline for range-based 3D modelling.*

more precise iterative closest point (ICP) technique that takes into consideration the different points' accuracy is required (for reviews of ICP, see Rodrigues *et al.*, 2002; Matabosch *et al.*, 2004). Details of geometric modelling and possible object representation methods can be found in Campbell and Flynn (2001). Texture acquired by independent digital cameras is registered with the geometric model using common points between the 2D image and the 3D model. In effect, this is finding the camera pose using the model points as control points. This must be done for every image unless the camera is fixed to the scanner, then it may only be done once. Scanners that acquire the red–green–blue (RGB) colour directly with the geometry have also been developed (Blais *et al.*, 2005). Details of the 3D modelling pipeline can be found in Soucy *et al.* (1996), Bernardini and Rushmeier (2002) and Brusco *et al.* (2005).

10.4.2 Classes of laser scanners

There are two main classes of active 3D systems:

- Triangulation-based systems (in theory up to 5 m range, but less in practice due to decrease in signal-to-noise ratio:
  - single-spot (flying spot scanner), Figure 10.7a (see colour section), or profile measurement (slit scanner), Figure 10.7b, coupled to mechanical motion devices, Figures 10.7c,d.
  - full-field (no scanning) such as multi-point projection, fringe projections, line projection by diffraction gratings, and moiré (Jähne *et al.*, 1999).
- Time delay systems (usually over 10 m range):
  - single-spot coupled to mirror-based scanning mechanisms known as time of flight (TOF), frequency modulated continuous wave (FMCW) or amplitude modulated continuous wave (AMCW) scanners (based on pulses, Figure 10.7e, and AM or FM modulation, Figure 10.7f); and
  - full-field using micro-channel plates or custom-built silicon chips with flood laser light projector (Jähne *et al.*, 1999).

Interestingly, for ranges between 5–10 m, there are very few laser scanners available commercially. This span of distances represents a transition between triangulation and time delay-based systems. Triangulation-based systems require a large baseline to operate in that span which is not always practical, and most time delay systems will not provide the accuracy desired at this range.

The choice of a 3D solution must consider that the geometric fidelity and visual quality depends mainly on the accuracy and data noise level (measurement uncertainty) and the level of surface details that can be sensed by the scanner (spatial resolution). Thus, we will give measurement uncertainty described by the standard deviation σ for each class of 3D laser scanners, as well as the parameters affecting the spatial resolution. The uncertainty of all the range capture methods presented hereafter can be expressed by a single equation that is a function of the signal-to-noise *(SNR)* (Skolnik, 1980; Poor, 1994). For a *SNR* >10, range uncertainty for any laser scanner discussed here has the following form:

$$\sigma_r \approx C \frac{1}{\sqrt{SNR}} \tag{10.1}$$

where C is a constant dependent on the range capture method. Table 10.1 lists C as a function of parameters that vary from one method to another. Other symbols in Table 10.1 are described in the following section. The effect of laser speckle is also given along with factors affecting the maximum range that can be measured with no ambiguity. We will see that for triangulation-based methods, for

Table 10.1   Range uncertainty as a function of *SNR* and speckle noise (see text for definitions)

Method	Constant C	Speckle noise δ_z	Max. range	Typical values
Triangulation	$\dfrac{Z^2}{f\,D}\dfrac{1}{BW}$	$\dfrac{\lambda}{\pi\sqrt{2}}\dfrac{Z^2}{D\,\Phi}$	Limited by optical geometry	0.05–1 mm range < 5 m @ 10 kHz
Pulsed wave (PW)	$\dfrac{c}{2}T_r$	Affects amplitude	Limited by pulse rate f_p	5–50 mm large depth of field (DOF) @ 10 kHz
AMCW	$\dfrac{\lambda_m}{4\pi}$	Affects amplitude	Limited by frequency f_m	0.05–5 mm DOF 1–100 m @ 100 kHz
FMCW	$\dfrac{\sqrt{3}}{2\pi}\dfrac{c}{\Delta f}$	Affects amplitude	Limited by chirp duration T_m	0.01–0.25 mm DOF < 10 m @ 0.10 kHz

large *SNR*, speckle noise becomes the dominant factor for the performance of the scanner. For time delay-based systems, speckle will affect the amplitude of the returned signal. Some details on the scanning methods and on how these parameters were obtained are given next.

Triangulation-based systems

The basic principle of optical triangulation for a single spot is shown in Figure 10.8a. The scanning of a laser spot in the *X-Y* lateral directions can be achieved by proper combination of mirrors, lenses and motors (galvanometers), such as the two orthogonal mirrors illustrated in Figure 10.8b. For line scanners, this can be achieved by mounting it on a rotation or translation platform (Figure 10.7).

A single-spot measurement works as follows. A laser source projects a beam on a surface of interest. The scattered light from that surface is collected from a vantage point distinct from the projected light beam (distance *D*). This light is focused onto a position sensitive detector (herein called spot sensor, usually a charge coupled devide (CCD). The knowledge of both projection and collection angles (α and β) relative to a baseline (*D*) determines the dimensions of a triangle. This laser spot sensor is in fact an *angle sensor*. For an incremental change in distance, ΔZ, one measures the resulting angle shift $\Delta\beta$, which is observed through a longitudinal shift in laser spot position. The errors with a triangulation-based laser scanner come mainly from the error in p (σ_p). Applying the error propagation principle to the range equation (Blais, 2004), gives the uncertainty estimate along the Z-axis as:

$$\sigma_z \approx \frac{Z^2}{f\,D}\delta_p \tag{10.2}$$

where *Z* is the distance to the surface, *f* is the effective position of laser spot sensor (effective focal length); and σ_p is the uncertainty in spot position. The value of σ_p depends on the type of sensor used, the peak detector algorithm, *SNR*, speckle noise, and the imaged spot shape. For *SNR* >10:

$$\sigma_p \approx \frac{1}{\sqrt{SNR}\;BW} \tag{10.3}$$

where *BW* is the rms signal bandwidth, the normalised second moment of the power spectral density of the signal in units of spatial frequencies (Poor, 1994). Substituting Equation (10.3) in Equation (10.2) gives the uncertainty in *Z* as:

$$\sigma_z \approx \frac{Z^2}{f\,D}\frac{1}{\sqrt{SNR}\;BW} \tag{10.4}$$

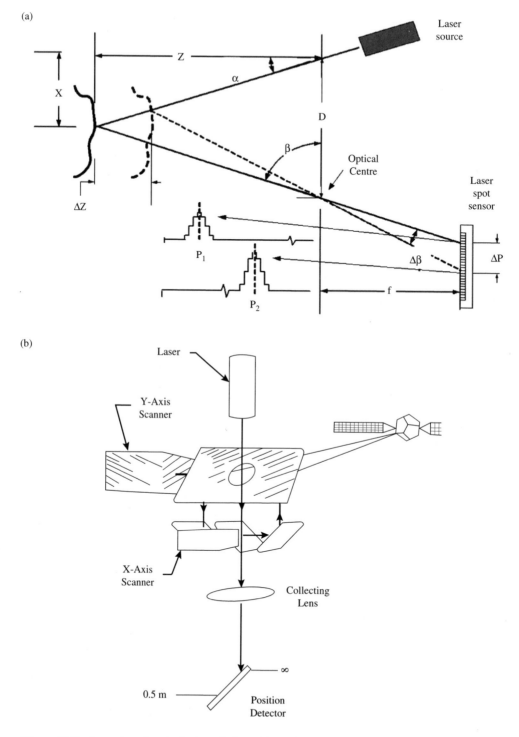

Figure 10.8 *Laser-based optical triangulation: (a) single spot; (b) full volume scanning (flying spot) based on two orthogonal mirrors.*

When the *SNR* is large, the dominant source of noise will be the speckle noise spot error (Baribeau and Rioux, 1991; Dorsch *et al.*, 1994). It is given by:

$$\sigma_p \approx \frac{1}{\pi \sqrt{2}} \lambda fn \tag{10.5}$$

where *fn* is the receiving lens f-number, equal to *f/Φ* where Φ is the lens aperture diameter, and λ is the laser wavelength. Substituting Equation (10.5) into Equation (10.2), *Z* uncertainty due to speckle noise is:

$$\sigma_z \approx \frac{\lambda}{\pi \sqrt{2}} \frac{Z^2}{D\,\Phi} \tag{10.6}$$

This shows that σ_z does not depend on *f* and a low uncertainty is achieved by the use of a short laser wavelength, e.g. blue laser as opposed to a near-infrared laser. Further, the last term in Equation 10.6 can be decomposed into two terms, one for the baseline-distance ratio *D/Z* (related to the triangulation angle) and one for the observation angle *Φ/Z*. So it is now clear that when these two are large, the better the accuracy (small σ_z). Assuming high *SNR* and a centroid-based method for peak detection, Equation (10.4) will be negligible compared to Equation (10.6). Figure 10.9 shows the range uncertainty as a function of measurement volume or the depth of field (DOF) for all laser systems (Rioux *et al.*, 1987). The solid line is for triangulation-based systems with *SNR* > 100 (resulting in neglecting Equation 10.4) and typical design values: λ = 0.633 μm (red-coloured laser beam), Φ = 25 mm, and *D/Z* = 0.35.

The collected signal strength depends also on the source laser power, the collecting lens aperture, and the reflectance and morphological properties of the surface (e.g. diffuse, translucent, transparent, specular, polarisation sensitive, roughness, colour, sharp edges and concavities). Since the square root of the *SNR* is inversely proportional to the square of the distance, therefore as the amount of light collected decreases, the *SNR* will deteriorate and will overwhelm the effect of the speckle noise and the range measurement uncertainty will be a function of the distance to the fourth power!

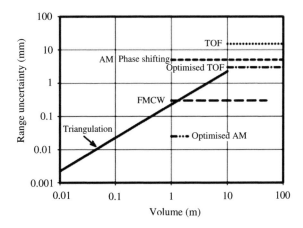

Figure 10.9 *Range uncertainty vs. the measurement volume (DOF), in logarithmic-scale, of typical range scanners (λ is 0.633 mm, Φ is 25 mm, SNR is large and constant, sampling rate >100 3D points per second).*

Time delay systems

Long-range sensors (range exceeding 10 m) are usually based on the TOF technology. Since, in a given medium, light waves travel with a finite constant velocity, the measurement of the light travelling time from a source to a reflective surface and back (round trip, τ) offers a convenient way to measure range; R is $c \times \tau/2$ (c is the speed of light, 299,792,458 m/s in vacuum). Different strategies have been devised to exploit this basic principle. There are TOF scanners using repetitive laser pulses (pulsed wave, PW), AMCW, not pulsed, using phase difference, and FMCW exploiting beat frequencies (Skolnik, 1980; Wehr and Lohr, 1999; Amann *et al.*, 2001). The range uncertainty for a single pulse is approximately:

$$\sigma_{r-p} \approx \frac{c}{2} \frac{T_r}{\sqrt{SNR}} \tag{10.7}$$

where T_r is the rise time of the laser pulse leading edge. Assuming SNR is 100 and T_r is 1 nanosecond, the range uncertainty is about 15 mm. Most high-speed (>1000 points per second) commercial systems based on TOF provide a range uncertainty between 5 mm and 50 mm as long as a high SNR is maintained (Figure 10.9). Averaging N independent pulse measurements will reduce σ_{r-p} by a factor proportional to the square root of N.

For all TOF systems, there is a maximum distance that can be measured without ambiguity. This distance is strictly limited by the ranging rate. In order to illustrate this, in Figure 10.10a, three range measurements are acquired at equally spaced intervals given by $1/f_p$, where f_p is the pulse repetition rate in Hz. For the first two measurements, A and B, the delay time τ is less than $1/f_p$ and can be correctly sensed, as shown in Figure 10.10b. For the third measurement, as τ is larger than $1/f_p$, the returned signal passes the end of the pulse cycle and the system fails to determine the correct start time since it cannot distinguish between the different cycles. This is sometimes called the aliasing effect. The distance after which this occurs, is referred to as the *ambiguity* range, or maximum range with *non-ambiguity*. It is measured by $1/f_p$ multiplied by $c/2$, and can increase by taking two or more measurement each with different frequency (Gokturk *et al.*, 2004). For an f_p of 10,000 Hz, the maximum distance that can be measured without ambiguity is 15 km, and the faster the measurement rate, the smaller this distance becomes.

Since PW systems have less precise timing, their accuracy is relatively poor at short range. CW modulation gets around the imprecision of short pulses by modulating the power, or the wavelength of the laser beam. For AM, the intensity of the laser beam is amplitude modulated,

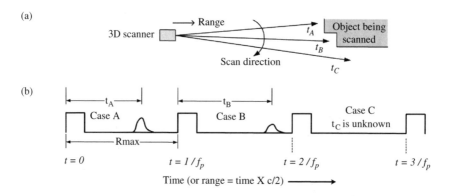

Figure 10.10 *TOF based on repetitive short pulse transmission: (a) laser scanner measuring three points; (b) waveforms illustrating the measurement principle.*

e.g. sine wave. The projected laser beam and collected laser light are compared and the phase difference ($\Delta\varphi$) between the two waveforms yields the time delay ($\tau = \Delta\varphi/2\pi \times \lambda_m/c$, where λ_m is the wavelength of the amplitude modulation). Again, the range, R, is found from the relationship $c \times \tau/2$.

The range uncertainty is approximately given by:

$$\sigma_{r-AM} \approx \frac{1}{4\pi} \frac{\lambda_m}{\sqrt{SNR}} \tag{10.8}$$

Typical commercial systems can scan a scene at rates of 10–500 kHz. Some optimised AM systems can achieve sub-100 µm uncertainty values but over very shallow depth of fields (see Figure 10.9).

The second most popular CW system is based on frequency modulated (FM) with either coherent or direct detection. In one implementation, the frequency (wavelength) of the laser beam is linearly modulated either directly at the laser diode or with an acousto-optic modulator. The linear modulation is usually shaped as a triangular or saw-tooth wave (chirp). The chirp duration, T_m ($1/f_m$), can last several milliseconds. In general, the range uncertainty is approximately given by:

$$\sigma_{r-FM} \approx \frac{\sqrt{3}}{2\pi} \frac{c}{\Delta f} \frac{1}{\sqrt{SNR}} \tag{10.9}$$

where Δf is the tuning range or frequency excursion (Skolnik, 1980; Stone *et al.*, 2004). The ambiguity range depends on the chirp duration and is given by $cT_m/4$. With coherent detection, a system can achieve a theoretical measurement uncertainty of about 2 µm on a cooperative surface located at 10 m (Stone *et al.*, 2004). Furthermore, because of the coherent detection, the dynamic range is typically very high. In practice, a typical commercial system can provide a measurement uncertainty of about 30 µm at a data rate of 40 points per second and 300 µm at about 250 points per second (Guidi *et al.*, 2005).

10.4.3 Laser propagation and spatial resolution

In the preceding sections, range uncertainty was presented for different measurement principals. This range uncertainty describes how well a method extracts longitudinal measurements, and does not affect the spatial resolution in X and Y (lateral resolution). For a flying spot scanner, the laser beam propagation properties or diffraction, and the DOF are the limiting factors for the lateral resolution. Even in the best laser emitting conditions, the beam does not maintain collimation with distance because of diffraction. Effects of the scanning mechanism and the number of pixels in the detector on the lateral resolution are smaller. Figure 10.11 shows the diffraction-limited resolution as a function of DOF or the measurement volume (Rioux *et al.*, 1987).

10.4.3 Objects material and surface texture effects

The underlying assumption of active optical geometric measurements is that the imaged surface is opaque, diffusely reflecting and uniform, like a vapour-blasted aluminium (VB-Al) surface. Since not all materials are like that, one should expect unreliable measurements on some types of surface. Marble departs from the ideal hypothesis by exhibiting two critical optical properties: translucency and non-homogeneity. This causes a bias in the measured distance and also increases the noise level (Godin *et al.*, 2001). Beraldin (2004) measured flat pieces of VB-Al and marbles using both a FMCW system and a close range triangulation

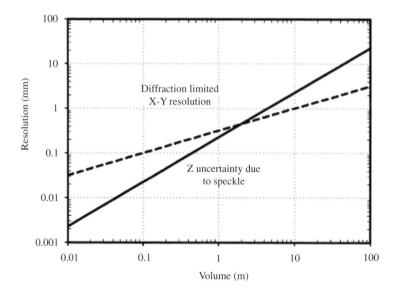

Figure 10.11 *Dashed line: diffraction-limited X-Y lateral spatial resolution as a function of measured volume (DOF), Solid line: Z uncertainty for triangulation-based systems.*

laser scanner. In both cases, a plane equation was fitted using the same algorithm. The FMCW system gave 14 μm on VB-Al but 87.6 μm on marble. The laser triangulation system gave 30 μm on VB-Al and 49 μm on marble. Dorsch *et al.* (1994) studied machined metal surfaces that show macroscopic scratches with locally varying properties and found that the uncertainty can worsen by a factor of two compared to cooperative surfaces. The accuracy also drops when measurements are performed on sharp discontinuities such as edges (El-Hakim and Beraldin, 1994), and even on flat surfaces at highly contrasting colour transitions like black and light colours. All such material type, surface, and edge effects should be expected regardless of the type of sensor used (Buzinski *et al.*, 1992; Hebert and Krotkov, 1992; Blais *et al.*, 2005).

10.4.5 3D data processing issues

In this section, we discuss a few examples that highlight some concerns linked to 3D modelling from range data. To generate a high-resolution model, one must take into account the unreliable data, which is not usually flagged by the scanner software. As an example, two scans were performed on a bronze surface, one taken at a stand-off distance to avoid saturation but yielded low resolution (Figure 10.12a) and one at closer range but with a saturation zone in the upper right corner of the image (Figure 10.12b). In this case, the operator must decide whether to proceed with a low-resolution 3D scan or keep the high-resolution data but manually remove the saturated zone. Saturations can increase systematic errors and should be removed from the processing pipeline, such as data registration; otherwise unwanted patterns can occur in the final model (Figure 10.12c).

Systematic errors are present in all 3D systems due to, for example, poor calibration, material properties, or motion during scanning. The latter produces waves in the raw 3D data, as shown in Figure 10.13 (see colour section). Either using a faster scanner or filtering the raw 3D scans can remove the waves. However, filtering can alter the *X–Y* spatial resolution.

Another issue is the operator's 3D modelling experience. A case in point is a scanned metope that was modelled by two different operators, average and expert. Figure 10.14a shows the result after alignment, merging and compression in the case of the average operator. This mesh contains about 36,000 polygons. The expert operator produced the result shown in Figure 10.14b starting from the same 3D images and the same 3D modelling software. This mesh contains only 20,000 polygons but exhibits much better quality. The difference between these models comes from the fact that the average operator did not remove the 3D artefacts originating from

(a)   (b)

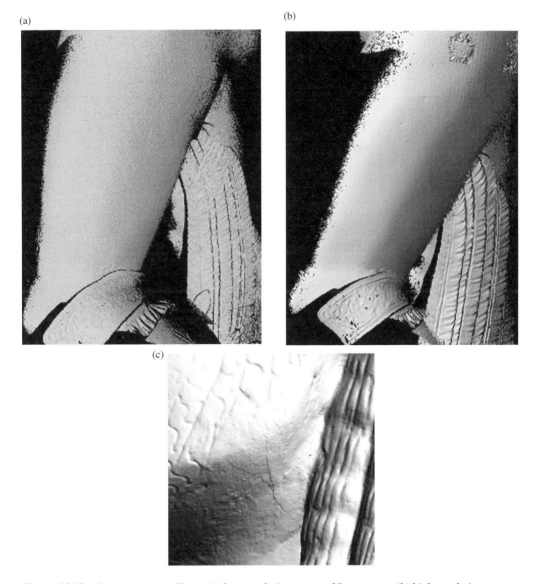

(c)

Figure 10.12   *Some scanning effects: (a) low-resolution scan at 30 cm range; (b) high-resolution scan at 15 cm range; (c) junction lines from alignment of saturated data in (b).*

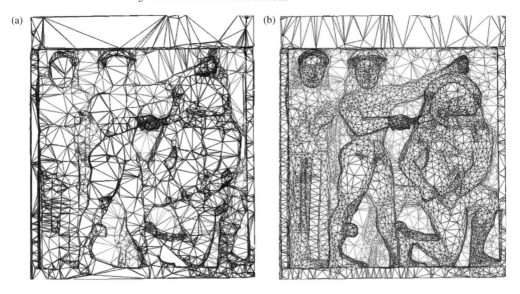

Figure 10.14 *(a) 36,000-polygon model prepared by an average operator; (b) only 20,000-polygon model, but better quality, prepared by an expert operator.*

abrupt shape variations or edges, as discussed in Section 10.4.4. This resulted in a large number of triangles trying to represent those erroneous spikes near the edges. The removal of such artefacts from the data allowed for a better redistribution of polygons where shape variations actually existed.

10.4.6 Performance evaluation and standards

A measurement is useful only if its uncertainty is known. The statement of uncertainty is usually based on comparisons with standards traceable to the national units (SI units), such as the standards available to manufacturers of theodolites and coordinate measurement machines (CMMs) for assessing their measuring systems. A guideline called VDI/VDE-2634 (2001) has been prepared in Germany for some triangulation-based optical 3D systems. It lists acceptance testing and monitoring procedures for evaluating the accuracy of fringe projection and moiré, as well as photogrammetric and scanning systems. Cheok and Stone (2004) conducted exploratory experiments to characterise the performance laser detection and ranging (LADAR) instruments. Their experience points to the need for a calibration and evaluation facility at the National Institute of Standards and Technology, USA (NIST), and to the need for uniform specifications and test procedures for characterising LADAR systems. Currently no internationally recognised standard or certification method exists to evaluate accuracy or measurement uncertainty and resolution, of laser range sensors. Hence, the user should devise methods to ensure confidence in what is being measured and perform periodic verifications even when the manufacturer provides a specification sheet. In practice, an object known with accuracy that is ten times better than that of the scanner should be employed as a reference. Studying the scientific literature published on testing range cameras (Johansson, 2002; Böhler *et al.*, 2003; Schulz and Ingensand, 2004) should help in preparing a verification methodology that best suits the application. Definitions of terms can be found in the VIM standard for metrology (VIM, 1993).

10.5 Integration of data from multiple sources

As discussed in Section 10.1.2, a single type of 3D data or technique is usually not enough to model relatively large or complex objects and sites with acceptable accuracy, level of details, and realism. Therefore, integration of multiple data types is an active research issue in 3D modelling. The concept was initially developed several years ago for mobile mapping systems (Schwarz and El-Sheimy, 2004), which combine digital cameras and perhaps laser scanning with GPS (global positioning system), INS (inertial navigation system), and other kinetic sensors. Many other methods for integrating different sensors and techniques have recently become available. Nagai *et al.* (2004) fitted an unmanned helicopter with a digital camera, a laser scanner, IMU (integrated motion unit) and GPS. Direct geo-referencing is achieved automatically using all the sensors. The scanning data determined the geometric 3D shape while the digital images provided the texture. Georgopoulos and Modatsos (2002) used tourist slides, engineering drawings and geodetic measurements to model the church of the Holy Sepulchre in Jerusalem. Bundle adjustment with self-calibration was applied with geodetic and engineering drawings measurements as the control. Grün *et al.* (2004) also integrated different images, including some from the internet, taken at different times with different types of camera to model the now-destroyed great Buddha of Bamiyan in Afghanistan. Partially existing sites have been reconstructed as they once were by digitising the remains using photogrammetry and laser scanning, and making use of available records and drawings (Stumpfel *et al.*, 2003; Mueller *et al.*, 2004; Beraldin *et al.*, 2005). Bernardini *et al.* (2002) combined structured light 3D sensing and photometric stereo to model Michelangelo's Florentine Pietà. Integrating laser scanning with image-based modelling and rendering (Sequeira *et al.*, 2001) and image-based modelling with image-based rendering (Debevec *et al.*, 1996) have also been reported. El-Hakim *et al.* (2004a) applied photogrammetry with aerial and terrestrial images on main shapes, and laser scanning on façade details to model the abbey of Pomposa in Italy. Similar ideas, except limiting photogrammetry to the global framework, edges and corners, and laser scanning for all other parts, have been described (Lamond and Watson, 2004; Alshawabkeh and Haala, 2004). Holes present in the scan data are filled from the photogrammetric model. Rönnholm *et al.* (2003) developed an interactive technique to register laser scanning data with panoramic images and applied it to large heritage sites. El-Hakim *et al.* (2004b) used a bundle adjustment to register laser scanning data with texture data, using common control points between the two types of data, to model cave rock art in Australia. Flack *et al.* (2001) developed tools specifically designed to assemble models created by various techniques.

10.5.1 Model assembly criteria

The following four issues must be addressed in order to turn models created by different data sets into a single representation appropriate for 3D reconstruction and visualisation:

(1) Relative scale and orientation must be correctly determined.
(2) Joint primitives, specifically surfaces, edges, and vertices, from adjacent models must match perfectly. This is often a problem if the exact same point locations in common parts between those models are not available.
(3) No gaps, redundant surfaces or intersecting edges are acceptable. This can easily happen when combining models acquired using different types of sensors and records which do not reflect the "as built" conditions.
(4) Textures on adjacent models should integrate seamlessly.

To satisfy most criteria, interactive post-processing of the integrated models is usually the obvious solution. However, some techniques such as mesh smoothing and mesh blending (Yu *et al.*,

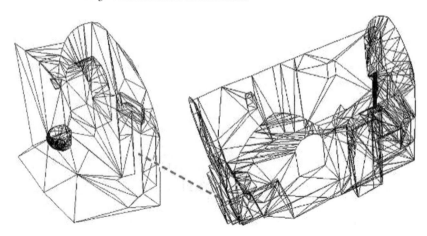

Figure 10.15 *Two adjacent models connected by a portal.*

2004; Liu *et al.*, 2005), have been recently developed to facilitate the procedure, particularly criteria (2) and (3), and to reduce the time-consuming trials and errors usually experienced with commercial software.

10.5.2 Model assembly procedure

The models can be registered (criterion (1) above) manually with available modelling software using common points between them (Figure 10.15). The process will be unnecessary if the individual models were directly created in the same coordinate system using control points or geo-referencing. But some models have no visible common points with other models or have no geo-referencing. In this case floor plans, if available, can be used for positioning those models (Figure 10.16). Once all models are registered, an integration procedure must be applied to satisfy criteria (2)–(4).

Based on the discussion in Sections 10.2–10.4, it is prudent to arrange the assorted models in a hierarchy, by the data source, where the top model contains the least detail as shown in Figure 10.17. The details, accuracy and reliability increase at each successive level. As a rule for selecting which data to use in overlapped areas, data in one level overrides the data in previous levels. Details, accuracy, and reliability are better with models from aerial images than from existing drawings or surveying, and increase for models from close-range images. A relative accuracy improvement and higher level of details can be achieved by TOF–PW laser scanners and further improvement by the TOF-CW and triangulation-based laser scanners if used at close range (less than 2 m). As an example, the initial general model may consist of an entire structure or site from floor plans or aerial images, or a combination of both. Detailed models from close-range images and/or laser scanner are then imported along common edges and vertices. The mesh of the general model is then adjusted to purge redundant surfaces by keeping those from the more detailed models, and to fill the gaps. In some cases, the models need to be manipulated to create holes for portals.

10.6 Visualisation

This section gives some details on visualisation issues affecting quality (photo-realism) and speed (interactivity) to provide realistic experience and smooth navigation through the model. This should be at a rate of at least 20 frames-per-second otherwise the human brain can detect latency or jitter, even complete loss of interactivity may occur.

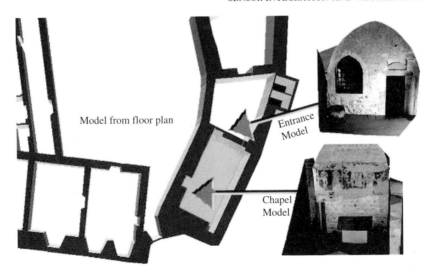

Figure 10.16   *Positioning detailed models using floor plans.*

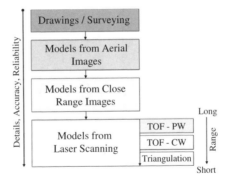

Figure 10.17   *Hierarchy of model assembly.*

10.6.1 Photo-realism

Traditional texture mapping techniques, which are simply draping static imagery over geometry, are not sufficient to represent reality. Several geometric and radiometric distortions, as well as image phenomena such as *aliasing* and other abnormalities can severely impede the visual quality (Yamauchi *et al.*, 2005). Also, without proper illumination that takes into account surface material, texture mapping alone does not give a truly realistic feel. Issues affecting photo-realism and available solutions to reduce their effect are discussed next.

Distortions from geometric sources

The textures on the models can be visibly affected by even slight geometric errors. Weinhaus and Devich (1999) cover geometric distortions, mostly those resulting from the various transformations between image space and object space. The sources of geometric errors affecting texture quality can be organised into three categories: first, mapping between polygon plane and image plane. The correct mapping is given by a projective transform. Since most 3D viewers do not use this transform, visible distortions arise (Figure 10.18b). The distortion will be less perceptible if the surface is parallel to the image plane and/or the polygon size is very small. To correct this distortion, parameters of a projective transform are determined, using the 3D coordinates of the polygon, and applied to re-sample, or warp, the texture within each polygon (Figure 10.18c). The easiest texture warping method is point sampling in which the texture value nearest to the desired sample

(a)   (b)

(c)

Figure 10.18 *Geometric mapping errors: (a) texture source image with oblique building side; (b) uncorrected texture map; (c) re-sampled texture map.*

point is used, although this usually results in aliasing artefacts. There is a need to apply sophisticated filters, such as space variant filters whose shape varies as they move across the image.

Secondly, camera calibration and orientation are important. Proper calibration with lens distortion must be applied otherwise visible errors, particularly line discontinuities, will be apparent at common edges of adjacent polygons mapped from different images (Figure 10.19a (see colour section)). It is also important that the internal calibration parameters remain valid for all images by either keeping the camera settings constant or applying self-calibration. External camera orientation must also be accurately determined with bundle adjustment.

Finally, there are the surface reconstruction assumptions. Deviations of the mesh of polygons from the underlying true object surface give rise to texture errors. Since a single polygon represents a plane, applying large polygons on surfaces with even slight curvature will result in features being projected on the wrong surface or disappearing altogether, particularly near surface intersections (Figures 10.19b,c).

Distortions from radiometric sources

The distortions from radiometric effects are usually visible along common edges of adjacent polygons mapped from different images. Due to variations in lighting, surface specularity and camera gain settings, sensed colour and intensity for a segment shown in images taken from separate positions will not match. When these images are stitched together on the model, seams and discernible patches will become visible. The digitised brightness value for a pixel is a measure of the amount of light energy reflected from the surface, known as radiance. There is an unknown non-linear mapping between the pixel value and the radiance. This non-linearity is largest near the saturated part of the response curve. Knowing the response function allows merging of images taken at different exposure settings, different angles, or even by different cameras. The inverse of the response function is used to recover the scene radiance, up to a scale (Debevec and Malik, 1997). The response function for a camera can be determined in a controlled environment, just like a laboratory calibration, and then applied to any image taken by this camera. When the response function is not available, seamless texture maps can be achieved by two different approaches: first, *texture blending*. This technique uses the weighted average of pixel values of polygons in the overlapping sections between adjacent textures. Pulli *et al.* (1998) applied a weighting scheme with three different weights: directional weight, sampling-quality weight, and feathering weight. The task of the directional weight is to favour rays originating from views whose viewing direction is close to that of the virtual camera. The sampling-quality weight reflects how well a ray samples the surface. Each pixel is assigned a weight that is defined as the cosine of the angle between local surface normal and the direction from the surface point to the camera. The feathering weight is used mostly to hide artefacts due to differences in lighting among images. Most blending techniques require a model with a uniform polygon or triangle size and shape and use this to select the size of the overlapped region. If the size of that region is too small the seams between regions will not be completely removed. On the other hand, if the region is too large important details may be blurred. To alleviate these problems, Baumberg (2002) developed an automatic multi-band approach to preserve the high-frequency detail in the overlapping region that does not require uniform triangle size. Other varieties of texture blending can be found in Bernardini *et al.* (2001), Rocchini *et al.* (2002) and Wang *et al.* (2001).

Secondly, *colour adaptation* can be used. El-Hakim *et al.* (1998) used a per-image correction called "global colour adaptation" and a per-polygon correction termed "local colour adaptation". The global colour adaptation estimates colour offsets between images. Colour differences along the border of adjacent regions of polygons are minimised by least-squares adjustment (Figure 10.20 (see colour section)). Weighted observations ensure non-singularity and prevent the off-set from drifting across larger scenes. The local colour adaptation modifies the colours of each polygon to ensure smooth transitions to all adjacent polygons. This is not straightforward since each edge of a polygon will have a different offset as it borders a different polygon. This requires an iterative solution to force a gradual change to the colours within a polygon. A plane is fitted to all colour offsets observed at the borders using least squares adjustment that minimises these differences. Then, after correcting the colours according to the plane parameters, the process is repeated until there are no noticeable changes.

The user has to decide which option is better for the case at hand: texture blending which better reduces artefacts at adjacent polygons but tends to blur the textures, or colour adaptation, which produces sharper texture maps but does not completely correct the texture differences, or both. With the latter, Guindon (1997) achieved seamless continuity by combining scene-wide statistical normalisation (matching means and variances of scene colours) with local operations such as feathering or blending in overlapped regions.

Other texture aberrations

When textured models are viewed from a long distance, aliasing, moiré effect, jagged lines, and other visible abnormalities may result. This is because when the surface is far away, a display pixel on that surface may be associated with several texture elements from the texture map. MIP-maps (from the Latin *multum in parvo*), a sequence of textures each of which is a progressively lower resolution image of the original (by a factor of two), are commonly used, especially in computer games, to solve this problem (Williams, 1983). The most appropriate texture is selected depending on the polygon size or distance. Texture memory caching and bi-linear or tri-linear filtering to remove aliasing from the created images are also needed. Since real-time re-sampling and filtering can be computationally expensive, MIP-maps may be pre-computed and stored. This will increase the total size of texture by a maximum of one-third. Most recent hardware supports MIP-mapping operations, but the actual managing and set-up of the caching scheme is up to the application.

Surfaces far away or at oblique angle will be covered by a small part of the image. Since this part has to be stretched to cover the 3D surface, blur and mark elongation will take place (Figure 10.21). View-dependent texture mapping, by taking sufficient images and selecting the most appropriate image depending on the location of the surface, can solve this problem (Debevec *et al.*, 1998; Ismert *et al.*, 2003). A logical choice is to select the image that is taken along the normal and in which the polygons appear largest. On the other hand, this may result in having too many different images assigned to adjacent batches of polygons. Accordingly, a balance must be struck between the best location and trying to have a large number of adjacent polygons with texture from the same image.

Illumination and reflectance

To truly achieve photo-realism, proper illumination must be applied. Lighting a computer-generated model entails specifying the illumination model and light position, intensity, direction and colour (Lee *et al.*, 2004). Global illumination techniques, where all objects in the scene affect light simulation, can greatly increase the level of realism compared to a single light source (Shirley and Morley, 2003; Damez *et al.*, 2003; Wald *et al.*, 2005). Also, the interaction between the incident illumination and the reflective properties of a surface should be taken into consideration since colour varies depending on how the light is reflected from the surface.

(a)

(b)

Figure 10.21   *Texture stretching: (a) texture image; (b) stretched texture on floor surfaces.*

The surface bi-directional reflectance distribution function (BRDF) describes its reflectance properties, and the interaction between illumination and BRDF is usually obtained by convolution (Nayar *et al.*, 1990; Ramamoorthi and Hanrahan, 2004). Since the incident illumination on a surface is scattered in several directions, the BRDF is a 4D function that describes which fraction of that light is reflected into a given direction. A direction is set with respect to the surface normal. This remains an active area of research in computer graphics and a variety of techniques are available (Rusinkiewicz, 1998; Debevec *et al.*, 2000; Shirley, 2002; Goesel *et al.*, 2003; Kautz *et al.*, 2003). Another problem is that even when proper texturing and lighting is applied, the rendered surface may look too smooth with no bumps or scratches, which is fairly unrealistic. An improvement can be achieved by applying bump mapping (Rushmeier *et al.*, 1997). This is done by introducing small variations in surface normal to add roughness and a more realistic look to the lighted surface without the need to model each wrinkle as a separate element.

10.6.2 Interactive visualisation

The ability to interact with 3D models is a continuing problem due to the fact that model sizes are increasing at a faster rate than computer hardware advances. The rendering algorithm should be capable of delivering images at real-time frame rates, even for a very large number of polygons and with high-resolution textures. Akenine-Moeller and Haines (2002) gives a good overview of the subject. Here the discussion relates to the main factors affecting the performance of a rendering system, mainly hardware limitations and model size, geometry and texture, and gives some basic solutions for each. This is followed by an outline of selected advanced systems designed for huge models.

Hardware processing and memory limitations

Specifications affecting processing speed on a given hardware are: the number of geometric transformations that can be performed per second, the number of triangles that can be rendered with texture per second, the available main and texture memory, the disk and network access speed (bandwidth), and data transfer rate from the main memory to the graphic card memory. Some of these hardware specifications will only affect start-up time. For example, significant computations are initially needed for the rendering software to build an internal representation of the scene graph. Also high bandwidth is crucial to rapidly load the model and texture in memory at start-up. Humphreys *et al.* (2001) and Strengert *et al.* (2004) used parallel processors or clusters to cope with large models. Lario *et al.* (2005) used on-board video memory to store geometry information to reduce data transfer from the main memory to the graphic card, which is the main bottleneck.

Size of the geometric model

Hierarchical level of detail (LOD), where objects far away from the viewpoint are rendered with pre-sampled lower resolution representations, is a standard technique that is supported by most scene graph libraries. Cohen *et al.* (2003) developed a low-level lightweight application program interface (API) system for geometric LOD that is integrated in the OpenGL library. Luebke *et al.* (2002) have surveyed mesh simplification techniques, such as progressive meshes (Hoppe, 1996), whose objective is to minimise the size of the output mesh given the allowed maximum error. Hoek and Damon (2004) have reviewed automatic model simplification and run-time LOD management techniques. They discuss LOD framework, LOD management, LOD simplification models and metrics for error evaluation. In addition to applying an LOD algorithm, there may be a need to apply *visibility culling*. This approach skips objects that are outside the viewing frustum, face away from the viewer, or are occluded by other objects or surfaces (Cohen-Or *et al.*, 2003; Zhang, 1998). Once visibility is determined, hidden surface removal is performed with a variant of the classic Z-buffer algorithm (Catmull, 1974).

Size of the texture

For many recent projects, the total texture size is several hundreds of megabytes. If all the texture fits into the on-board texture memory then there is usually no problem, however, if the texture memory is less than the active texture (which must be in the texture memory), swapping will occur, resulting in noticeable performance degradation. In this case, the viewing application must concoct a texture caching routine that decides on which texture to keep or move into memory and which to move to RAM, or even to the system virtual memory (Cline and Egbert, 1998; Dumont *et al.*, 2001). Fortunately it is rare that all the textures are needed for any one frame. To take advantage of this, two techniques are suitable: *view frustum culling*, which skips rendering and loading into memory parts of the model that are outside the view frustum; and *MIP-mapping*, which was originally developed to reduce aliasing (Section 10.6.1). To make texture caching more efficient it is recommended that the scene be divided into smaller uniform groups, each with a small texture with width and height that are to the power of two. This makes it possible to control the file size and the rendering speed more precisely. Wei (2004) proposed a tile-based texture-mapping algorithm by which it only physically stores a small set of texture tiles instead of a large texture. The algorithm generates an arbitrarily large and non-periodic virtual texture map from the small set of stored texture tiles. Consequently, only a small constant storage is used regardless of the size of texture. Another approach is *adaptive texture maps* (Kraus and Ertl, 2002; Hwa *et al.*, 2004). These are texture maps with locally adaptive resolution and an adaptive boundary of the map's domain instead of a rectilinear domain. The locally adaptive resolution permits continuous simplification, and allows efficient representation of fine details rather than increasing the resolution of the whole texture map. Adaptive boundaries considerably reduce the memory demands when the texture data contains large colour-uniform regions. Besides textures, global illumination techniques (see Section 10.6.1) are still time consuming and must be optimised for interactive visualisation.

Advanced systems for rendering huge models

Advanced systems to handle models containing hundreds of millions, or more, polygons are being developed, and some examples are given here. Aliaga *et al.* (1999) has presented a system for rendering very complex 3D models at interactive rates. It selects a subset of the model as preferred viewpoints and partitions the space into virtual cells. Each cell contains near geometry rendered using LOD and visibility culling, and far geometry rendered as a textured depth mesh. GigaWalk (Baxter *et al.*, 2002) is a system for interactive walkthroughs of complex, gigabyte-sized environments. It combines occlusion culling and LOD and uses two graphics pipelines with one or more processors. Cignoni *et al.* (2004) developed the Adaptive Tetra-Puzzles system where, for each frame, the best-fit variable resolution LOD is selected according to the view frustum and the requested visualisation accuracy. The selection is taken from a pre-constructed multi-resolution hierarchy of geometric model patches. GoLD (Geo-morphing of LOD) was developed for view-dependent real-time visualisation of huge multi-resolution models (Borgeat *et al.*, 2005). It uses geo-morphing to smoothly interpolate between geometric patches in a hierarchical LOD structure. The same approach also applies to the texture patches.

10.7   Applications examples

Examples using data from multiple sensors and techniques are presented in this section. Projects selected include evaluation and validation of the results. The projects show what can be achieved and what accuracy should be expected in practice.

10.7.1 Temple "C" of Selinunte

Selinunte in Sicily is the site of some historically important Greek ruins such as the sixth-century BC Temple "C", or Apollo's Temple (Figure 10.22a). This is an example of a site that needed a mixture of technologies to digitally reconstruct it. The project was divided into two phases. In the first phase, a room at the archaeology museum in Palermo containing artefacts from the site in Selinunte (Figure 10.22b) was completely modelled using multiple techniques. Two types of laser scanner on different elements in the room were needed for the details. One scanner was used for 0.5 mm resolution at close range. It was applied to the three museum metopes associated with Temple C, each has an

(a)   (b)

Figure 10.22   *(a) Temple C now; (b) Selinunte room at the Archaeology museum, Palermo, Italy.*

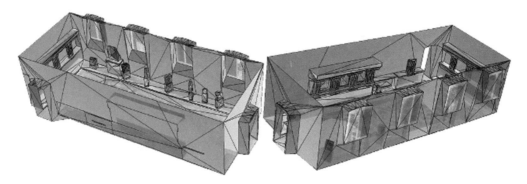

Figure 10.24   *Two views of the photogrammetry model.*

Figure 10.25   *Textured model.*

average size of 1 m × 1 m × 0.4 m, and other small sculptures from the temple. Figure 10.23 (see colour section) shows the results of the modelling of a metope known as "Perseo and Medusa". The second scanner captured 2–10 mm details at longer range and was applied to larger objects with small surface variations and some room surfaces such as the vaulted ceiling. Most of the 3D model of the museum room was created using photogrammetry (Figures 10.24 and 10.25) and CAD modelling. A rendering of the complete room is shown in Figure 10.26 (see colour section).

The second phase of the project will be the reconstruction of Temple C of the Acropolis of Selinunte, as it once was, using CAD tools, scanning of remains, and its modelled parts at the museum from the first phase. The model will be based on historical information available at the University of Lecce, Apulia, Italy and at the "Museo Archeologico Regionale" of Palermo, Sicily. See Beraldin *et al.* (2005) for details.

10.7.2 The Stenico Castle

This castle in Trentino is one of the oldest and most important medieval castles in Northern Italy, and is an interesting mixture of styles of buildings added over several centuries (Figure 10.27). The structures extend over an area of about 100 m × 64 m and are up to 35 m in height. The castle grounds are elevated by over 30 m above the road approaching it. The site consists of several components, mainly a number of buildings and towers organised around four courtyards, a Renaissance loggia, a thirteenth-century chapel with medieval frescos, and tall thick gated walls enclosing everything.

The castle was completely modelled using an image-based semi-automatic technique (see Section 10.3.3) and existing floor plans. Several sets of images were taken from ground level and from a helicopter with a high-resolution digital camera. The imaging took a total of four hours in two visits to the site. Two camera settings were used during the project, one at the shortest and one at the longest focal length depending on the available space around the object. The aerial images were taken from an average range of 120 m and average vertical distance of 65 m above the castle. The field of view of each image usually covered all buildings with a field of view of 150–170 m. One pixel in the image corresponded to about 8 cm. Surveying of 115 points with a total station was also needed to help register all models in one global coordinate system. A general model was created from the aerial images (Figure 10.28), and 17 detailed models were created using ground images, such as the loggia shown in Figure 10.29 (see colour section).

Figure 10.27   *Aerial view of Stenico Castle.*

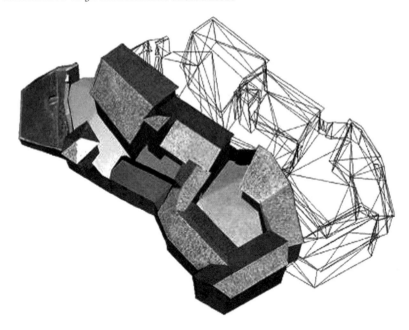

Figure 10.28   *General model of castle buildings.*

The accuracy was assessed by the variance–covariance matrix of the computed 3D points, and validated by surveyed checkpoints. The achieved accuracy for the points from the aerial images was fairly homogeneous in the three coordinates and averaged: 17 mm (X), 15 mm (Y) and 16 mm (Z). This is one part in 10,000 and represents 0.2 pixels. For the ground-level models, the accuracy ranged from 1 mm to 2 mm, which gives about one part in 6000 and represents 0.3 pixels. The superior relative accuracy reached from aerial images, even though they were taken from much longer ranges, is attributed to the better geometric configuration compared to the ground images. The latter had less than ideal locations due to obstructions and tight spaces around most buildings. Although much higher relative accuracy has been reported from photogrammetric measurements on targets for industrial applications, the numbers given here are what can be expected on unmarked large structures in an uncontrolled environment. Using the floor plans, surveyed points and points along common regions, detailed models were integrated with the general model. The latter was re-meshed to account for point changes and sometimes to create a hole where detailed models were inserted. Sections still missing from the general model were added from the floor plans using wall heights known from the image-based models. See Gonzo *et al.* (2004) for details.

10.7.3 Donatello's Maddalena

The objective of the 3D acquisition and modelling of the 180 cm tall Maddalena statue, sculpted by Donatello in 1455, was to develop a non-invasive methodology for monitoring fragile wooden sculptures over the years. The main requirement of the work is high model accuracy, therefore new quality control procedures based on both the self-check of 3D data and the use of complementary 3D sensors, were developed for testing the model. Comparing the 3D model created from a range sensor with bundle adjustment on digital images of the statue demonstrated that the interactive closest point (ICP) method implemented to register the multiple scans caused a loss of metric accuracy even though the single-range images were highly

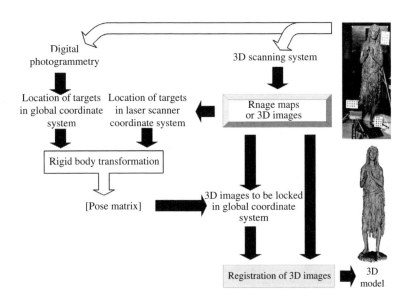

Figure 10.30 *Outline of the 3D modelling procedure with multiple techniques.*

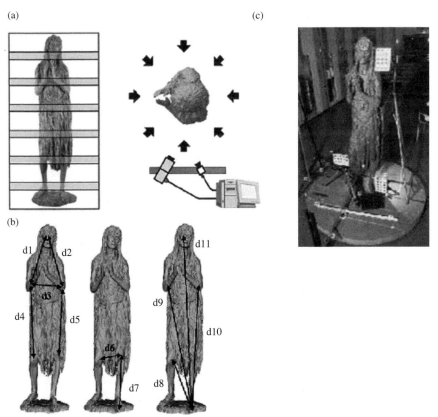

Figure 10.31 *(a) Range sensing set up; (b) distances used for verification; (c) targets for image registration and accuracy evaluation.*

accurate. Sensor fusion was then used to correct the alignment errors. This was achieved by combining the range data with photogrammetric data to take advantage of the good local accuracy and point density of the former and the high overall accuracy of the latter. The procedure is outlined in Figure 10.30.

The custom range system used for this project involves the projection of special patterns over the area of interest and capturing it with a CCD camera (Figure 10.31a). Lateral resolutions ranging from 0.1 mm on critical parts of the statue such as cracked areas to 0.5 mm on smoother areas were required. More than 170 range images were needed to cover the object with this resolution. Due to the nature of the wood surface, significant noise was encountered. Taking multiple-range images from the same location and averaging them succeeded in reducing the noise level. The optimum number of repeated images was four, which resulted in range uncertainty (standard deviation of depth) of 0.05 mm (single-range image).

Bundle adjustment provided dimensional verification on the 3D model (Figure 10.31b) to check the ICP method. The ICP alone produced errors up to 4 mm, which was reduced to below 0.5 mm when integrated with photogrammetry. Targets placed around the statue (Figure 10.31c) were used for the bundle adjustment and accuracy verification. See Guidi *et al.* (2003) for details.

10.8 Selecting the right approach

This chapter has examined most of the different ways to acquire, process, and visualise 3D information about real objects and environments. From the above discussions and examples, one can see that there is no single approach or sensor that can work effectively in all situations. Establishing which approach is the optimum for an application is difficult due to the shortage of system performance specification standards for most optical 3D capture and modelling systems, such as the ones available for surveying equipment and CMMs. Another difficulty with the non-contact 3D systems is that their performance depends significantly on the object being measured, which may require testing different systems on the actual intended object before determining the most suitable one. A 3D modelling system also consists of several integrated components and specialised software that must all be evaluated within controlled experiments. Since designing and setting up such experiments are still in an early stage of development, the analysis in this chapter is based on the authors' as well as others' reported experience.

Figure 10.32 provides rough guidelines for selecting which technique to use to acquire the different types of feature, object or site. Automated image-based techniques have been included, although they are still of limited practical use and there are no accepted accuracy or reliability figures available. This may ultimately change due to the efforts currently being invested in those techniques, but until then semi-automated techniques are strongly recommended. It is important to mention that not only geometric accuracy or uncertainly is to be considered when choosing a system, but also the amount of details it is capable of capturing, the photo-realism, and the ability to visualise the output in an interactive manner or in real time. Laser scanners must be used based on the knowledge of the capability of each different technology at the desired stand-off distance. Currently, the 5–10 m range creates many challenges as no one system provides optimal performance. We can foresee in the future that FMCW systems could become affordable and deliver data rates above 1000 points per second. Figure 10.9 displayed the measurement uncertainty described by the standard deviation σ for each class of 3D laser scanners and should be consulted when deciding upon the appropriate scanner. However, that information is based on the uncertainty of a single-point range measurement on an ideal surface. One needs to take into account the uncertainly of the full scan or range image, i.e. the *XYZ* space, the registration of multiple scans, surface type, and the way the data is processed by the user with a particular

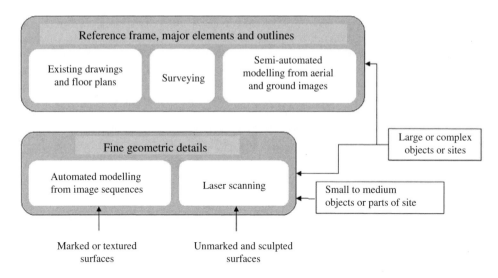

Figure 10.32   *The appropriate approaches for given tasks.*

modelling software. The issue becomes even more complicated when it involves integrating models created from different sensors and techniques since the final model quality will be a function of how well the different data is registered and integrated.

Finally, we can undoubtedly appreciate that selecting and applying the appropriate 3D capture and modelling technique for a given application still requires considerable training and experience. Nevertheless, we hope that the information presented in this chapter, as well as the cited references, will lead to a more manageable task.

10.9 References

Akenine-Moeller T. and Haines, E., 2002. *Real-time Rendering* (2nd edn.), A.K. Peters, Wellesley, Massachusetts, USA.

Aliaga, D., Cohen, J., Wilson, A., Baker, E., Zhang, H., Erikson, C., Hoff, K., Hudson, T., Stürzlinger, W., Bastos, R., Whitton, M., Brooks, F. and Manocha, D., 1999. MMR: an interactive massive model rendering system using geometric and image-based acceleration. *Symposium on Interactive 3D Graphics*, Atlanta, Georgia, USA, 199–206.

Allen, P.K., Stamos, I., Troccoli, A., Smith, B., Leordeanu, M. and Murray, S., 2003. New methods for digital modelling of historic sites. *IEEE Computer Graphics and Applications*, 23(6), 32–41.

Alshawabkeh, Y. and Haala, N., 2004. Integration of digital photogrammetry and laser scanning for heritage documentation. *International Archives of Photogrammetry and Remote Sensing*, (V), 424–429.

Amann, M.-C., Bosch, T., Lescure, M., Myllylä, R. and Rioux, M., 2001. Laser ranging: a critical review of usual techniques for distance measurement. *Optical Engineering*, 40(1), 10–19.

Bacigalupo, C. and Cessari, L., 2003. Survey techniques and virtual reality for recovery plan of a fortified Mediterranean town. *International Workshop on Vision Techniques for Digital Architectural and Archaeological Archives*, Ancona, Italy, July, 40–42.

Baribeau, R. and Rioux, M., 1991. Influence of speckle on laser range finders. *Applied Optics,* 30, 2873–2878.

Baumberg, A., 2002. Blending images for texturing 3D models. *British Machine Vision Conference (BMVC)*, Cardiff, UK, 404–413.

Baxter III, W.V., Sud, A., Govindaraju, N.K. and Manocha, D., 2002. *GigaWalk: Interactive walkthrough of complex environments*. University of North Carolina at Chapel Hill Technical Report TR02-013, Chapel Hill, North Carolina, USA.

Beraldin, J.-A., 2004. Integration of laser scanning and close-range photogrammetry – the last decade and beyond. *International Archives of Photogrammetry, Remote Sensing and Spatial Information Sciences*, 35(B5), 972–983.

Beraldin, J.-A., Picard, M., El-Hakim, S., Godin, G., Paquet, E., Peters, S., Rious, M., Valzano, V. and Bandiera, A., 2005. Combining 3D technologies for cultural heritage interpretation and entertainment. *Videometrics VIII, SPIE Proceedings* 5665, San Jose, California, USA, 108–118.

Beraldin, J.-A., Picard, M., El-Hakim, S.F., Godin, G., Valzano, V., Bandiera, A. and Latouche, C. 2002. Virtualizing a Byzantine crypt by combining high-resolution textures with laser scanner 3D data. *International Conference on Virtual Systems and Multimedia (VSMM'02)*, Gyeongju, Korea, 3–14.

Bernardini, F. and Rushmeier, H., 2002. The 3D model acquisition pipeline. *Computer Graphics Forum*, 21(2), 149–172.

Bernardini, F., Martin, I.M. and Rushmeier, H., 2001. High-quality texture reconstruction from multiple scans. *IEEE Transactions on Visualization and Computer Graphics*, 7(4), 318–332.

Bernardini, F., Rushmeier, H., Martin, I.M., Mittleman, J. and Taubin, G., 2002. Building a digital model of Michelangelo's Florentine Pieta. *IEEE Computer Graphics and Applications*, 22(1), 59–67.

Blais, F., 2004. Review of 20 years of range sensor development. *Journal of Electronic Imaging*, 13(1), 232–240.

Blais, F., Taylor, J., Cournoyer, L., Picard, M., Borgeat, L., Dicaire, L.-G., Rioux, M., Beraldin, J.-A., Godin, G., Lahnanier, C. and Aitken, G., 2005. Ultra-high resolution imaging at 50µm using a portable XYZ-RGB color laser scanner. *International Workshop on Recording, Modelling and Visualization of Cultural Heritage*, Ascona, Switzerland.

Böhler, W., 2005. Comparison of 3D laser scanning and other 3D measurement techniques. *International Workshop on Recording, Modelling and Visualization of Cultural Heritage*, Ascona, Switzerland. Baltsavias, E.P., Grün, A., Van Goel, L. and Patesaki, M. (Eds.), Taylor and Francis, London, UK, 89–99.

Böhler, W., Bordas Vicent, M. and Marbs, A., 2003. Investigating laser scanner accuracy. *18th CIPA International Symposium*, Antalya, Turkey, 30 September–4 October. *International Archives of Photogrammetry, Remote Sensing and Spatial Information Sciences*, 34(5/C15), 474–479.

Borgeat, L., Godin, G., Blais, F., Massicotte, P. and Lahanier, C., 2005. GoLD: Interactive display of huge coloured and textured models. *ACM Transactions on Graphics (SIGGRAPH'05)*, 24(3), 869–877.

Born, M. and Wolf, E., 1999. *Principles of Optics*, 7th edn., Cambridge University Press, Cambridge, UK.

Boulanger, P., Godin, G. and Rioux, M., 1992. Application of 3-D active vision to rapid prototyping reverse engineering. *Proceedings of the Third International Conference on Rapid Prototyping*, Dayton, Ohio, USA, 7–10 June, 213–223.

Brusco, N., Andreetto, M., Giorgi, A. and Cortelazzo, G.M., 2005. 3D registration by textured spin-images. *Fifth International Conference 3D Digital Imaging and Modelling (3DIM'2005)* Ottawa, Ontario, Canada, 13–16 June, 262–269.

Buzinski, M., Levine, A. and Stevenson, W.H., 1992. Performance characteristics of range sensors utilizing optical triangulation. *IEEE National Aerospace and Electronics Conference*, Dayton, Ohio, USA, 18–22 May, 1230–1236.

Campbell, R.J. and Flynn, P.J., 2001. A survey of free-form object representation and recognition techniques. *Computer Vision and Image Understanding*, 81(2), 166–210.

Cantzler, H., 2003. *Improving Architectural 3D Reconstruction by Constrained Modelling*. Ph.D. thesis, School of Informatics, University of Edinburgh, UK.

Catmull, E., 1974. *A Subdivision Algorithm for Computer Display of Curved Surfaces*. Ph.D. thesis, Department of Computer Science, University of Utah, Utah, USA.

Cheok, G.S. and Stone, W.C., 2004. Performance evaluation facility for ladars. In *Proceedings of Laser Radar Technology and Applications IX*, SPIE Vol. 5412, 54–65.

Cignoni, P., Ganovelli, F., Gobbetti, E., Marton, F., Ponchio, F. and Scopigno, R., 2004. Adaptive tetra-Puzzles: efficient out-of-core construction and visualisation of gigantic multi-resolution polygonal models, *ACM Transactions on Graphics (SIGGRAPH'04)*, 23(3), 796–803.

Cline, D. and Egbert, P.K., 1998. Interactive display of very large textures. *IEEE Visualization'98*, Research Triangle Park, North Carolina, USA, 18–23 October, 343–350.

Cohen, J., Luebke, D., Duca, N. and Schubert, B., 2003. *GLOD: Level of Detail for the Masses*. Technical Report, JHU-CS-GL03-4, Johns Hopkins University, Baltimore, Maryland, USA.

Cohen-Or, D., Chrysanthou, Y., Silva, C.T. and Durand, F., 2003. A survey of visibility for walkthrough applications. *IEEE Transactions on Visualization and Computer Graphics*, 9(3), 412–431.

Damez, C., Dmitriev, K. and Myszkowski, K., 2003. State of the art in global illumination for interactive applications and high-quality animation. *Computer Graphics Forum*, 21(4), 55–77.

Debevec, P.E. and Malik, J., 1997. Recovering high dynamic range radiance maps from photographs. *SIGGRAPH'97*, Los Angeles, California, 3–8 August, 369–378.

Debevec, P.E., Hawkins, T., Tchou, C., Duiker, H.-P., Sarokin, W. and Sagar, M., 2000. Acquiring the reflectance field of a human face. *SIGGRAPH'2000*, New Orleans, Louisiana, USA, 23–28 July, 145–156.

Debevec, P.E., Taylor, C.J. and Malik, J., 1996. Modelling and rendering architecture from photographs: A hybrid geometry and image-based approach. *SIGGRAPH'96*, New Orleans, Louisiana, USA, 4–9 August, 11–20.

Debevec, P.E., Yu, Y. and Borshukov, G., 1998. Efficient view-dependent image-based rendering with projective texture-mapping. *9th Eurographics Rendering Workshop*, Vienna, Austria, 29 June–1 July, 105–116.

DeLeon, V., 1999. VRND: Notre-Dame cathedral, a globally accessible multi-user real-time virtual reconstruction. *International Conference of Virtual Systems and Multimedia (VSMM'99)*, Dundee, Scotland, 1–3 September, 484–491.

Dick, A.R., Torr, P.H., Ruffle, S.J. and Cipolla, R., 2001. Combining single view recognition and multiple view stereo for architectural scenes. *8th IEEE International Conference on Computer Vision (ICCV'01)*, Vancouver, Canada, 9–12 July, Vol. 1, 268–274.

Dikaiakou, M., Efthymiou, A. and Chrysanthou, Y., 2003. Modeling the walled city of Nicosia. *4th International Symposium on Virtual Reality, Archaeology and Intelligent Cultural Heritage (VAST'2003)*, 57–66.

Dorsch, R.G., Hausler, G. and Herrmann, J.M., 1994. Laser triangulation: fundamental uncertainty in distance measurement. *Applied Optics* 33, 1306–1314.

Dumont, R., Pellacini, F. and Ferwerda, J.A., 2001. A perceptually-based texture caching algorithm for hardware-based rendering. *12th Eurographics Workshop on Rendering*, London, UK, 25–27 June, 246–256.

Eid, A. and Farag, A., 2004. A unified framework for performance evaluation of 3-D reconstruction techniques. *IEEE Computer Vision and Pattern Recognition Workshop (CVPRW'04)*, Washington, DC, USA, 27 June–2 July, 41–48.

El-Hakim, S.F., 2002. Semi-automatic 3d reconstruction of occluded and unmarked surfaces from widely separated views. *ISPRS Com. V Symposium*, Corfu, Greece, 2–6 September, 143–148.

El-Hakim, S.F. and Beraldin, J.-A., 1994. On the integration of range and intensity data to improve vision-based three-dimensional measurements. *Videometrics III, SPIE Proceedings* 2350, 306–321.

El-Hakim, S.F., Brenner, C. and Roth, G., 1998. A multi-sensor approach to creating accurate virtual environments. *ISPRS Journal of Photogrammetry and Remote Sensing*, 53(6), 379–391.

El-Hakim, S.F., Beraldin, J.-A., Picard, M. and Godin, G., 2004a. Detailed 3D reconstruction of large-scale heritage sites with integrated techniques. *IEEE Computer Graphics and Applications,* 24(3), 21–29.

El-Hakim, S.F., Fryer, J.G., Picard, M. and Whiting, E., 2004b. Digital recording of aboriginal rock art. *International Conference on Virtual Systems and Multimedia (VSMM'04)*, Ogaki, Japan, 17–19 November, 344–353.

Evers-Senne, J.-F. and Koch, R., 2005. Image-based rendering of complex scenes from a multi-camera rig. *IEEE Proceedings Vision, Image, and Signal Processing*, 152(4), 470–480.

Faugeras, O., Luong, Q.-T. and Papadopoulo, T., 2001. *The Geometry of Multiple Images*, MIT Press, Cambridge, Massachusetts, USA.

Fleming, W., 1998. *3D Photorealism Toolkit*, Wiley, Chichester, UK.

Flack, P.A., Willmott, J., Browne, S.P., Arnold, D.B. and Day, A.M., 2001. Scene assembly for large scale urban reconstructions. *International Symposium on Virtual Reality, Archaeology and Intelligent Cultural Heritage (VAST'2001)*, Glyfada, Greece, 28–30 November, 227–234.

Foni, A.E., Papagiannakis, G. and Magnenat-Thalmann, N., 2002. Virtual Hagia Sophia: restitution, virtualisation and virtual life simulation. *UNESCO World Heritage Congress*, Venice, Italy, 14–16 November.

Fraser, C.S., 2005. Network orientation models for image-based 3D measurement. *ISPRS International Workshop 3D Virtual Reconstruction and Visualisation of Complex Architectures (3D-Arch'2005)*, Mestre-Venice, Italy.

Fruh, C. and Zakhor, A., 2003. Constructing 3D city models by merging aerial and ground views. *Proceedings of the IEEE Computer Graphics and Applications: Special Issue on 3D Reconstruction and Visualization of Large-scale Environments*, 23(6), 52–61.

Fusiello, A., 2000. Uncalibrated Euclidean reconstruction: a review. *Image and Vision Computing*, 18, 555–563.

Georgopoulos, A. and Modatsos, M., 2002. Non-metric bird's eye view. *ISPRS Com. V Symposium*, Corfu, Greece, 359–362.

Gibson, S., Hubbold, R.J., Cook, J. and Howard, T.L.J., 2003. Interactive reconstruction of virtual environments from video sequences. *Computers and Graphics*, 27, 293–301.

Godin, G., Beraldin, J.-A., Taylor, J., Cournoyer, L., Rioux, M., El-Hakim, S., Baribeau, R., Blais, F., Boulanger, P., Picard, M. and Domey, J., 2002. Active optical 3-D imaging for heritage applications. *Proceedings of the IEEE Computer Graphics and Applications: Special Issue on Computer Graphics in Art History and Archaeology*, 22(5), 24–36.

Godin, G., Rioux, M., Beraldin, J.-A., Levoy, M., Cournoyer, L. and Blais, F., 2001. An assessment of laser range measurement on marble surfaces. *5th Conference on Optical 3D Measurement Techniques*, Vienna, Austria, Wichmann Verlag, Heidelberg, Germany, 49–56.

Goesele, M., Granier, X., Heidrich, W. and Seidel, H.-P., 2003. Accurate light source acquisition and rendering. *ACM Transactions on Graphics (Proceedings of SIGGRAPH'03)*, 22(3), 621–630.

Gokturk, S.B., Yalcin, H. and Bamji, S., 2004. A time-of-flight depth sensor – system description, issues and solutions. *Computer Vision and Pattern Recognition Workshop (CVPRW'04)*, Washington, DC, USA, 27 June–2 July, 35.

Gonzo, L., El-Hakim, S.F., Girardi, S., Picard, M. and Whiting, E., 2004. Photo-realistic 3D reconstruction of castles with multiple-sources image-based techniques. *International Archives of Photogrammetry and Remote Sensing*, 35(5), 120–125.

Grossmann, E. and Santos-Victor, J., 2005. Least-squares 3D reconstruction from one or more views and geometric clues. *Computer Vision and Image Understanding*, 99, 151–174.

Grün, A., Remondino, F. and Zhang, L., 2004. Photogrammetric reconstruction of the Great Buddha of Bamiyan. *Photogrammetric Record*, 19(107), 177–199.

Guidi, G., Cioci, A., Atzeni, C. and Beraldin, J.-A., 2003. Accuracy verification and enhancement in 3D modeling: application to Donatello's Maddalena. *International Conference on 3D Imaging and Modelling (3DIM)*, 334–342.

Guidi, G., Frischer, B., De Simone, M., Cioci, A., Spinetti, A., Carosso, L., Micoli, L.L., Russo, M. and Grasso, T., 2005. Virtualizing ancient Rome: 3D acquisition and modelling of a large plaster-of-Paris model of imperial Rome. *Videometrics VIII*, San Jose, California, USA, 18–20 January, *SPIE Proceedings* 5665, 119–133.

Guindon, B., 1997. Assessing the radiometric fidelity of high resolution satellite image mosaics. *ISPRS Journal of Photogrammetry and Remote Sensing*, 52(5), 229–243.

Hanke, K. and Oberschneider, M., 2002. The medieval fortress Kufstein, Austria – an example for the restitution and visualization of cultural heritage. *ISPRS Comm. V Symposium*, Corfu, Greece, 2–6 September, 530–533.

Haval, N., 2000. Three-dimensional documentation of complex heritage structures. *IEEE Multimedia*, 7(2): 52–56.

Hebert, M. and Krotkov, E., 1992. 3-D measurements from imaging laser radars: how good are they? *Image and Vision Computing*, 10(3), 170–178.

Heok, T.K. and Damen, D., 2004. A review of level of detail. *IEEE International Conference on Computer Graphics, Imaging and Visualization (CGIV'04)*, Penang, Malaysia, 26–29 July, 70–75.

Hilton, A., 2005. Scene modelling from sparse 3D data. *Image and Vision Computing*, 23, 900–920.

Hoppe, H., 1996. Progressive meshes. *ACM Computer Graphics (SIGGRAPH'96)*, New Orleans, Louisiana, USA, 4–9 August, 99–108.

Humphreys, M., Eldridge, M., Buck, I., Everett, M., Hanrahan, P. and Stoll, G., 2001. WireGL: a scalable graphics system for clusters. *SIGGRAPH'01*, Los Angeles, California, USA, 12–17 August, 129–140.

Hwa, L.M., Duchaineau, M.A. and Joy, K.I., 2004. Adaptive 4-8 texture hierarchies. *IEEE Visualization*, Austin, Texas, USA, 10–15 October, 219–226.

Ikeuchi, K. and Sato, Y. (Eds.), 2001. *Modelling from Reality*, Kluwer, Norwell, Maine, USA.

Ikeuchi, K., Nakazawa, A., Hasegawa, K. and Ohishi, T., 2003. The Great Buddha Project: modelling cultural heritage for VR systems through observations. *IEEE/ACM International Symposium on Mixed and Augmented Reality (ISMAR 2003)*, Tokyo, Japan, 7–10 October, 7–16.

Ismert, R.M., Bala, K. and Greenberg, D.P., 2003. Detail synthesis for image-based texturing. *Interactive 3D Graphics (I3D)*, Monterey, California, USA, 27–30 April, 171–175.

Jähne, B., Haußecker, H. and Geißler, P., 1999. *Handbook of Computer Vision and Applications*, Academic Press, San Diego, California, USA.

Johansson, M., 2002. Exploration into the behaviour of three different high-resolution ground-based laser scanners in the built environment. *Workshop on Scanning for Cultural Heritage Recording – Complementing or Replacing Photogrammetry*, Corfu, Greece, 1–2 September, 33–38.

Kautz, J., Lensch, H.P.A., Goesele, M., Lang, J. and Seidel, H.-P., 2003. Modeling the world: the virtualization pipeline. *12th International Conference on Image Analysis and Processing (ICIAP' 03)* Mantova, Italy, 12–19 September, 166–175.

Kernighan, B.W. and Van Wyk, C.J., 1996. Extracting geometric information from architectural drawings. *Workshop on Applied Computational Geometry (WACG)*, Philadelphia, Pennsylvania, USA, 27–28 May, 82–87.

Kersten, T., Acevedoi Pardo, C. and Lindstaedt, M., 2004. 3D acquisition modelling and visualization of north German castles by digital architectural photogrammetry. *International Archives of Photogrammetry, Remote Sensing and Spatial Information Sciences*, 35(B2), 126–131.

Kraus, M. and Ertl, T., 2002. Adaptive texture maps. *Eurographics Conference on Graphics Hardware*, Saarbruken, Germany, 1–2 September, 7–15.

Lamond, B. and Watson, G., 2004. Hybrid rendering – a new integration of photogrammetry and laser scanning for image based rendering. *IEEE Theory and Practice of Computer Graphics, (TPCG 2004)*, Birmingham, UK, 8–10 June, 179–186.

Lario, R., Pajarola, R. and Tirado, F., 2005. Cached geometry manager for view-dependent LOD rendering. *13th International Conference in Central Europe on Computer Graphics, Visualization and Computer Vision (WSCG'2005)*, Plzen, Czech Republic, 31 January–4 February, 9–16.

Lee, C.H., Hao, X. and Varshney, A., 2004. Light collages: lighting design for effective visualization. *IEEE Visualisation'2004*, Austin, Texas, USA, 10–15 October.

Lee, S.C. and Nevatia, R., 2003. Interactive 3D building modelling using a hierarchical representation. *IEEE Workshop Higher-level Knowledge in 3D Modelling and Motion Analysis (HLK'03)*, with ICCV'03, Nice, France, 13–16 October, 58–65.

Levoy, M., Pulli, K., Curless, B., Rusinkiewicz, S., Koller, D., Pereira, L., Ginzton, M., Anderson, S., Davis, J., Ginsberg, J., Shade, J. and Fulke, D., 2000. The digital Michelangelo project: 3D scanning of large statues, *SIGGRAPH'00*, New Orleans, Louisiana, USA, 23–28 July, 131–144.

Lewis, R. and Sequin, C., 1998. Generation of 3D building models from 2D architectural plans. *Computer-Aided Design*, 30(10), 765–779.

Liebowitz, D., Criminisi, A. and Zisserman, A., 1999. Creating architectural models from images. *EUROGRAPHICS '99*, 18(3), 39–50.

Liu, Y.-S., Zhang, H., Yong, J.-H., Yu, P.-Q. and Sun, J.-G., 2005. Mesh blending. *The Visual Computer*, 21(11), 915–927.

Luebke, D., Reddy, M., Cohne, J., Varshney, A., Watson, B. and Huebner, R., 2002. *Level of detail for 3D graphics*, Morgan-Kaufmann, San Francisco, California, USA.

Matabosch, C., Salvi, J., Pinsach, X. and Carcia, R., 2004. Surface registration from range image fusion. *IEEE International Conference on Robotics and Automation*, New Orleans, Louisianna, USA, 26 April–1 May, 678–683.

Mueller, P., Vereenooghe, T., Vergauwen, M., Van Gool, L. and Waelkens, M., 2004. Photo-realistic and detailed 3D modelling: the Antonine nymphaeum at Sagalassos (Turkey). *Computer Applications and Quantitative Methods in Archaeology (CAA2004)*, Prato, Italy, 13–17 April, 221–230.

Nagai, M., Shibasaki, R., Kumagai, H., Mizukami, S., Manandhar, D. and Zhao, H., 2004. Construction of digital surface model by multi-sensor integration from an unmanned helicopter. *International Workshop on Processing and Visualization using High-Resolution Imagery*, Pitsanulok, Thailand, 18–20 November (on CD).

Nayar, S., Ikeuchi, K. and Kanade, T., 1990. Determining shape and reflectance of hybrid surfaces by photometric sampling. *IEEE Transactions on Robotics and Automation*, 6(4), 418–430.

Nguyen, T.-H., Oloufa, A.A. and Nassar, K., 2005. Algorithms for automated deduction of topological information. *Automation in Construction*, 14, 59–70.

Pollefeys, M., Van Gool, L., Vergauwen, M., Verbiest, F., Cornelis, K., Tops, J. and Koch, R., 2004. Visual modelling with a hand-held camera. *International Journal of Computer Vision* 59(3), 207–232.

Poor, H.V., 1994. *An introduction to signal detection and estimation*, 2nd edn., Springer-Verlag, New York, NY, USA, Chap. VII.

Pulli, K., Abi-Rached, H., Duchamp, T., Shapiro, L.G. and Stuetzle, W., 1998. Acquisition and visualization of colored 3-D objects. *International Conference on Pattern Recognition*, Brisbane, Australia, 16–20 August, Vol. 1, 11–15.

Ramamoorthi, R. and Hanrahan, P., 2004. A signal-processing framework for reflection. *ACM Transactions on Graphics*, 23(4), 1004–1042.

Rioux, M., Bechthold, G., Taylor, D. and Duggan, M., 1987. Design of a large depth of view three-dimensional camera for robot vision. *Optical Engineering*, 26(12), 1245–1250.

Rocchini, C., Cignoni, P., Montani, C. and Scopigno, R., 2002. Acquiring, stitching and blending diffuse appearance attributes on 3D models. *The Visual Computer*, 18, 186–204.

Rodrigues, M., Fisher, R. and Liu, Y., 2002. Special issue on registration and fusion of range images. *Computer Vision and Image Understanding*, 87(1–3), 1–131.

Rodriguez, T., Sturm, P., Gargallo, P., Guilbert, N., Heyden, A., Jauregizar, Menéndez, F.J.M. and Ronda, J.I., 2005. Photorealistic 3D reconstruction from handheld cameras. *Machine Vision and Applications*, 16(4), 246–257.

Rönnholm, P., Hyyppä, H., Pöntinen, P., Haggrén, H. and Hyyppä, J., 2003. Interactive orientation of digital images. *Photogrammetric Journal of Finland*, 18(2), 58–69.

Roy-Chowdhury, A.K. and Chellappa, R., 2002. Towards a criterion for evaluating the quality of 3D reconstructions. *IEEE International Conference on Acoustics, Speech, and Signal Processing (ICASSP '02)*, Vol. 4: IV-3321– IV-3324.

Ruggeri, M., Dias, P., Sequeira, V. and Gonçalves, J.G.M., 2001. Interactive tools for quality enhancement in 3D modelling from reality. *9th Symposium on Intelligent Robotics Systems, SIRS*. LAAS-CNRS, Toulouse, France, 18–20 July, 157–165.

Rushmeier, H., Taubin, G. and Gueziec, A., 1997. Applying shape from lighting variation to bump map capture. *Eurographics Rendering Workshop*, St. Etienne, France, 16–18 June, 35–44.

Rusinkiewicz, S., 1998. A new change of variables for efficient BRDF representation. *Eurographics Rendering Workshop*, Vienna, Austria, 29 June–1 July, 11–22.

Schindler, K. and Bauer, J., 2003. A model-based method for building reconstruction. *ICCV Workshop on Higher-level Knowledge in 3D Modelling and Motion (HLK'03)*, Nice, France, 17 October, 74–82.

Schulz, T. and Ingensand, H., 2004. Terrestrial laser scanning – investigations and applications for high precision scanning, *FIG Working Week*, Athens, Greece, 22–27 May.

Schwarz, K.P. and El-Sheimy, N., 2004. Mobile mapping systems – state of the art and future trend: *ISPRS XXth Congress, Vol. XXXV*, Istanbul, Turkey, 12–23 July, 759–768.

Sequeira, V., Wolfart, E., Bovisio, E., Biotti, E. and Goncalves, J., 2001. Hybrid 3D reconstruction and image-based rendering techniques for reality modelling. *Videometrics and Optical Methods for 3D Shape Measurements, SPIE Proceedings* 4309, 126–136.

Shirley, P., 2002. *Fundamentals of Computer Graphics*, A.K. Peters, Wellesley, Massachusetts, USA.

Shirley, P. and Morley, R.K., 2003. *Realistic Ray Tracing*, 2nd edn., A.K. Peters, Wellesley, Massachusetts, USA.

Shum, H.-Y., Petajan, E. and Osterman, J. (Eds.), 2003. Special issue on image-based modelling, rendering and animation. *IEEE Transactions on Circuits and Systems for Video Technology*, 13(11).

Skolnik, M.L., 1980. *Introduction to Radar Systems*, 2nd edn., McGraw-Hill, New York, NY, USA.

Soucy, M., Godin, G., Baribeau, R., Blais, F. and Rioux, M., 1996. Sensors and algorithms for the construction of digital 3-D colour models of real objects. *International Conference on Image Processing (ICIP'96)*, Lausanne, Switzerland, 16–19 September, Vol. II, 409–412.

Stone, W.C., Juberts, M., Dagalakis, N., Stone, J. and Gorman, J., 2004. *Performance analysis of next-generation LADAR for manufacturing, construction, and mobility*, National Institute of Standards and Technology Gaithersburg, Maryland, USA, NISTIR 7117.

Strengert, M., Magallon, M., Weiskopf, D., Gothe, S. and Ertl, T., 2004. Hierarchical visualization and compression of large volume datasets using GPU clusters. *Eurographics Symposium on Parallel Graphics and Visualization*, Grenoble, France, 10–11 June, 1–7.

Stumpfel, J., Tchou, C., Yun, N., Martinez, P., Hawkins, T., Jones, A., Emerson, B. and Debevec, P., 2003. Digital reunification of the Parthenon and its sculptures. *4th International Symposium on Virtual Reality, Archaeology and Intelligent Cultural Heritage, VAST'2003*, Brighton, UK, 5–7 November, 41–50.

van den Heuvel, F.A., 1998. 3D reconstruction from a single image using geometric constraints. *ISPRS Journal of Photogrammetry and Remote Sensing*, 53(6), 354–368.

Varady, T., Martin, R.R. and Cox, J., 1997. Reverse engineering of geometric models – an introduction. *Computer-Aided Design*, 29(4), 255–268.

VDI/VDE 2634, 2001. Optical 3-D measuring systems – Imaging systems with point-by-point probing. VDI, Düsseldorf, Germany.

VIM, 1993. *ISO International Vocabulary of Basic and General Terms in Metrology*, 2nd edn., International Organization for Standardization, Geneva, Switzerland.

Wald, I., Friedrich, H., Marmitt, G., Slusallek, P. and Seidel, H.-P., 2005. Faster isosurface ray tracing using implicit KD-trees. *IEEE Transactions on Visualization and Computer Graphics*, 11(5), 562–572.

Wang, L., Kang, S.B., Szeliski, R. and Shum, H.-Y., 2001. Optimal texture map reconstruction from multiple views. *Computer Vision and Pattern Recognition (CVPR01)*, Kauai, Hawaii, USA, 9–14 December, 347–354.

Wei, L.-Y., 2004. Tile-based texture mapping on graphics hardware. *Eurographics Conference on Graphics Hardware*, Grenoble, France, 29–30 August, 55–64.

Weinhaus, F.M. and Devich, R.N., 1999. Photogrammetric texture mapping onto planar polygons. *Graphical Models and Image Processing*, 61(1), 61–83.

Wehr, A. and Lohr, U., 1999. Airborne laser scanning – an introduction and overview. *ISPRS Journal of Photogrammetry and Remote Sensing*, 54, 68–82.

Werner, T. and Zisserman, A., 2002. New technique for automated architectural reconstruction from photographs. *7th European Conference on Computer Vision*, Part I, Copenhagen, Denmark, 28–31 May 2002, Springer, Berlin, Lecture Notes in Computer Science, Vol. 2351, 541–555.

Wilczkowiak, M., Trombettoni, G., Jermann, C., Sturm, P. and Boyer, E., 2003. Scene modelling based on constraint system decomposition techniques. *9th IEEE International Conference on Computer Vision (ICCV'03)*, Nice, France, 14–17 October, Vol. 2, 1004–1010.

Williams, L., 1983. Pyramidal parametrics. *Computer Graphics (Proc. SIGGRAPH '83)*, 17(3), 1–11.

Yamauchi, H., Lensch, H.P.A., Haber, J. and Seidel, H.-P., 2005. Textures revisited. *The Visual Computer*, 21(4), 217–241.

Yu, Y., Zhou, K., Xu, D., Shi, X., Bao, H., Guo, B. and Shum, H.Y., 2004. Mesh editing with Poisson-based gradient field manipulation. *SIGGRAPH'04*, Los Angeles, California, USA, 8–12 August, 644–651.

Zhang, H., 1998. *Effective Occlusion Culling for the Interactive Display of Arbitrary Models*. Ph.D. thesis, Department of Computer Science, University of North Carolina-Chapel Hill, Chapel Hill, North Carolina, USA.

Index